# 4th International Conference on Applied Geophysics (AG 2018)



E3S Web of Conferences Volume 66 (2018)

Cracow, Poland
28-29 June 2018

**Editors:**

**Z. Pilecki**
**H. Marcak**
**J. Dubinski**
**J. Dec**

ISBN: 978-1-5108-7556-2

**Printed from e-media with permission by:**

Curran Associates, Inc.
57 Morehouse Lane
Red Hook, NY 12571

**Some format issues inherent in the e-media version may also appear in this print version.**



Printed by Curran Associates, Inc. (2019)

For additional information, please contact EDP Sciences – Web of Conferences at the address below.

EDP Sciences – Web of Conferences
17, Avenue du Hoggar
Parc d'Activité de Courtabœuf
BP 112
F-91944 Les Ulis Cedex A
France

Phone: +33 (0) 1 69 18 75 75
Fax: +33 (0) 1 69 28 84 91

contact-edps@webofconferences.org

**Additional copies of this publication are available from:**

Curran Associates, Inc.
57 Morehouse Lane
Red Hook, NY 12571 USA
Phone: 845-758-0400
Fax: 845-758-2633
Email: curran@proceedings.com
Web: www.proceedings.com

# TABLE OF CONTENTS

# Geophysical identification of voids and loosened zones in the shallow subsurface of post-mining areas

*Zenon* Pilecki*

Mineral and Energy Economy Research Institute of the Polish Academy of Sciences, Cracow, Poland

**Abstract.** The shallow historic exploitation of Zn-Pb/Fe ore deposits as well as hard coal has generated many discontinuous deformations on the terrain surface in the Upper Silesian Coal Basin/Poland. Discontinuous deformations occur in different forms as sinkholes, synclines, cracks, faults or ditches. The basic cause of their occurrence is the presence of void and loosened zones in the shallow subsurface. If the appropriate conditions arise, the sinkhole process begins to move upwards and may cause a discontinuous deformation on the terrain surface. Typically, geophysical methods are used for void and loosened zone identification. The most effective methods are gravimetric, seismic, electric resistivity and ground penetrating radar (GPR). Geophysical testing, requires distinct changes in the physical properties in the rock mass. The identified geophysical anomalies should be verified by control borehole and borehole tests to confirm the presence of the void and loosened zones in the rock mass. The results of control drilling and borehole tests determine the need to apply treatment works. In order to assess the threat of the occurrence of discontinuous deformations in the areas of historical shallow mining in Upper Silesia, a classification system based on geophysical tests has also been developed.

**Keywords:** shallow exploitation, voids and loosened zones, discontinuous deformations, geophysical methods

## 1 Introduction

Shallow subsurface voids and loosened zones of mining origin may be the cause of the activation of sinkhole processes [1–11] (Fig. 1). In unfavorable geological and mining conditions, sinkhole processes may appear on the terrain surface creating various kinds of discontinuous deformations. In the conditions of shallow exploitation of deposits of Zn-Pb/Fe ores and hard coal in the northern part of the Upper Silesian Coal Basin (USCB), these are most often the forms of the sinkholes (Fig. 2) [9]. In general, most sinkholes have dimensions of up to 3.0 m. The arithmetic mean of a sinkhole surface length amounts to 5.0 m, the minimum is 0.5 m and the maximum 17 m. Also, the mean surface width amounts to 3.6 m, the minimum width is 0.3 m and the maximum width 12 m [9].

* Corresponding author: pilecki@meeri.pl

https://doi.org/10.1051/e3sconf/20186601001

**Fig. 1.** (a) A ditch on post mining terrain of shallow Zn-Pb exploitation near Bytom town (phot. Z. Pilecki); (b) A cuboid sinkhole over an old shaft near Chorzów town (phot. M. Rotkegel).

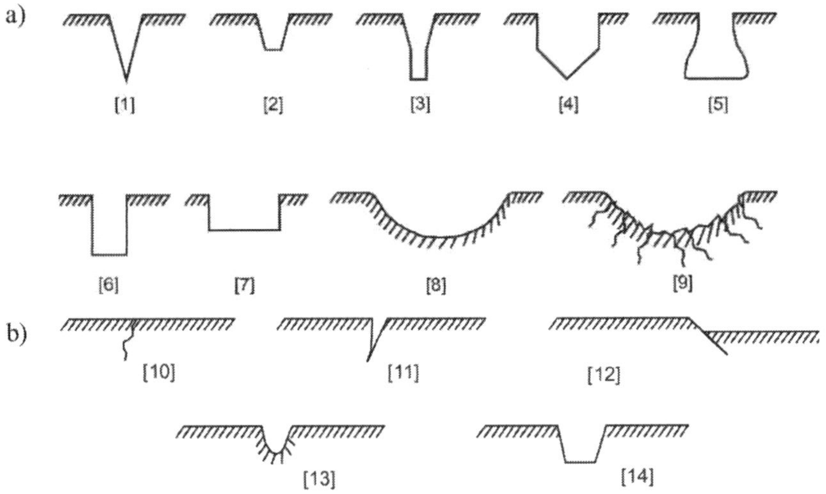


**Fig. 2.** Kinds of discontinuous deformations on the terrain surface in the vertical section observed in the area of Upper Silesia Coal Basin – (a) surface and (b) linear deformations (on the basis [1, 6]); [1] – conical sinkhole, [2] – truncated sinkhole, [3] – conical-cylindrical sinkhole, [4] – cylindrical – conical sinkhole, [5] – bell sinkhole, [6] – cylindrical sinkhole, [7] – cuboid sinkhole, [8] – continuous syncline, [9] – discontinuous syncline, [10] – crack, [11] – fissure, [12] – fault deformation, [13] – ditch, [14] – smoothed ditch.

## 2 Model of the formation sinkhole process

The model of the sinkhole formation process induced by shallow exploitation of various deposits has been discussed, among others, by [2–7, 10–11]. The identification of voids and loosened zones from the point of view of geophysical investigations have been undertaken, among others, by [6, 8–9, 12–20].

In a very general sense, the sinkhole formation process responsible for propagating the void towards the terrain surface may be presented using the "pressure arch" theory [21]. The model assumes that the void has the shape of an ellipse with a height of 2c and a width of 2a.

According to the theory of elasticity, radial stresses $\sigma_r$ on the outline of the void are described by the following dependencies [21]:

a) in the roof and floor of the void

$$\sigma_r^v = \sigma_x^0(1+2\frac{c}{a}) - \sigma_z^0 \tag{1}$$

b) in the sidewalls of the void

$$\sigma_r^h = \sigma_z^0(1+2\frac{a}{c}) - \sigma_x^0 \tag{2}$$

where:
$\sigma_z^0$ *and* $\sigma_x^0$ are the primary stresses in the vertical and horizontal direction respectively in the rock mass.

From dependencies (1) and (2) it follows that the ratio of the void height to its width is related to the distribution of stress at the border with the rock mass :

$$\frac{1}{2}(\frac{\sigma_z^0}{\sigma_x^0} - 1) = \frac{c}{a} \tag{3}$$

The ratio $\sigma_z^0 / \sigma_x^0$ increases from a value of 1 for hydrostatic conditions at large depths, to large values for voids in very shallow locations. If the void is located closer to the terrain surface, its shape should be more elongated in the vertical direction to maintain the force equilibrium at the border with the surrounding medium. Therefore, the different shape of the void and the unbalance of the relevant forces may cause the development of gravitional shear forces leading to the fracturing of rock mass. The fracture zone will grow until a new forces equilibrium is established related to the new shape of the void. The image of the void and its surroundings may be presented as a configuration of four zones (Fig. 3) [5, 6]:
(I) – **void**, often filled with water, containing rock fragments and soil material deposited in the floor;
(II) – **intense fractures**, mainly in the roof and sidewalls of the void, with the density of the fractures decreasing with greater distance from the void boundary,
(III) – **fractures surrounding the void**, to the border with the undisturbed rock mass,
(IV) – **undisturbed rock mass** by the propagating void.

A geophysical image of void with the surrounding fractured rock mass can be simplified to two zones, the most important from the point of view of geophysical identification are as follows:

- **central zone** - void with rock fragments and soil material deposited on its floor and possibly filled with water, characterized by a very high contrast of physical properties in comparison with the undisturbed rock mass.
- **external zone** - it is a fracture zone from the border of the void to the border with the undisturbed rock mass. The properties of this zone depend on the density of the fractures. In comparison to undisturbed rock mass, it is characterized by lower seismic wave velocity, greater anisotropy depending on the orientation of the fracture systems, increased electric resistance of the medium, lower volumetric density, or changes to electromagnetic parameters. This zone strengthens the anomalous effect of the central zone in the geophysical identification of the void.

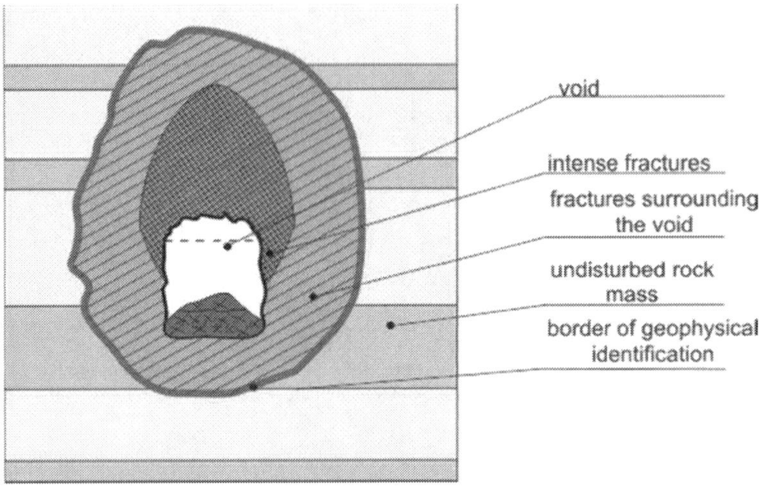


**Fig. 3.** Geophysical model of a void of mining origin [5].

Void propagation towards the terrain surface may be illustrated by numerical simulation, assuming that tensile stress resulting from the gravity force has the decisive influence on the growth of fractures in the void roof (Fig. 4).

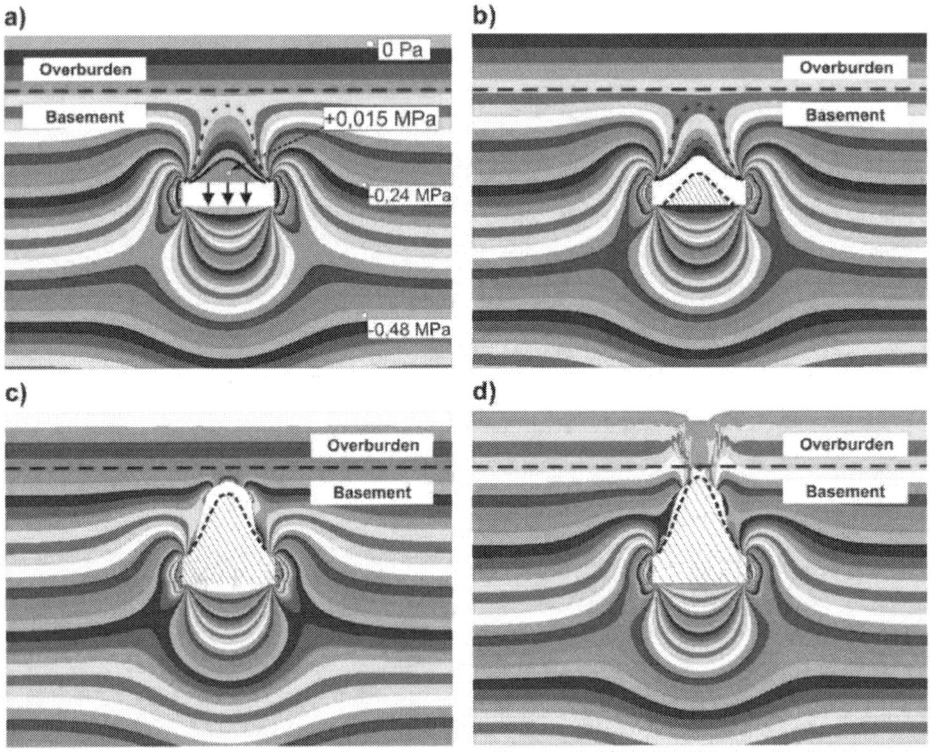


**Fig. 4.** (a) Redistribution of primary stress and the development of the fracture zone in the roof of the void; (b) the collapsing process of the roof of the void comprising a tensile stress zone; (c) secondary collapsing of the roof of the void; (d) reaching the border with loose overburden formations and the occurrence of a sinkhole on the terrain surface [11].

## 3 Methodology of the identification of voids and loosened zones in areas of shallow exploitation using geophysical methods

The geophysical identification of voids and loosened zones, referred to as weak zones, requires distinct changes in the physical properties of the medium [5, 6]. These changes are resulted from a comparison of the properties of the void zone together with the fracture zone to the properties of the surrounding, undisturbed medium. The external boundary of this system is not clearly determinable, since it is most often of a transitional nature related to a change in the fracture density. It should also be emphasized that the geophysical image of weak zones changes over time and depends on the impact of many factors activating the sinkhole process in the medium.

For the geophysical identification of weak zones in USCB conditions it was assumed that the basic research method is microgravimetry. The results of gravimetric tests were correlated with the results of seismic measurements or in case of further ambiguity with the results of the GPR or resistance method. In particular, the proposed methodology includes the following stages [9]:

Stage I - determination of weak zones using the gravimetric method, seismic refraction profiling and mapping of mining terrain also using aerial, satellite or drone images. Recently, terrain laser scanning has proven to be useful for mapping terrain prone to discontinous deformation [20].

Stage II - verification of gravimetric and seismic anomalies with other seismic techniques, or using a georadar or electric resistance method. The selection of the method should result from the measurement conditions - in the areas of occurrence of metal ore deposits the use of the electric resistance method is not recommended.

Stage III - verification of geophysical anomalies by means of cored control drilling along with recognition using a borehole camera, or borehole georadar, or other method.

Stage IV - in the case of liquidation of the weak zone, for instance by backfilling, the quality of the filling should be checked with one of the geophysical methods used at the stage of anomaly identification.

The use of the gravimetric method as the basic technique results from the fact that the volumetric density is a clear contrasting physical parameter in the tested basement disturbed by shallow exploitation. Figure 5 shows an example of such gravity surveys supplemented by seismic surveys. The aim of the seismic refraction profiling presented in Figure 5 was to determine the course of the boundary between overburden and basement, and the velocity changes of the refraction wave in the basement. Seismic refraction measurements were supported by the vertical profiling of S-wave velocity changes.

The GPR or electric resistance method are used as complementary methods depending on the measurement conditions. These methods use other properties of the medium, which may allow for a more reliable image of the weak zones in rock mass.

The verification of the results of geophysical identification usually consist of making a control borehole along with penetration by a camera or a borehole georadar. The results of drilling and borehole testing allow for a decision to be made about applying treatment works, and protecting the terrain surface from deformation.

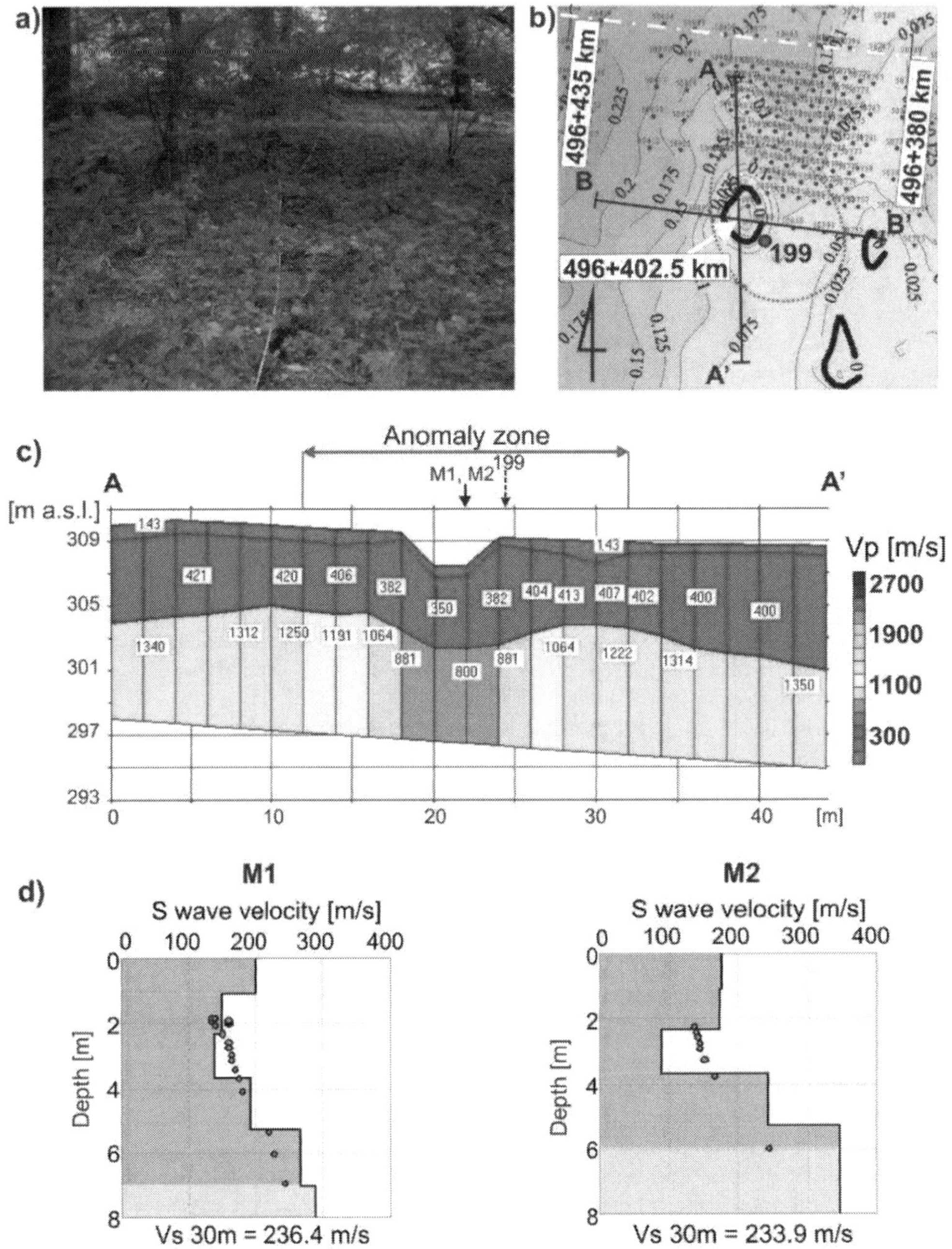


**Fig. 5.** (a) View of the sinkhole above the liquidated shaft 796 in an area of historical, shallow exploitation of zinc and lead ores deposit near Piekary Śl.; (b) a gravimetric map with an anomaly marked (source: J. Madej with the team) and the location of seismic profiles; (c) refractive cross-section with a visible seismic anomaly; (d) vertical profiling of the S wave velocity at the site of the seismic anomaly [6].

## 4 Classification of the threat assessment of the occurrence of discontinuous deformations on the terrain surface

The assessment of the threat of the occurrence of discontinuous deformations in the post-mining areas in Upper Silesia based on analytical methods is very difficult. This is mainly due to the unknown degree of rock mass distruction folowing the shallow exploitation of metal ore deposits or hard coal seams and often deep exploitation of coal seams. Under these conditions, information concerning the sinkhole threat is provided by geophysical identification. In the case of areas designated for building, classifications based on the likelihood of discontinuous deformations are used. Table 1 presents a modified classification of Pilecki and Kotyrba [22] to assess the likelihood of discontinuous deformations occurring in the areas of shallow exploitation of metal ore deposits or hard coal seams in USCB. The classification assumes that the probability of discontinuous deformations occurrence in conditions of overburden of loose soils with a thickness above 5m, is low due to the self-filling of excavations. In rock mass disturbed by exploitation, with the increase in weak zone depth, the probability of the occurrence of discontinous deformations on the terrain surface clearly decreases.

**Table 1.** Classification of the assessment of discontinuous deformation occurrence with the use of geophysical surveys in conditions of shallow exploitation of metal ore deposits and hard coal seams in the USCB [9].

| Threat category | **Threat criteria** |
|---|---|
| **A Negligible threat**[1] | no clear geophysical anomalies |
| | no discontinuous deformations and other terrain surface deformations |
| | a possible shallow exploitation of metal ore (or coal seams) deposits at a depth greater than 30 m, thickness of roof rocks at least five times greater than the height of excavations, or the presence of excavations at a depth greater than 30m |
| | no excavations connected to the terrain surface |
| | the mining area could be treated |
| | no outcrops of faults at a depth less than 15 m |
| | thickness of overburden of loose soils greater than 5 m, or stable level of water in the soil |
| **B Small threat**[1] | weak geophysical anomaly determined by at least one method |
| | no discontinuous deformations and other deformations of the terrain surface, or the mining area treated and documentation of the liquidation of excavations is available |
| | shallow exploitation of metal ore (or coal seams) deposits at a depth of more than 20 m, thickness of roof rocks at least five times greater than the height of excavations, or exploitation of iron ore, or excavations at a depth greater than 20m |
| | thickness of overburden of loose soils greater than 5 m |
| | no stable water level in the soil |
| | outcrops of faults at a depth greater than 5 m |
| **C Medium threat**[2] | a clear geophysical anomaly determined by at least one method |
| | symptoms indicating a single discontinuous deformation of the terrain surface or documented, liquidated discontinuous deformation |
| | the exploitation of ore deposits (or coal seams) at a depth greater than 20 m, the thickness of the overlying rocks is between three and five times the height of the excavations, or the presence of excavations at a depth greater than 30m, or the excavations having a connection to the terrain surface with documented proof of their liquidation |
| | thickness of overburden of loose soil less than 5 m, lack of water or local water occurrence in the loose soil |

| | |
|---|---|
| | outcrops of faults at a depth less than 5 m |
| **D Large threat[2]** | a strong geophysical anomaly determined by at least two methods |
| | occurrence of discontinuous deformations in the past or clear symptoms indicating numerous deformations of the discontinuous terrain surface |
| | shallow exploitation of metal ore (or coal seam) deposits at a depth less than 20 m, and the thickness of the overlying rock layers less than three times the height of excavations |
| | shallow exploitation of metal ore (or coal seam) deposits at a depth less than 20 m, and the thickness of the overburden of loose soil less than 5 m |
| | the presence of excavations liqudated without documentation having a connection with the terrain surface, or the presence of excavations of unknown liquidation at a depth of less than 20 m |
| | conducted mining works below shallow excavations |
| **E Very large threat[1]** | a strong geophysical anomaly determined by at least two methods |
| | occurrence of discontinuous deformations in the past or clear symptoms indicating numerous discontinuous deformations of the terrain surface |
| | shallow exploitation of metal ore (or coal seam) deposits, or the occurrence of excavations at a depth less than 30 m |

[1]with all of the listed criteria
[2]with one of the listed criteria

## 5 Summary

Discontinuous deformations on the mining and post-mining terrain are most often the result of the presence of voids and loosened zones at shallow depths created as a result of shallow exploitation. If the appropriate conditions arise, the sinkhole process begins to move upwards and may cause a discontinuous deformation on the terrain surface. The dynamics of this process depends on local conditions. Goszcz [4] formulated the basic criterion to assess the discontinuous deformations threat: *a post-mining void may cause a discontinuous deformation on the terrain surface, if the amount of rock material filling in this void does not cause self-filling as it moves towards the terrain surface.*

Changes in physical properties associated with the void and the surrounding fracture zone increases the electrical resistance of the medium, and decreases the seismic waves velocity, reduces the volumetric density of the medium and changes the electromagnetic parameters giving an anomalous effect which is measurable using geophysical methods.

From previous experience in the areas of shallow deposits of hard coal and metal ores in USCB, it may be assumed that microgravimetric studies are the basic geophysical method. The results of gravimetric tests are most often correlated with the results of refraction seismic profiling as well as depending on the measurement conditions with the results of the GPR and electrical resistivity method. The identified geophysical anomalies should be verified by control borehole and borehole tests, e.g. the borehole georadar is used to confirm the presence of weak zones in the rock mass. The results of control drilling and borehole tests determine the need to apply treatment works.

In order to assess the probability of the occurrence of discontinuous deformations in the areas of historical shallow mining in Upper Silesia, a classification based on geophysical tests has been developed.

## References

1. M. Chudek, W. Janusz, J. Zych, *Studium dotyczące stanu rozpoznania tworzenia się i prognozowania deformacji nieciągłych pod wpływem podziemnej eksploatacji złóż (Study on recognition of formation and forecasting of discontinuous deformations under*

*the influence of underground exploitation)*. Zeszyty Naukowe Politechniki Śląskiej **141**, Gliwice (1988) (in Polish)

2. F. G. Bell, Bulletin of the International Association of Engineering Geology **37**, 91–98 (1988) doi: 10.1007/BF02590374
3. B. N. Whittaker, D. J. Reddish, *Subsidence behaviour of rock structure, w: Comprehensive rock engineering – principles, practice and projects*, J.A. Hudson ed., Pergamon Press, Oxford, New York (1993)
4. A. Goszcz, *Powstawanie zapadlisk i innych deformacji nieciągłych powierzchni na obszarach płytkiej eksploatacji górniczej (Formation of sinkholes and other discontinuous deformations of terrain surfaces on areas of shallow exploitation)*. Mat. Konf.: Szkoła Eksploatacji Podziemnej, Wyd. CPPGSMiE PAN, Kraków, 119–137 (1996) (in Polish)
5. H. Marcak, *Powstawanie zapadlisk i innych form deformacji nieciągłych powierzchni spowodowanych występowaniem pustek (Formation of sinkholes and other forms of deformation of discontinuous surfaces caused by voids)*. Mat. Konf.: Warsztaty Górnicze, Wyd. IGSMiE PAN, Kraków, 71–84 (1999)
6. E. Popiołek, Z. Pilecki (eds.), *Ocena przydatności do zabudowy terenów zagrożonych deformacjami nieciągłymi za pomocą metod geofizycznych (Assessment of areas threatened with discontinuous deformations for building suitability using geophysical methods)* Wyd. IGSMiE PAN, Kraków (2005)
7. Z. Pilecki, A. Baranowski, *Estimation of dimension of a regular-type sinkhole activated by abandoned shafts*, Publications of the Institute of Geophysics, Polish Academy of Sciences **M-29**, 395, 271–279 (2006)
8. Z. Pilecki, Gospod. Surowcami Min. **25**, 3, 319–331 (2009)
9. Z. Pilecki, *Uzdatnienie podłoża autostrady A-1 na terenach pogórniczych płytkiej eksploatacji rud metali (Treatment of the A-1 motorway basement in post-mining areas of shallow mining of metal ore deposits)* Studia, Rozprawy Monografie **184,** Wyd. IGSMiE PAN, Kraków (2014)
10. P. Strzałkowski, Arch. Min. Sci. **60**, 1, 63–71 (2015) doi: 10.1515/amsc-2015-0005
11. K. Krawiec, Z. Pilecki, *Numeryczna symulacja procesu zapadliskowego w warunkach geologicznych i górniczych niecki bytomskiej na terenie pogórniczym płytkiej eksploatacji złóż rud metali (Numerical simulation of the sinkhole process in geological and mining conditions of the Bytom basin in the post-mining area of shallow mining of metal ore deposits)* Technika Poszukiwań Geologicznych **1**, 47–62 (2012)
12. Z. Fajklewicz, *Znaczenie badań geofizycznych w procesie przywracania wartości uzytkowych terenom naruszonym dokonaną eksploatacją górniczą (Significance of geophysical research in the process of restoring utility values to sites violated by mining)* Mat. Konf.: Warsztaty Górnicze, Wyd. IGSMiE PAN, Kraków, 45–86 (2001)
13. H. Marcak, Z. Pilecki, *Some Geophysical and Geomechanical Remarks on Recognition Sinkhole Processes in Post-Mining Areas*, Proc. 12th European Meeting of Environmental and Engineering Geophysics, 4–6 September 2006, Helsinki, Finland (2006)
14. Z. Pilecki, *The role of geophysical methods in the estimation of sinkhole threat in the post-mining areas of shallow exploitation in the Upper Silesian Coal Basin, Poland*, Gospod. Surowcami Min. **24**, 3/1, 27–40 (2008)
15. Z. Fajklewicz, *Origin of the anomalies of gravity and its vertical gradient over cavities in brittle rock*, Geophys. Prospect. **34**, 6, 1233–1254 (1986)

16. G. C. Colley, *The detection of caves by gravity measurements*, Geophys. Prospect. **11**, 1–10 (1963)
17. D. Lyness, *The gravimetric detection of mining subsidence*, Geophys. Prospect. **33**, 567–576 (1985)
18. J. C. Cook, *Seismic mapping of underground cavities using reflection amplitudes*, Geophysics **30**, 4, 527–538 (1965)
19. R. Neumann, *La gravimétrie de haute précision-application aux recherches de cavities*, Geophys. Prospect. **15**, 1, 1–2 (1967)
20. E. Pilecka, D. Szwarkowski, E3S Web Conf. **24**, 01006 (2017) doi: 10.1051/e3sconf/20172401006
21. A. Sałustowicz, *Zarys mechaniki górotworu (Outline of rock mass mechanics)* Wyd. Śląsk, Katowice (1968) (in Polish)
22. Z. Pilecki, A. Kotyrba, *Problematyka rozpoznania deformacji nieciągłych dla potrzeb projektowania konstrukcji drogowych na terenie płytkiej eksploatacji rud metali (Problems of recognizing discontinuous deformations for the purpose of designing road structures in shallow metal ore mining*) Prace Naukowe GIG **III**, 379–392 (2007)

# Evaluation of the forecast reliability of site effect for regional mining tremors

*Józef* Dubiński, *Krystyna* Stec*, and *Grażyna* Holeczek

Central Mining Institute, Katowice, Poland

**Abstract.** It is obligatory for Polish underground mines of hard coal and copper to prepare long-term exploitation projects for 2 – 5 years. All the natural hazards, including seismic ones, that may occur have to be taken into account in such projects. In the case of mining seismicity the forecast of site effect plays an important role. To recognize the reliability of this forecast for strong regional mining tremors special methodology was used. The peak ground velocity and peak ground acceleration, recorded after the occurrence of a regional tremor by surface seismometric network stations, were compared with the distribution of the forecast values of these parameters determined for similar seismic energy. Differences between these values are the basis for the evaluation of the reliability of the forecast site effect. The practical implementation was presented with the tremor of seismic energy $E = 5{\cdot}10^8$ J ($M_L = 3.63$) which occurred on 18th January 2018. The small value $\Delta PGV_{Hmax}$ (4%) of the difference between the recorded $PGV_{Hmax}{}^{rec}$ and the forecast values $PGV_{Hmax}{}^{pr}$ confirms that the measurement values correlate well with the forecast values. The difference value $\Delta PGA_{H10}$ for the peak ground acceleration equals 31% showing a worse correlation with the forecast data.

**Keywords:** mining tremors, site effect, velocity, acceleration

## 1 Introduction

Mining tremors are dynamic phenomena caused by underground mining operations. They occur as a result of violently displaced, fractured or caved-in rock mass layers. In the Upper Silesian Coal Basin (USCB) there is a distinct group of tremors which are directly associated with mining face advance, so-called exploitation tremors, of seismic energy $10^2 \leq E < 10^7$ J (local magnitude $M_L$ $0.1 \leq M_L < 3.2$) and a group of strong tremors of $E \geq 10^8$ J ($M_L \geq 3.2$) so-called regional tremors [1]. Regional tremors are associated with mining operations over a greater area, covering one or several mines, and the occurrence of tectonic zones or other geological structures in the area of mining operations (folds, synclines, overlaps) [2]. Continuous seismographic measurements, conducted by the Upper Silesian Seismological Network (USSN) of the Central Mining Institute and mine networks, enable the observation

* Corresponding author: kstec@gig.eu

of all the mining tremors occurring in the USCB, the location of their foci and the determination of their energy [3].

The USSN database of 1977 – 2017 contains 62,650 tremors of seismic energy $E \geq 10^5$ J. The collected data concerning the energy and the number of mining tremors in given energy classes is presented in Table 1. In the USCB, to determine the value of local magnitude, the dependence of $\log E = 1.8 + 1.9\log M_L$ (where: $E$ – seismic energy in J, $M_L$ –local magnitude) is applied.

**Table 1.** List of mining tremors in the Upper Silesian Coal Basin, 1977 – 2017.

| Energy $E$, J / Local magnitude $M_L$ | | No. of tremors |
|---|---|---|
| $10^5 \leq E < 10^6$ | $1.7 \leq M_L < 2.2$ | 53.260 |
| $10^6 \leq E < 10^7$ | $2.2 \leq M_L < 2.7$ | 8.433 |
| $10^7 \leq E < 10^8$ | $2.7 \leq M_L < 3.3$ | 851 |
| $10^8 \leq E < 10^9$ | $3.3 \leq M_L < 3.8$ | 88 |
| $10^9 \leq E < 4 \cdot 10^9$ | $3.8 \leq M_L < 4.2$ | 18 |
| | Total | **62.650** |

Regional tremors, despite their rarity (approx. 1 – 3 every year), resemble weak earthquakes. They have a strong influence on surface infrastructure and people's well-being. They often cause fear and significant discomfort. The phenomena often result in damaged buildings and property (Photos 1). In the USCB, as a result of mining tremors, a few hundred buildings were damaged [4].

**Photo 1.** Damage in buildings after a tremor of seismic energy $E = 1 \cdot 10^9$ J, $M_L = 3.8$).

The aforementioned seismicity means that protecting surface infrastructure is a significant problem. When strong tremors of regional nature occur, the intensity of ground motion caused by a tremor becomes an important parameter. Basic parameters employed to determine the intensity of tremors are amplitudes of peak ground velocity (*PGV*), peak ground acceleration (*PGA*) and the duration of the tremors. The exploitation tremors usually have relatively higher vibration velocity and acceleration, yet only in the epicentral zone, and short duration of their vibrations, thus their intensity is sensed only within the epicentral zone. Regional tremors have a distinctly longer seismic wave (lower frequencies of the dominant vibrations), longer duration and a greater area of high vibration intensity. Developing seismic forecasts for Polish mining basins is one of the significant elements of mining projects made for 2 to 5 year horizons. They include forecasting the places of the occurrence of mining tremors and, first of all, strong regional seismic phenomena together with their forecasted seismic energy and the distribution of ground motion parameters. Moreover, selected places on the surface, which require special protection, are constantly monitored providing data on the actual values of the ground motion parameters. This enables the evaluation of the reliability of these forecasts in reference to the parameters of surface vibrations determining their intensity. It has to be emphasised that the reliability of the forecasts is crucial for construction work planned in areas subjected to the dynamic influence of mining operations. On their basis local authorities determine the conditions for planned buildings, including additional measures to protect the objects against damage caused by seismic phenomena.

## 2 Forecasting the seismic energy of regional tremors

Forecasts of the parameters of site effect induced by regional rock mass tremors, first of all contain the evaluation/forecast of the maximum seismic energy which may occur when the mining project is realised, at a given probability of tremor occurrence. The energy is determined with the use of statistical distributions based on the catalogues of tremors which occurred in similar geological and mining conditions. The mathematical models applied here are asymptotic extreme distributions called Gumbel distribution functions. Their significant advantage is the fact that they make relatively general assumptions of the type of data which can be described in such a way. For the calculations made for the article, concerning forecasts of the occurrence of regional tremors, of the foci located in local tectonic zones, Gumbel distribution function III was assumed.
Gumbel distribution function III is expressed with the following equation:

$$G(x) = Exp\left[-\left(\frac{W-x}{W-u}\right)^{B}\right] \tag{1}$$

where:
$x = log\,E$ – random variable describing the seismic energy logarithm,
$W, B, u$ – distribution parameters.

Parameter $W$ is the upper limit in which random variable $x$ can assume. Hence it is limited to the top values, which follow the seismometric observations and the physics of the phenomenon (level of stress in the observed area). The most important parameter characterising seismic risk, when statistical distribution function $G(x)$ is assumed to be known, is seismic hazard $RD(x)$ which determines the occurrence probability of the tremors with logarithm of seismic energy $x$ or greater during $D$ time intervals. Seismic hazard is expressed with the equation:

$$RD(x) = 1 - G(x)^{D} \tag{2}$$

The conducted research into the evaluation of the repeatability of strong regional rock mass tremors were based on the catalogue of tremors of seismic energies $1\,10^2 \leq E < 6{\cdot}10^8$ J ($1.7 \leq M_L < 3.7$), which occurred between 1 January 2007 and 8 November 2017. The catalogue contained 80,652 events. The monthly distribution of the maximum seismic energy of the tremors is presented in Fig. 1, while Fig. 2 presents the Gumbel distribution function III and the measurement data. The calculated parameters of the distribution function $G(x)$ were applied to assess seismic risk $RD(x)$, i.e. the occurrence probability of a tremor of energy $E \geq 10^x$ J during the 3-year period of the planned mining exploitation (2018 – 2020). The data is presented in Table 2 and Figure 3.

To forecast the parameters of site effect, following the USCB rules, the maximum seismic energy is determined which, with a 10% probability, will not be exceeded. As the table and the figure show, in the analysed 10% probability area for 3 years, seismic energy of $E = 5 \cdot 10^8$ J ($M_L = 3.63$) will not be exceeded. The locations of potential foci of regional tremors were associated with greater faults and pillars for mine workings in the vicinity of planned coal longwalls.

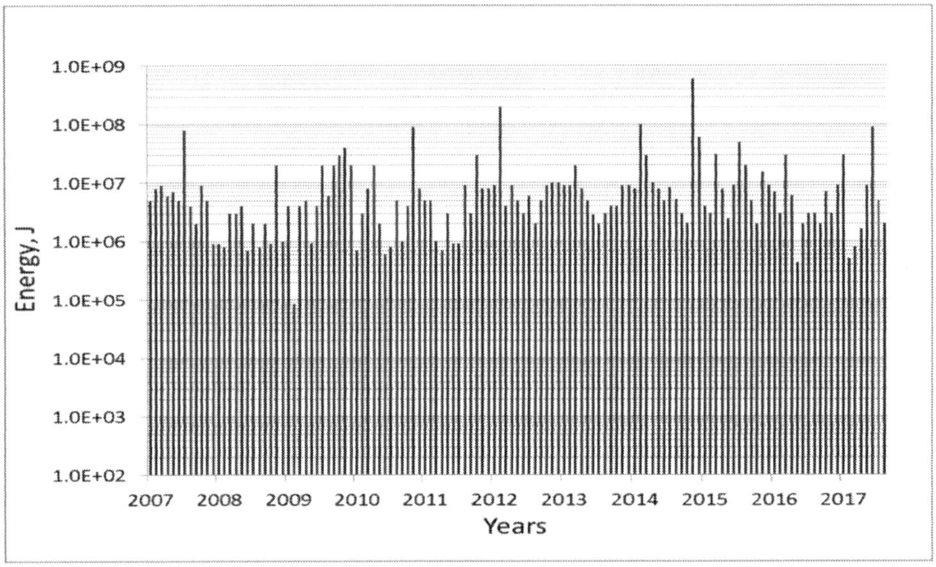


**Fig. 1.** Monthly maximum energy tremors distribution for the period from 1 Jan 2007 to 8 Nov 2017.

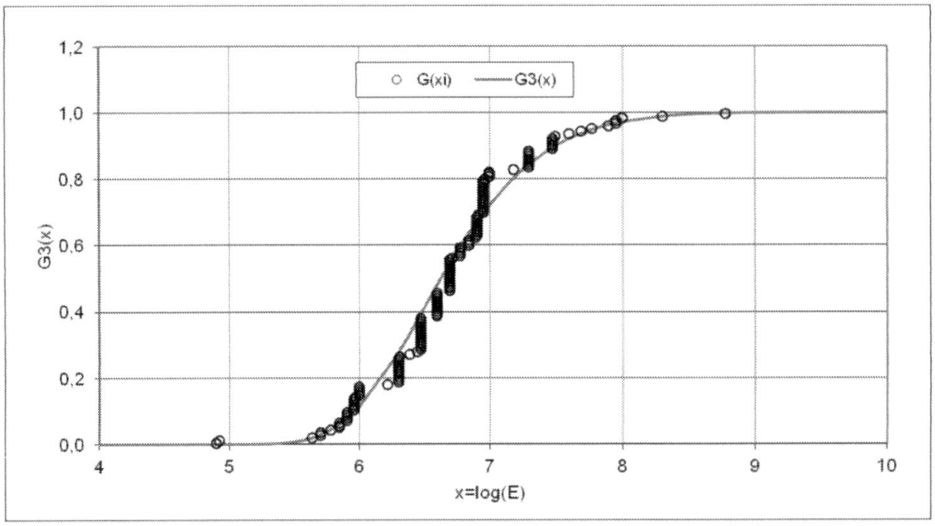


**Fig. 2.** Gumbel distribution function III (red line) and measurement data (circles) for the period from 1 Jan 2007 to 8 Nov 2017.

**Table 2.** Occurrence probability the forecast seismicity over a 3-year period of the planned mining exploitation (2018 – 2020).

| Seismic energy, J | $1\cdot10^7$ | $5\cdot10^7$ | $6\cdot10^7$ | $7\cdot10^7$ | $8\cdot10^7$ | $9\cdot10^7$ | $1\cdot10^8$ | $2\cdot10^8$ | $3\cdot10^8$ | $4\cdot10^8$ | $5\cdot10^8$ |
|---|---|---|---|---|---|---|---|---|---|---|---|
| Probability % | 100.0 | 90.00 | 84.63 | 79.12 | 73.74 | 68.66 | 63.93 | 33.86 | 20.70 | 13.93 | 9.99 |

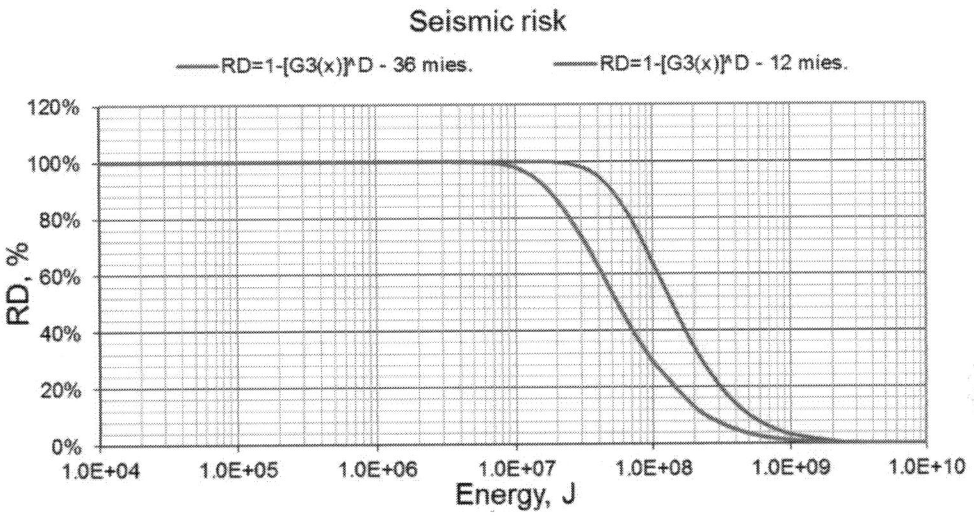


**Fig. 3.** Seismic risk of regional tremors for the exploitation planned for 3 years (red line) and one year (blue line) in coal mine.

## 3 Evaluation of the reliability of the forecast isoseism of ground motion velocity and acceleration based on measurement data

Regional scale empirical dependences were employed to forecast parameters of ground motion for the planned mining operations [5, 6]. They are based on records from numerous seismometric stations located in the USCB. They determine the dependence of the resultant peak ground velocity $PGV_{Hmax}$ or peak ground acceleration $PGA_{H10}$ of up to 10 Hz on the seismic energy of the tremors and epicentral distance, considering the types of ground defined in EUROCODE 8 (PN-EN 1998-1).

The research employed attenuation relations for peak ground velocity amplitude $PGV_{Hmax}$ and peak ground acceleration amplitude $PGA_{H10}$ in the ground rock [5]. According to the EUROCODE 8 standard it is ground type A.

$$\log (PGV_{Hmax}) = 0{,}_{209log}(E) - 0{,}33R - \log(R) - 0{,}814 \quad (3)$$

$$\log (PGA_{H10}) = 0{,}223\log(E) - 0{,}041R - \log(R) + 0{,}586 \quad (4)$$

where:
$PGV_{Hmax}$ – peak ground velocity, mm/s,
$PGA_{H10}$ – peak ground acceleration of frequency up to 10Hz, mm/s$^2$
$E$ – seismic energy of tremor, J,
$R$ – epicentral distance, m.

For the forecast, in this instance, the maximum seismic energy of tremors of $E = 5\cdot10^8$ J ($M_L = 3.63$), which may occur in the analysed area in 2018-2020 with a 10% probability, maps of distribution of predicted peak ground velocity $PGV_{Hmax}^{pr}$ isoline (Fig. 4) and predicted peak ground acceleration $PGA_{H10}^{pr}$ isoline (Fig. 5) were drawn.

In January 2018, in the analysed area, a regional tremor of seismic energy of $E = 5{\cdot}10^8$ J ($M_L = 3.63$) occurred and this was recorded by four surface seismometric stations. Their location with recorded value is shown in Figures 4 and 5. The distributions isolines consider the trend of tectonic zones and the distribution of subsoil amplification coefficient in this area presented in Fig. 6.

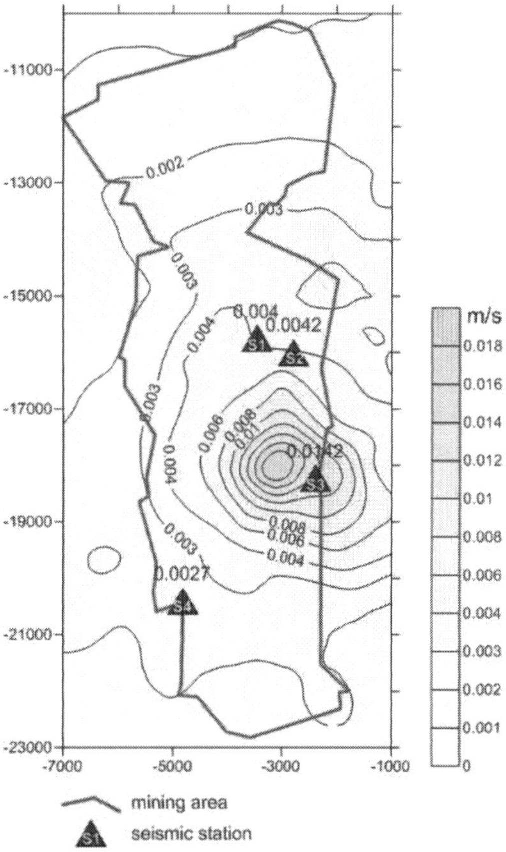


**Fig. 4.** Distribution of the forecast $PGV_{Hmax}^{rec}$ and recorded peak ground velocity $PGV_{Hmax}^{pr}$.

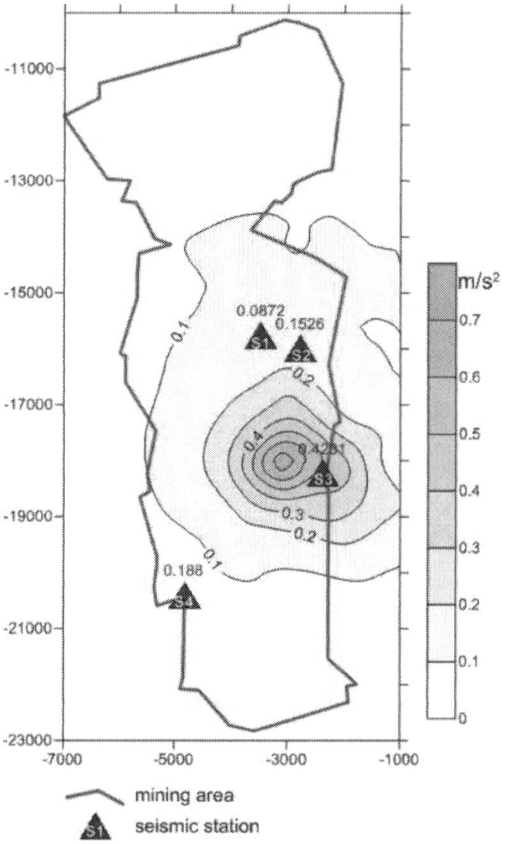


**Fig. 5.** Distribution of the forecast $PGA_{H10}^{rec}$ and recorded peak ground acceleration $PGA_{H10}^{pr}$.

The reliability of the forecast site effect for regional mining tremors was analysed through comparing the values of the peak ground velocity $PGV_{Hmax}^{rec}$ and the peak ground acceleration $PGA_{H10}^{rec}$, recorded by the stations of the surface seismometric network after a strong regional tremor of seismic energy of $5{\cdot}10^8$ J ($M_L = 3.63$), with forecast values of the distribution of velocity $PGV_{Hmax}^{pr}$ and acceleration $PGA_{H10}^{pr}$ for the same seismic energy.

$$\Delta PGV_{Hmax} = PGV_{Hmax}^{rec} - PGV_{Hmax}^{pr} \tag{5}$$

$$\Delta PGA_{H10} = PGA_{H10}^{rec} - PGA_{H10}^{pr} \tag{6}$$

where:

$\Delta PGV_{Hmax}$, $\Delta PGA_{H10}$ – difference which the forecast reliability is based on.

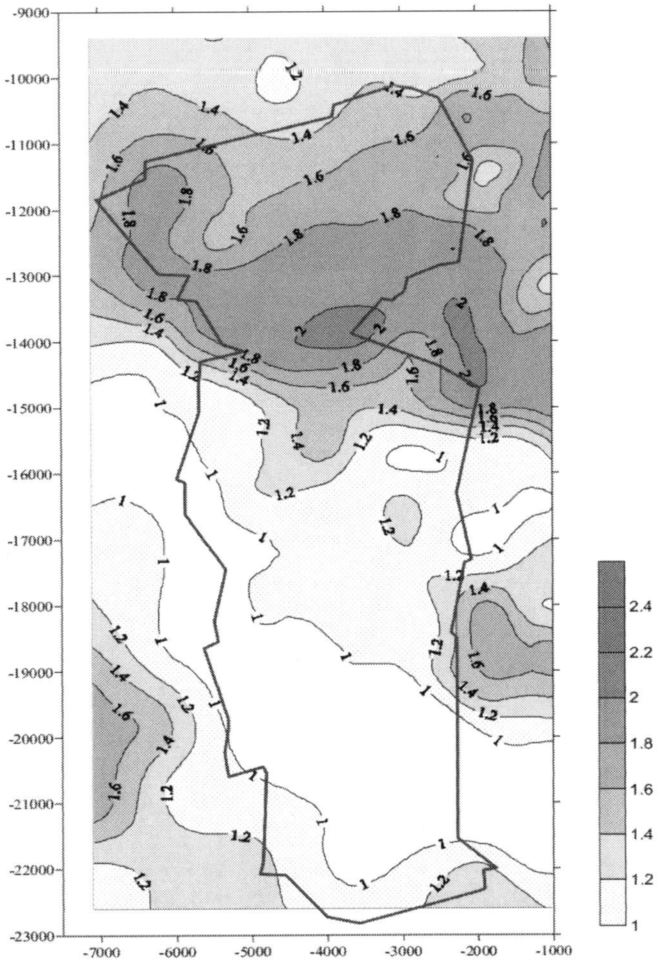


**Fig. 6.** Distribution of subsoil amplification coefficient.

In Table 3, columns 5 and 8 show the percentage of the difference between the recorded and the forecast ground motion velocity and accelerations, which indicates the quality and reliability of the forecast side effect, caused by strong mining tremors of regional origin. It shows that, the average difference between the forecast and the really recorded values for the ground motion velocity $\Delta PGV_{Hmax}$ is 4%. For the ground motion acceleration $\Delta PGA_{H10}$ the difference is much higher and equals 31%. This means that ground motion velocity is a much better and much more reliable forecast parameter.

**Table 3.** Recorded and forecast ground motions velocity and acceleration for a tremor of seismic energy $E = 5{\cdot}10^8$ J ($M_L = 3.63$).

| Station no. | Epicentral distance, m | $PGV_{Hmax}{}^{rec}$ m/s | $PGV_{Hmax}{}^{pr}$ m/s | $\Delta PGV_{Hmax}$ % | $PGA_{H10}{}^{rec}$ m/s$^2$ | $PGA_{H10}{}^{pr}$ m/s$^2$ | $\Delta PGA_{H10}$ % |
|---|---|---|---|---|---|---|---|
| *1* | *2* | *3* | *4* | *5* | *6* | *7* | *8* |
| S1 | 2282 | 0.004 | 0.0039 | 3.0 | 0.087 | 0.127 | 46 |
| S2 | 2032 | 0.0042 | 0.0041 | 3.2 | 0.153 | 0.132 | 13 |
| S3 | 807 | 0.0142 | 0.0138 | 2.9 | 0.423 | 0.463 | 9 |
| S4 | 2965 | 0.0027 | 0.0025 | 6.5 | 0.188 | 0.080 | 57 |

## 4 Conclusion

In Polish underground mines of hard coal and copper ore, it is compulsory to prepare long-term exploitation projects that determine their spatial distribution in a 2 – to 5-year time horizon. These projects take into account all natural hazards, including seismic ones.

The seismic forecast is important both because of the threat of rockbursts and the impact of ground motion on the surface environment. It is particularly important in reference to strong seismic events of a regional nature.

The small value $\Delta PGV_{Hmax}$ (4%) of the difference between the recorded $PGV_{Hmax}{}^{rec}$ and the forecast values $PGV_{Hmax}{}^{pr}$ parameters confirms that the measured values correlate well with the values of the forecast ground motion velocity; this strong correlation proves the accuracy of the methodology used to forecast site effect for strong regional tremors and the reliability of the forecast results.

The greater value $\Delta PGA_{H10}$ (31%) of the difference between the recorded $PGA_{H10}{}^{rec}$ and the forecast values of $PGA_{H10}{}^{pr}$ parameters shows worse correlation and distinctly little reliability of the forecast results obtained with this parameter.

## References

1. K. Stec, *Characteristic of seismic activity of the Upper Silesian Coal Basin in Poland,* GEOPHYS J INT **168**, 2, 757-768 (2007) doi.org/10.1111/j.1365-246X.2006.03227.x
2. M. Kozłowska, B. Orlecka-Sikora, Ł. Rudziński, S. Cielesta, G. Mutke *Atypical evolution of seismicity patterns resulting from the coupled natural, human-induced and coseismic stresses in a longwall coal mining environment,* INT J ROCK MECH MIN **86**, 5-15 (2016) doi: 10.1016/j.ijrmms.2016.03.024
3. R. Patyńska, K. Stec, *Regional Rockburst Indicator for structural units of Upper Silesian Coal Basin*, STUD GEOTECH MECH **39**, 3, 27-37 (2017) doi:10.1515/sgem-2017-0027
4. G. Mutke, J. Chodacki, L. Muszyński, S. Kremers, R. Fritschen, *Mining Seismic Instrumental Intensity Scale MSIIS-15 – verification in coal basins*. AIMS 2015 - Fith Int. Symp.: Mineral Resources and Mine Development. RWTH Aachen University. **14**, 551-560. (ISBN 978-3-941277-22-9) (2015)
5. J. Chodacki, *New Ground Motion Prediction Equation for Peak Ground Velocity and Duration of Ground Motion for Mining Tremors in Upper Silesia*, ACTA GEOPHYS **64**, 6, 2449-2470 (2016) doi.org/10.1515/acgeo-2016-0109

# GPR ground-wave parameters changes due to variation of soil moisture

*Henryk* Marcak[1], *Sylwia* Tomecka-Suchoń[1], *Rafał* Czarny[2,*], *Paweł* Pysz[1], *Akinniyi* Akinsunmade[3], and *Tetiana* Kril[4]

[1]AGH, University of Science and Technology, Kraków, Poland
[2]MEERI PAS, Department of Geoengineering and Environmental Engineering, Kraków, Poland
[3]Nigerian Geological Survey Agency, Nigeria
[4]NAS, Institute of Geological Sciences, Kyiv, Ukraine

**Abstract.** The soil moisture distribution is important in terms of assessment of agricultural conditions for plant cultivations. The aim of this study is to develop methodology for monitoring soil moisture by the use of gound-penerating radar (GPR). This non-invasive geophysical method has been widely introduced for this purpose for the last decade. However, there is still lack of routinly application in agriculture. The main reasons are time-consuming data processing and aqusition, particulary for multi-offset measurments. In order to fill this gap we tried to adapt single-offset measurment. Our field study contained several measurments for different time span after ground irrigation. We used 800 MHz shielded and 200 MHz unshelded antennas. We focused on ground wave which propagates just beneath the surface. We observed relative velocity and amplitude spectrum changes of air and ground waves after water irrigation. These changes has an explanation in electromagnetic wave propagation theory. Water irrigation causes the increase of ground wave time arrivals and shift of amplitude spectrum towards lower frequencies.

**Keywords:** ground penetrating radar, GPR ground-wave, soil moisture, soil water content

## 1 Introduction

Obtaining information about the spatial and temporal distribution of water in ground is important for precision agricultural programs. Nowadays, the most common electromagnetic method for the soil water content estimation is TDR (time-domain reflectometry) [10]. However, this method can monitor water content at one location, therefore it is difficult and expensive to apply for wide range areas. This technical gap can be fulfilled by non-invasive GPR method which allows to probe soil water content from any area with high resolution.

GPR has made significant progress for last 25 years in water content estimation [8]. A various methodologies have been introduced and summarized by Huisman et al. [5] and by

* Corresponding author: czarny@min-pan.krakow.pl

Klotzsche et al. [8]. Those methods are based on single- or multi-offset (called wide angle reflection and refraction – WAAR) GPR measurements which can be done at surface or in the borehole.

At the beginning of this paper, theoretical background of GPR waves parameters changes in terms of water content estimation was described. Then, case study was demonstrated. Field experiment contained GPR measurements by 800 MHz shielded and 200 MHz unshielded antennas. Measurements were carried out in two series: before and after water irrigation. Relative velocity and power spectrum density (PSD) changes were measured with reference to soil moisture. Time delays and shifts of maximum energy at PSD for GPR ground wave after water irrigation were observed.

## 2 Theoretical background

The most common method for assessing soil water content with use of GPR is to measure the velocity of electormagnetic waves. The velocity $V$ depends on the relative dielectric permittivity $\varepsilon_r$ with relation described by formula:

$$\varepsilon_r = \frac{c^2}{V^2}, \tag{1}$$

where $c$ – free space electromagnetic wave propagation velocity (3 x $10^8$ m/s). The higher values of permittivity due to e.g. water content, the lower velocity of electromagnetic waves. The apparent $\varepsilon_r$, i.e. permittivity determined from GPR measurment can be directly connected to volumetric soil water content by use of empirical equation introduced by Jacobsen and Schrjoning [5]:

$$\theta = 5.3 \times 10^{-2} + 2.97 \times 10^{-2}\varepsilon_r - 5.5 \times 10^{-4}{\varepsilon_r}^2 + 4.3 \times 10^{-6}{\varepsilon_r}^3. \tag{2}$$

The changes in soil water content can be aslo observed in a fequency domain. The dielectical losses due to water saturation can shift maximum amplitude spectra towards lower frequencies [1,2]. From theoretical point of view GPR waves propagation depends on complex dilectric premitivity $\varepsilon^*$. According to Debye [3] the dipolar response of the medium to the electrical field for single relaxation time $\tau$ can be written as:

$$\varepsilon^*(\omega) - \varepsilon_\infty \approx \frac{1}{1-i\omega\tau}, \omega_p = \frac{1}{\tau}, \tag{3}$$

where $\omega_p$ - peak loss frequency, $\varepsilon_\infty$ - real part of $\varepsilon^*(\omega)$ at infine frequency. The consequence of electric dispersion is the dielectric loss in the medium reveals as a imaginary part of electrical permittivity:

$$\varepsilon^*(\omega) = \varepsilon' + i\varepsilon'', \tag{4}$$

where $\varepsilon'$ and $\varepsilon''$ are real and imaginary permittivity, respectively. However, this description isn't useful in practice. The more general approach was proposed by Jonscher [7], where imaginary and real parts of permittivity are proportional to the frequency:

$$\varepsilon'(\omega) \approx \varepsilon''(\omega) \approx \omega^{n-1} \text{ for } \omega > \omega_p. \tag{5}$$

For this assumption for positive frequencies the following formula can be introduced [1]:

$$\varepsilon^*(\omega) = \varepsilon_r \left(\frac{\omega}{\omega_r}\right)^{n-1} \left[\sin\left(\frac{n\pi}{2}\right) + i\cos\left(\frac{n\pi}{2}\right)\right], \tag{6}$$

for $1 > n > 0$.
The complex spectrum of the electrical signal has form [1]:

$$E(\omega, z) = E_9(\omega)e^{\left[i\frac{\omega}{V(\omega)}z\right]}e^{-\alpha z} \text{ for } \alpha = \frac{\omega}{2V_0 Q}. \tag{7}$$

$E_9(\omega)$ - signal transmitted at distance z, Q - quality factor of attenuation, $V_0$ – phase velocity for selected frequency. Periodical part in formula (7) is relativly small and it can be assumed that the spectrum depends only on $E_9(\omega)$, $V_0$ and $Q$ for a chosen frequency range. The examples of $E(\omega, z)$ for various Q, for velocity $V_0$ equal 10 cm/ns and $E_9(\omega)$ as a gaussian curve with maximum at 600 MHz is presented in Figure 1. The periodical component in the frequency range 400-800 MHz was neglected. Frequency as well as $E(\omega, z)$ decreases for small Q i.e. for high attenuation.

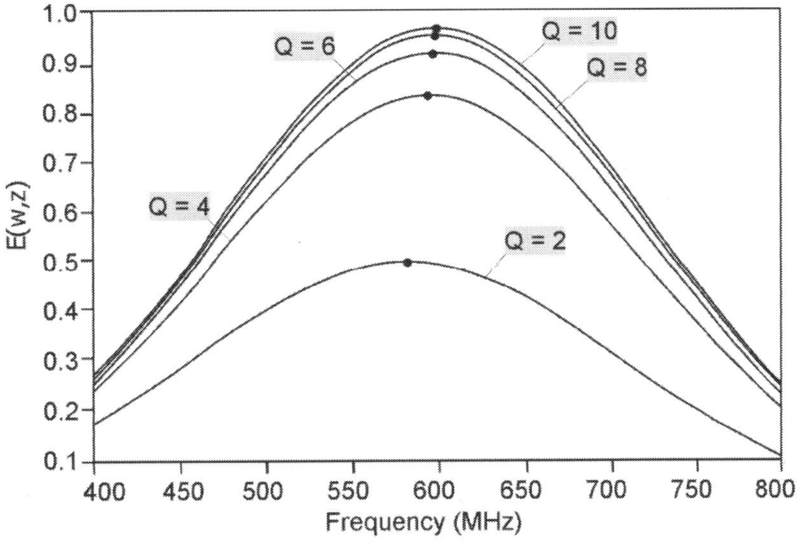


**Fig. 1.** The complex spectrum of the electrical signal transmitted at a distance $z$ in the frequency range 400-800 MHz, for velocity 10 cm/nsec and $E_9(\omega)$ in the form of gaussian curve.

## 3 Data and methodology

The aim of case study was to observe GPR waves parameters changes due to soil moisture. In order to track these changes dozens litres of water were poured down on the surface to mimic irrigation process. GPR method was applied before and after irrigation. The area of field study was dominated by sand formation up to 5 m depth. The measurements were contained 14 standard GPR single-offset profiles by 800 MHz shielded antenna and 2 GPR profiles with multi-offset technique (WAAR) by 200 MHz unshielded antenna. All profiles had 4 m long and were carried out through the same path.

The examples of radargrams before and after irrigation for 200 MHz and 800 MHz antennas are shown in Figure 2 and 3, respectively. In order to preserve as much information from gathered data as possible, only necessary processing steps like static correction and mean amplitude subtraction were done. In Figure 3 echograms from multi-offset measurements are presented. Air and ground waves are visible. Velocity of these waves decreases for measurement about 30 minutes after water irrigation. For air wave it is reduction from about 30 to about 28 m/ns and for ground wave from about 14 to about 12 m/ns. In Figure 2 it can be seen that traces are stretched after irrigation, particularly in the canter of the profile where irrigation was the highest (Fig. 2, box with black dashed line). Moreover, radargrams before irrigation and the day after irrigation are similar. Probably, water infiltrated to the deeper layers or/and evaporated the day after.

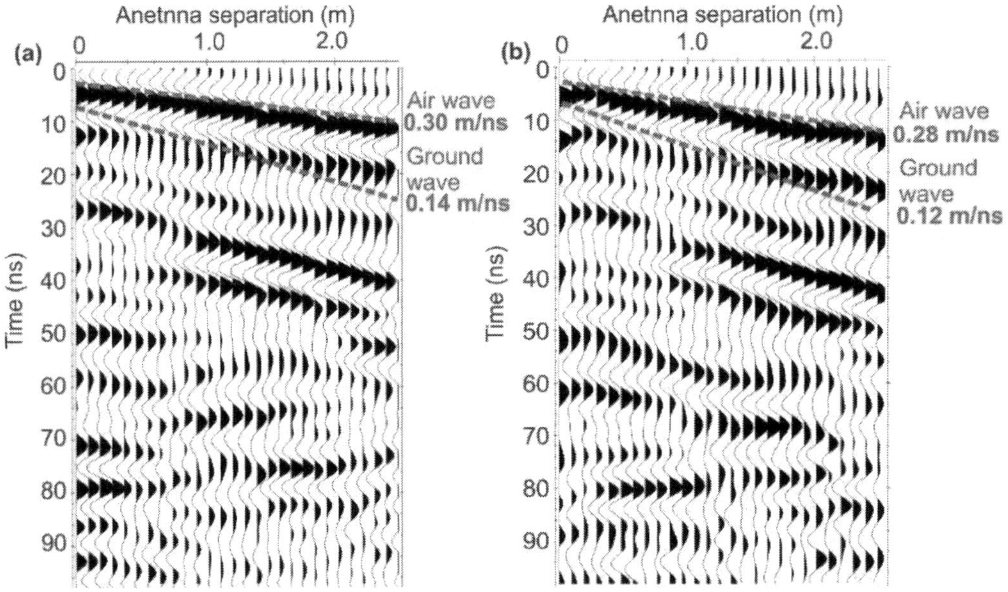


**Fig. 2.** Wide angle reflection and reflection (WARR) radargrams with the use of 200 MHz antenna. Velocity decreases for air and ground waves after water irrigation.

Even if WARR technique gives robust estimation of GPR waves velocity that can be transformed to volumetric water content (equation (2)), this method is time consuming, hence difficult to apply for agricultural purposes. In our study, we put much more effort to adapt single-offset GPR measurement to monitor some relative changes of GPR waves parameters. We focused on GPR ground wave acquired by 800 MHz single-offset antenna. Ground wave propagates just close to the surface, therefore it is suitable for agriculture. As described in paragraph 2 soil moisture can reduce velocity and frequency of GPR waves. Therefore, to find how the water influences velocity and frequency in our data we decided to average all traces within the single profile. It helped us to compare all the measurements. Eventually, each of 14 echograms, for different time span after irrigation, was represented by the single trace (Fig. 4). After averaging, each trace was shifted according to the time of the first maximum corresponding to GPR air wave. In Figure 4 two different time-windows are marked: for air wave and for ground wave by the use of velocities from WAAR (Fig. 2). Due to small distance between transmitter and receiver both waves overlap.

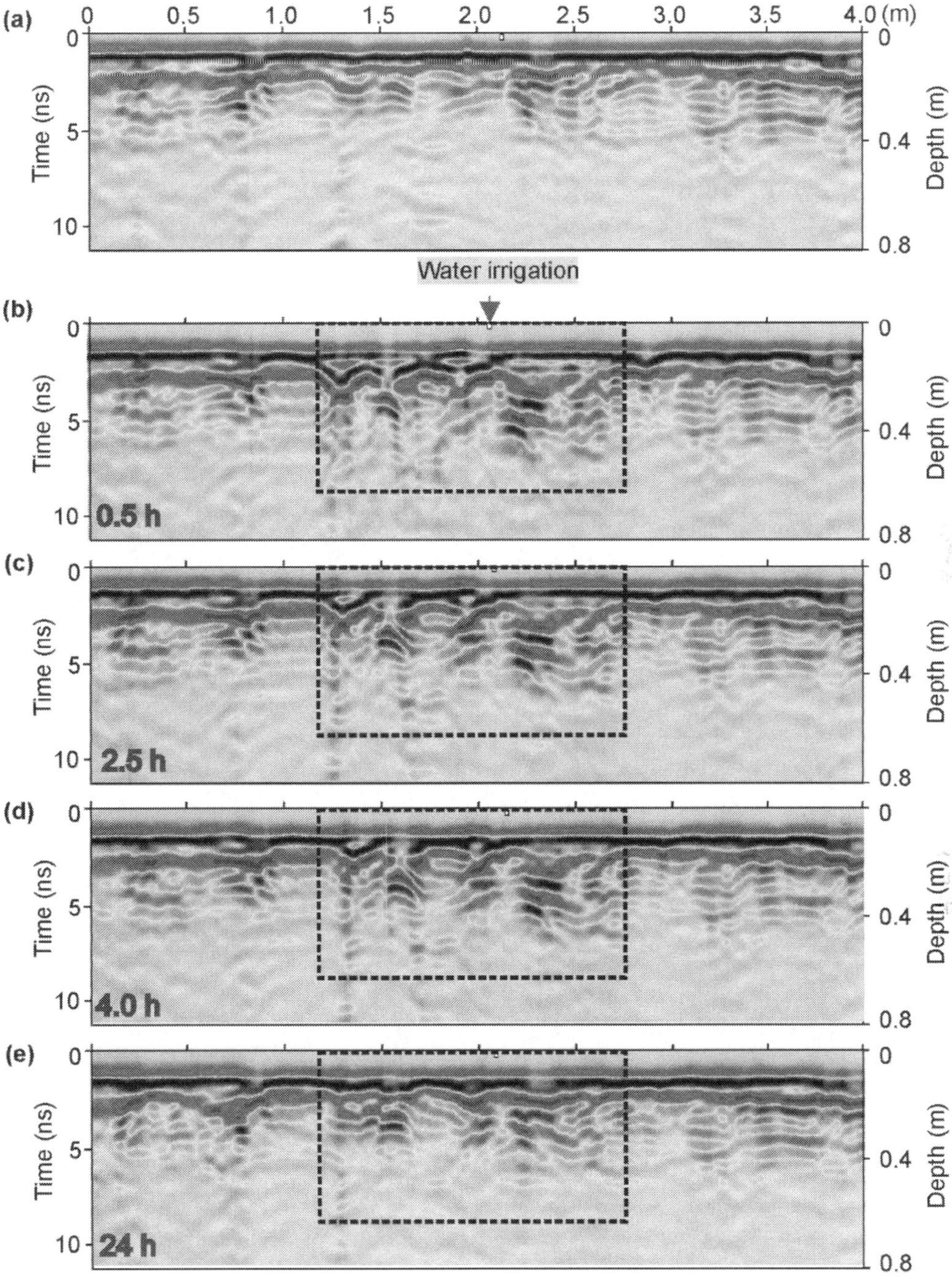


**Fig. 3.** Radargrams before (a) and after (b-e) water irrigation for different time span. Changes due to soil moisture are visible inside the box marked by black dashed line. Radargram after 24 h is similar to that one before water irrigation.

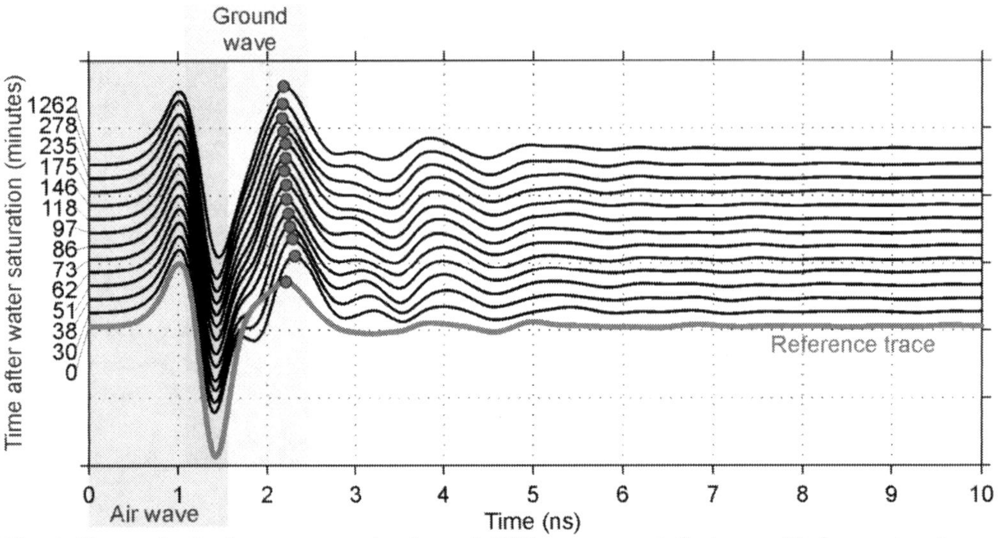


**Fig. 4.** The result of radargram averaging for each GPR measurment. Red curve (Reference trace) indicates measurement for dry soil. Red dot indicates the time of maximum amplitude observed in ground wave time-window.

## 4 Results and analysis

In Figure 5 and 6 GPR ground wave parameters changes are presented for different time span after irrigation. In Figure 5 arrival times of maximum amplitude observed in ground wave time-window are presented. The highest delay time is noticed for first measurement i.e. 30 minutes after irrigation. This delay progressively vanishes after next 40 minutes. Probably water infiltrated to the ground and/or evaporated. From observation at 73 minute to the last measurement, delay times have similar values.

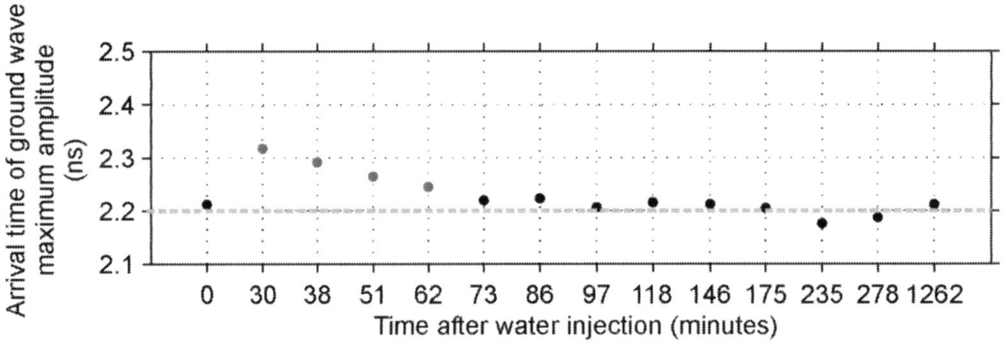


**Fig. 5.** Arrival times of maximum amplitude for ground wave time window. Time delays after water irrigation are noticed (red dots).

In figure 6 PSD for 14 measurements are shown. PSD was computed for ground wave time window. Frequency and PSD values decrease after water irrigation. These parameters are similar to those for dry soil after measurement at 118 minute.

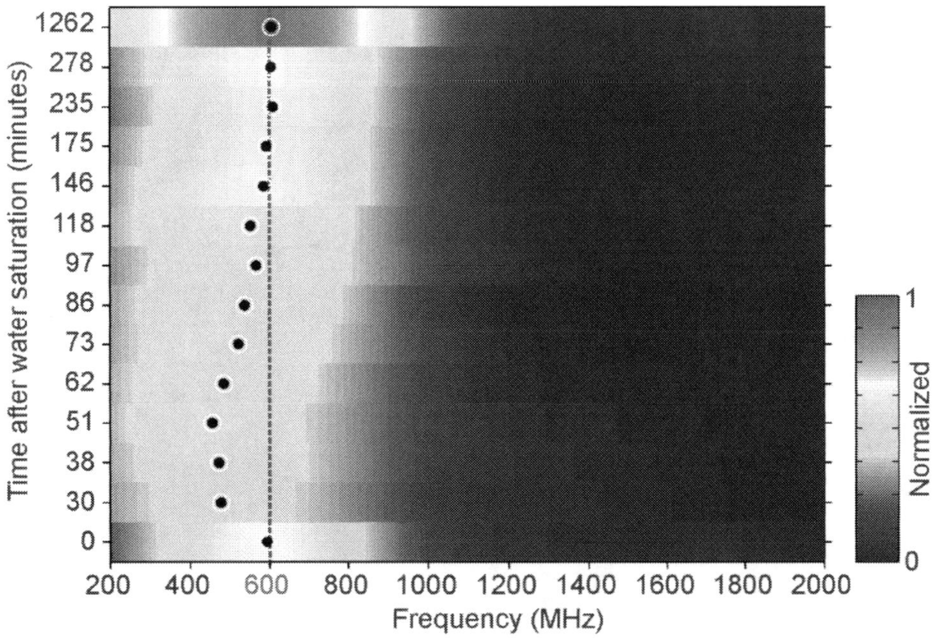


**Fig. 6.** Power spectrum density for every averaged trace. Black dot denotes maximum value on the PSD.

## 5 Conclusion

The paper has shown usefulness of using single-offset GPR antenna to monitor soil moisture. Measured parameters are relative and can't be directly connected with water content. However, in case of farmland where soil ingredients don't change significantly, one can determine empirical relation between these parameters and water content by using laboratory test. Future work on this project will attempt to further use single-offset GPR in determination of water content for agricultural purposes.

*This research was supported by AGH University of Science and Technology in Krakow no. 11.11.140.645.*

## Reference

1. M. Bano, *Constant dielectric losses of ground-penetrating radar waves*, Geophys. J. Int. **124**, 1, 279–288 (1996)
2. A. Benedetto, *Water content evaluation in unsaturated soil using GPR signal analysis in the frequency domain*, J. Appl. Geophys. **71**, 1, 26–35 (2010)
3. P. *Debye*, *Polar molecules*, Chemical Catalogue Co., Dover, New York (1929)
4. L. W. Galagedara, G. W. Parkin, J. D. Redman, *An analysis of the ground-penetrating radar direct ground wave method for soil water content measurement*, Hydrol. Process. **17**, 18, 3615–3628 (2003)
5. J. A. Huisman, S. S. Hubbard, J. D. Redman, A. P. Annan, *Measuring soil water content with ground penetrating radar*, Vadose Zone J. **2**, 476–491 (2003)

6. O. H. Jacobsen, P. Schjonning. *Comparison of TDR calibration functions for soil water determination*, Proceedings of the Symposium: Time-Domain Reflectometry Applications in Soil Science, 25–33, (1995)
7. A. K. Jonscher, *The 'universal' dielectric response*, Nature **267**, 5613, 673 (1977)
8. A. Klotzsche, F. Jonard, M. C. Looms, J. van der Kruk, J. A. Huisman, Measuring Soil Water Content with Ground Penetrating Radar: A Decade of Progress, Vadose Zone J. **17**, 1 (2018)
9. C. Paz, F. J. Alcalá, J. M. Carvalho, L. Ribeiro, *Current uses of ground penetrating radar in groundwater-dependent ecosystems research*, Sci. Total Environ. **595**, 868–885 (2017)
10. G. C. Topp, J.L. Davis, A. P. Annan, *Electromagnetic determination of soil water content: Measurements in coaxial transmission lines,* Water Resour. Res. **16**, 3, 574–582 (1980)

# Formation evaluation challenges in Pliocene gas-bearing reservoirs from the Romanian Western Black Sea shelf

*Bogdan Mihai* Niculescu*, and *Gina* Andrei

University of Bucharest, Faculty of Geology and Geophysics, Department of Geophysics, Bucharest, Romania

**Abstract.** This paper addresses some formation evaluation challenges and petrophysical particularities regarding two gas fields of Early Pliocene age, belonging to the biogenic hydrocarbon system of Western Black Sea Basin - Romanian continental shelf. Although these structures are located at the same depth and only 15 km apart, the wells that intercepted the sands and silts gas-bearing reservoirs indicate an important lateral facies variation and different reservoir qualities. We analyzed and interpreted data from exploration and appraisal wells that targeted these reservoirs, showing that: (1) there is a limited radioactivity contrast between the reservoir and non-reservoir intervals, so a clay volume determination based solely on the gamma ray log is not practical; (2) the reservoirs are characterized by high capillary-bound water contents, leading sometimes to abnormally low resistivity readings; (3) an additional resistivity suppression might be caused by the limited vertical resolution of the electrical logging tools, in the presence of thinly laminated sand - shale intervals; (4) the identification of gas-water contacts based exclusively on pressure gradients may be inaccurate and should be checked against the results of conventional geophysical logs interpretation and of nuclear magnetic resonance logs, for delineating the intervals with bound water or with movable fluids.

**Keywords:** biogenic gas, Black Sea, formation evaluation, Pliocene, well logging

## 1 Introduction

For oil and gas exploration wells, formation evaluation consists of the analysis and interpretation of geophysical well logging data and other types of subsurface information (e.g. well testing, formation testing and coring) in order to ascertain if commercially producible hydrocarbons are present and to determine the best means for their recovery. The key properties of reservoir rocks derived through formation evaluation (lithology, porosity, clay content, water and hydrocarbon saturations, hydrocarbons type, permeability, positions

* Corresponding author: bogdan.niculescu@gg.unibuc.ro

of initial fluid contacts, petrophysical cut-offs) are used in further exploration, field development and resource estimates.

This paper addresses some of the formation evaluation challenges and petrophysical particularities regarding two gas fields (conventionally named Field A and Field B) of Early Pliocene (Dacian) age, belonging to the biogenic hydrocarbon system of Western Black Sea Basin - Romanian continental shelf. The geology, tectonics, evolution, hydrocarbon systems and hydrocarbon potential of this region have been studied and presented, among other authors, by Ionescu et al. [1], Dinu et al. [2], Moroşanu [3, 4] and Georgiev [5]. The gas discoveries are situated to the east of Constanţa harbour, in a shallow water (45 – 125 m depth) exploration block which overlies the Tertiary-filled Histria Depression. They were identified and imaged using 2D and 3D seismic reflection surveys, started in 1992 and carried out until recent years. Although these dome structures are located at the same depth and only 15 km apart, in similar four-way dip closures, the exploration and appraisal wells that intercepted the sands and silts gas-bearing reservoirs show an important lateral facies variation between the two structures and clearly different reservoir qualities. Such variability cannot be detected on the seismic sections (both structures have quasi-identical reflection signatures), but is evident from the openhole geophysical logs responses and the drill cores taken from the reservoir intervals. In the same time, the reservoir intervals for both fields are characterized by a vertical variability of the grain size, with sandy units overlying silty units, in a more or less gradual transition.

The petrophysical analysis of the gas fields taken into consideration has presented a number of challenges since their discovery. The sands are highly unconsolidated, fine to very fine grained, trending to silts, with abundant small pores that can trap large volumes of bound water. Consequently, the main challenges have been in the determination of realistic water and gas saturations, clay content and porosity, gas-water contacts and the net contributing reservoirs. We show and discuss particular issues related to the log interpretation and formation evaluation for these fields, together with the implications of depositional facies variability upon the resource estimates.

## 2 Geological settings

The Western Black Sea Basin is an important hydrocarbon-bearing region in SE Europe. Up to now, on the Romanian continental shelf, the main area of interest from the standpoint of hydrocarbons potential is the Histria Depression (Fig. 1), a post-tectonic cover superimposed over the North Dobrogea Orogen, formed by extension beginning with Aptian – Albian and until Eocene, then followed by subsidence [2, 4]. This sedimentary basin gradually expands and deepens towards SE and merges with Western Black Sea Basin floor [5]. In this region, oil-bearing reservoirs have been identified in Albian, Late Cretaceous (Turonian, Coniacian, Santonian), Eocene and Oligocene formations, whereas gas shows or commercial gas accumulations are located in Cretaceous, Eocene and especially Late Miocene – Early Pliocene (Pontian and Dacian) formations [1, 4, 5]. In the pre-Oligocene and Oligocene formations, the traps are either tectonic (anticlines, faulted anticlines) or stratigraphic (pinchouts, drape anticlines). In the post-Oligocene formations, the traps are of stratigraphic type, such as depositional fans or anticlines of differential setting, extended over wide areas and sometimes affected by gravity faults and growth faults.

According to Moroşanu [4], four thermogenic petroleum systems and one biogenic gas system are present in the Romanian continental shelf area of the Western Black Sea. The biogenic system comprises the Midia, Ovidiu, Cobălcescu, Ana, Doina and Domino gas fields.

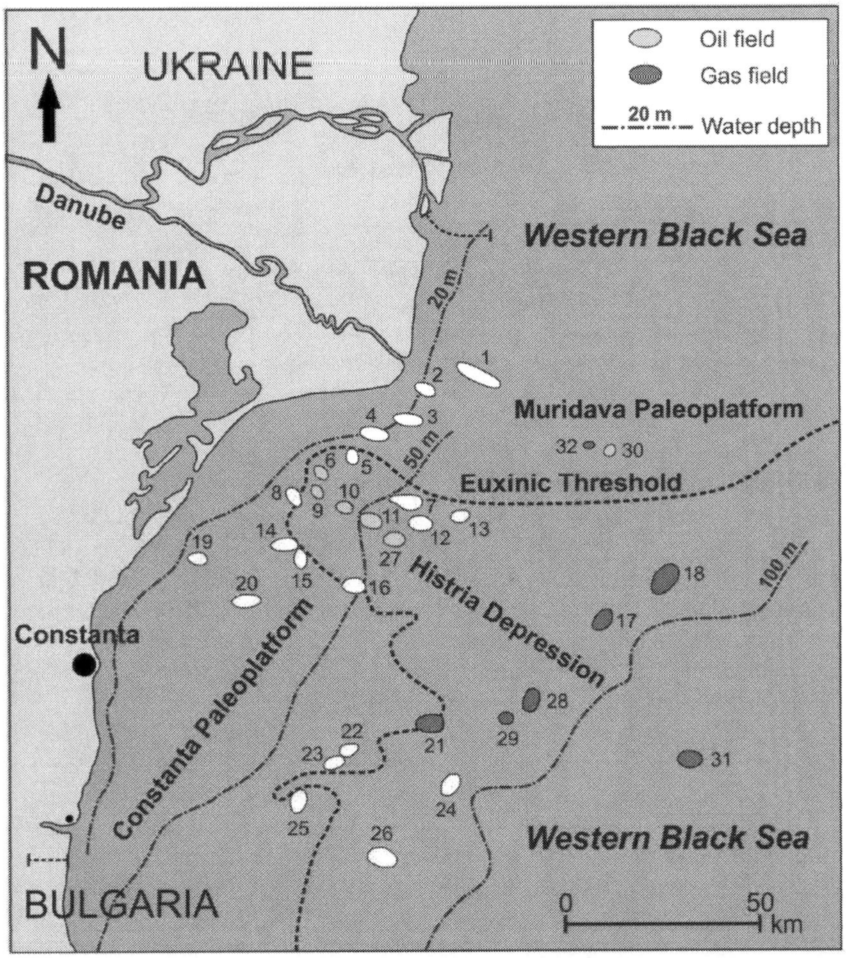


**Fig. 1.** Location map of the Romanian Black Sea shelf showing the main plays and leads. 1 – Sf. Gheorghe, 2 – Pelican, 3 – Sacalin, 4 – Sturion, 5 – Egreta, 6 – Portiţa, 7 – Heracleea, 8 – Venus, 9 – Sinoe, 10 – Lebăda-W, 11 – Lebăda-E, 12 – Minerva, 13 – Albatros, 14 – Iris, 15 – Lotus, 16 – Tomis, 17 – Ovidiu, 18 – Cobălcescu, 19 – Vadu, 20 – Corbu, 21 – Midia, 22 – Meduza, 23 – Neptun, 24 – Neptun-E, 25 – Delfin, 26 – Jupiter, 27 – Pescăruş, 28 – Doina, 29 – Ana, 30 – Muridava (Olimpyska), 31 – Domino, 32 – Eugenia (adapted from [4]).

Gas is reservoired in Pontian – Dacian sands; the source rocks are considered to be the Middle-Late Miocene (Sarmatian or Early Pontian) pelitic deposits and the seals can be represented by all pelitic Pliocene intervals. The traps for the biogenic gas system include drape anticlines, roll-over anticlines, monocline beds affected by listric faults and stratigraphic traps (pinch-outs, depositional fans).

The gas discoveries considered in this study are hosted in four-way closure dome structures by Dacian sands, the reservoirs tops being located at 1094 – 1120 m true vertical depth subsea. The sands are immature, poorly consolidated (little or no authigenic or diagenetic cements), fine to very fine grained, muddy to silty, sometimes thinly bedded. The overall depositional environment is most likely shallow marine, in water depths between 25 and 100 m (lower shoreface to upper offshore regime), with frequent sediment influx being provided from a delta system situated approximately to the NW. The wells drilled on these structures reveal a vertical variability of the reservoir intervals grain size, allowing their separation into a "Sand" upper facies (good reservoir quality, fine sands) of 15 – 28 m thickness and a "Silt" bottom facies (poor reservoir quality, silty sands, silts and muddy silts) of 14 – 47 m thickness. This vertical distinction is more clear in Field A (located towards SW in the exploration block taken into consideration) than in Field B (located towards NE in the exploration block), where the reservoir quality is lower, with higher silt and clay contents and fewer thick sand intervals.

## 3 Data and methodology

We analyzed and interpreted complex data made available by the Romanian oil and gas industry from several exploration and appraisal wells that targeted the Dacian gas-bearing reservoirs. All the wells were drilled with a water-based KCl mud (mud density: 9.6 to 10 lb/gal, equivalent to 1.15 – 1.20 g/cm$^3$) and generally show good hole conditions. The data consisted of wireline geophysical logs, formation tester pressure measurements and fluid samples (RFT – *Repeat Formation Tester* and MDT – *Modular Formation Dynamics Tester*), conventional and special laboratory petrophysical measurements on cores and X-ray diffraction (XRD) analyses.

A total of 105 successful pressure measurements were carried out with wireline formation testers in the wells from Field A and Field B. Reservoir pressures range from 1686 to 1719 psia (116.2 – 118.5 bar) in Field A and from 1600 to 1716 psia (110.3 – 118.3 bar) in Field B. The gas samples collected during wireline testing operations in three wells drilled in these fields show it has very high methane content (99.7 - 99.9 %) and does not contain $H_2S$. At the reservoirs pressure and temperature (38 – 40 °C, measured in the wireline logging and formation tester runs), gas density is approximately 0.08 g/cm$^3$.

XRD analyses performed on 16 core samples extracted from the "Sand" reservoir facies of Field A show a total clay content ranging from 13% to 53% by weight (29% on average). The clays present are predominantly illite + mica, mixed-layer illite/smectite and chlorite, with only subordinate kaolinite observed. The additional electrical conductivity associated with the presence of clay minerals in the reservoirs requires a "shaly sands"-type interpretation of the geophysical well logs, to compute realistic water and gas saturations.

The wireline well logging suites recorded over the 8.5 inch diameter final sections of the wells included total gamma ray, apparent resistivity (*Dual Laterolog*, *Laterolog Array* and focused microresistivity tools), neutron porosity, litho-density (bulk density and photoelectric factor) and compressional sonic transit time. For some of the recently drilled wells, electrical imaging (FMI – *Fullbore Formation Microimager*) and nuclear magnetic resonance (CMR – *Combinable Magnetic Resonance Tool*) logs were also available. Figures 2 and 3 show the geophysical well logs recorded in two exploration wells from Field A and Field B, over the main reservoir intervals (Fig. 2: track 1 – measured depth (MD) and borehole's temperature; track 2 – true vertical depth subsea (TVDSS); track 3 – zonation of the reservoir intervals; track 4 – total gamma ray, caliper and bit size; track 5 – apparent resistivities (*Laterolog Array* curves and flushed zone microresistivity); track 6 – neutron porosity, bulk density and photoelectric factor curves overlay on a limestone compatible scale; track 7 – neutron porosity and bulk density curves overlay on a sandstone compatible scale; track 8 – compressional sonic transit time (slowness or reciprocal velocity); track 9 – nuclear magnetic resonance (CMR) $T_2$ transverse relaxation time distribution; track 10 – formation tester pressure readings).

In the wells drilled in Field A (Fig. 2), the main reservoir unit consists of a 26 – 27 m thick upper interval with significant gas crossover on the density-neutron overlay, deep resistivities reaching 170 – 384 Ω·m, free fluid indication on the CMR's $T_2$ distribution (large $T_2$ values), and a strong compressional sonic $\Delta t$ response of 170 – 200 μs/ft (compared to the 130-140 μs/ft slowness recorded in the adjacent non-reservoir intervals), indicating gas-bearing formations.

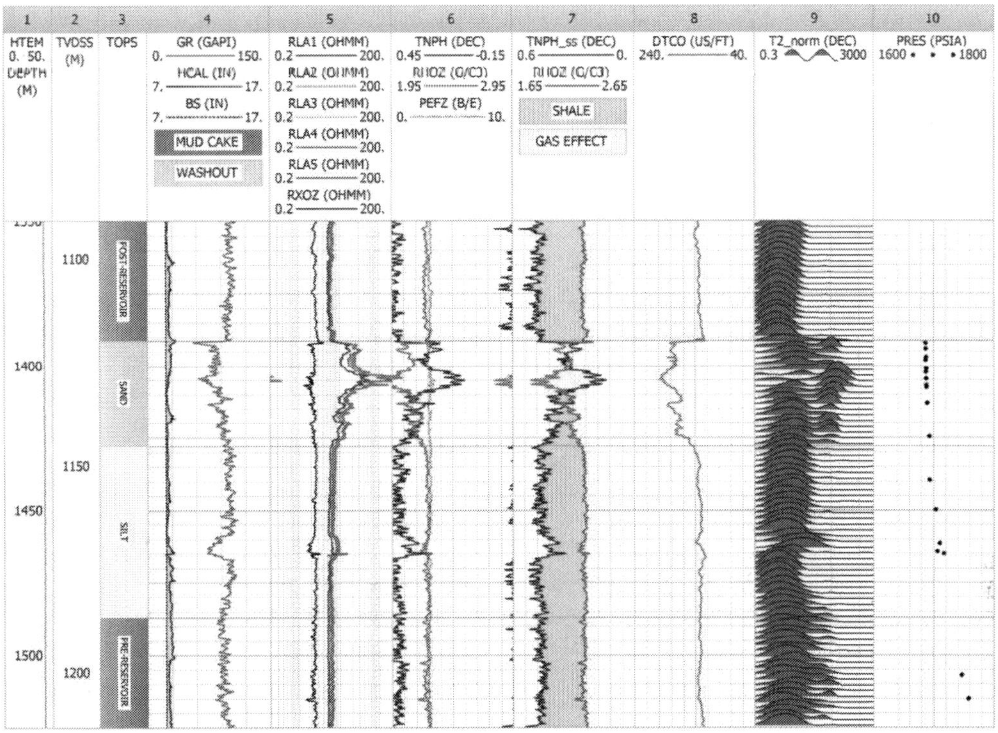


**Fig. 2.** Wireline geophysical well logs recorded in an exploration well from Field A.

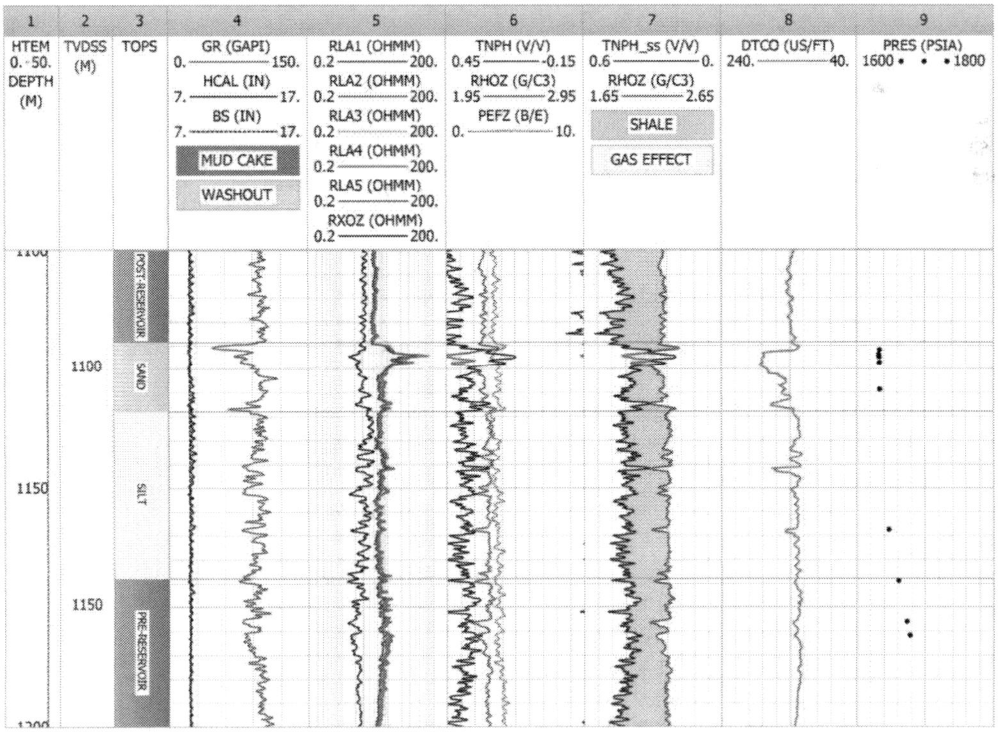


**Fig. 3.** Wireline geophysical well logs recorded in an exploration well from Field B.

Total gamma ray logs (*GR*) recorded in the wells shows an unusually low contrast between reservoir and non-reservoir intervals. The increased radioactivity of the reservoir rocks may be explained by a higher adsorption of radionuclides in the very fine sands and silts and, also, by the presence of potassium-bearing mica. The lower ("Silt") part of the reservoir unit, 38 – 41 m thick, displays no density-neutron crossover, slight and local resistivity and $T_2$ time increases (little free fluid), possibly indicating small gas saturations at particular depth levels. Generally, the *GR* log subtle trends observed along the reservoir intervals are of coarsening-upward type (prograding sequences).

In the wells drilled in Field B (Fig. 3) the "Sand" main reservoir unit, 15 – 28 m thick, appears with the same general well log signature as in Field A, i.e. with a suppressed *GR* response, few and thinner intervals with density-neutron gas crossover and much lower deep resistivities (maximum readings of 16 – 82 Ω·m), but again with a distinct sonic response suggesting gas-bearing formations ($\Delta t$ reaching 180 – 190 μs/ft in the main reservoir compared to 125 – 140 μs/ft in the non-reservoir intervals).

The wireline logging data processing and interpretation consisted of:

1) Clay volume ($V_{clay}$) estimation by using a combination of gamma ray and density-neutron clay indicators. A linear clay volume estimated from the *GR* log ($V_{clay,GR}$) was calibrated against a clay volume estimated from the density-neutron crossplot ($V_{clay,DN}$) in order to obtain $V_{clay,GR} \approx V_{clay,DN} \approx 1$ (100%) at the depth levels with good hole conditions and a maximum and consistent separation between the neutron and density curves ($\phi_N$ - $\phi_D$) on a sandstone compatible porosity scale. The corresponding expressions are

$$V_{clay,GR} = (GR - GR_{clean}) / (GR_{clay} - GR_{clean}) \quad (1)$$

$$V_{clay,ND} = (\phi_N - \phi_D) / (\phi_{Nclay} - \phi_{Dclay}) \quad (2)$$

$$\phi_D = (\delta_{ma} - \delta) / (\delta_{ma} - \delta_f) \quad (3)$$

$$\phi_{Dclay} = (\delta_{ma} - \delta_{clay}) / (\delta_{ma} - \delta_f) \quad (4)$$

where: $GR$ – measured gamma ray radioactivity (API), $GR_{clean}$ – gamma ray radioactivity of clean reservoirs intervals (API), $GR_{clay}$ – gamma ray radioactivity of clays (API), $\phi_N$ – measured neutron porosity, corrected for sandstone lithology (V/V), $\phi_{Nclay}$ – clay neutron porosity (V/V), $\phi_D$ – density porosity computed for a sandstone matrix (V/V), $\phi_{Dclay}$ – clay density porosity (V/V), $\delta$ – measured bulk density (g/cm$^3$), $\delta_{ma}$ – sandstone matrix density (g/cm$^3$), $\delta_{clay}$ – clay density (g/cm$^3$), $\delta_f$ – pore fluid (mud filtrate) density (g/cm$^3$). The $V_{clay,GR}$ calibration with respect to the reference $V_{clay,DN}$ can be achieved via an adequate selection of the clean and clay *GR* responses. The density-neutron combination has the advantage of not depending on formations radioactivity, but $V_{clay,DN}$ from Eq. (2) will not work in gas-bearing intervals. Once calibrated against $V_{clay,DN}$ on clay/shale or water-bearing intervals, $V_{clay,GR}$ can be used as a valid clay volume estimator for the entire interpreted intervals, because it is not affected by light hydrocarbon (gas) effects.

2) Effective porosity ($\phi$) estimation by using a density-neutron crossplot porosity model appropriate for shaly sands. The matrix (grain) densities $\delta_{ma}$ of 2.67 – 2.69 g/cm$^3$ used in the calculations were obtained from measurements on core plugs. The clay parameters were statistically selected from the logs, resulting neutron porosities $\phi_{Nclay}$ = 0.45 – 0.50 and densities $\delta_{clay}$ = 2.25 – 2.26 g/cm$^3$.

3) Estimation of formation water resistivity ($\rho_w$) by using effective porosity vs. true resistivity Pickett crossplots [6], i.e. $log(\phi) = f(log(\rho_t))$, and approximating $\rho_t$ with the deepest investigation apparent resistivity curves *RLA5* (Fig. 4). For the wells of Field A we obtained $\rho_w$ = 0.5 – 0.6 Ω·m at formations temperature, corresponding to salinities of 6600 – 9100 ppm equivalent NaCl.

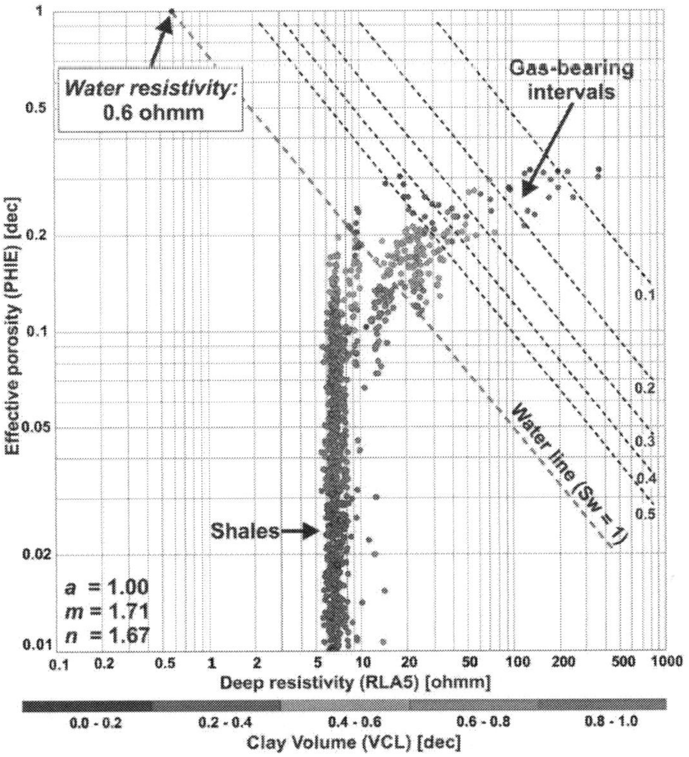


**Fig. 4.** Effective porosity (log-derived) vs. deep resistivity crossplot for an exploration well from Field A. The $S_w = 1$ "water line" is defined as passing through the change of trend of the resistivity – porosity data (shales vs. reservoir rocks). The very limited effective porosity variation range of the reservoir rocks does not allow a clear linear trend to be observed for the water-bearing levels.

For the wells of Field B we obtained $\rho_w = 0.9 - 1.2$ Ω·m at formations temperature, corresponding to salinities of 3400 – 4700 ppm equivalent NaCl. The relatively fresh, low salinity, formation waters suggest that the clay content has a significant effect upon the bulk rock resistivities, by providing an additional conduction path for the electrical current.

4) Estimation of water saturations ($S_w$, $S_{xo}$) and total, movable and residual hydrocarbon saturations ($S_h$, $S_{hm}$, $S_{hr}$) in the reservoir intervals by using a shaly sands saturation model [7], in the effective porosity system. Clay resistivities were statistically selected from the logs ($\rho_{clay} = 5.2 - 8.5$ Ω·m) and the water saturations were computed using Archie parameters [8] measured on core plugs, at a confining pressure representative for reservoir conditions. For Field A the average values of these parameters were $a = 1.0$ (the so-called tortuosity factor), $m = 1.71$ (cementation exponent), $n = 1.67$ (saturation exponent) and for Field B the average parameters were $a = 1.0$, $m = 1.54$, $n = 1.33$.
5) Processing the CMR $T_2$ transverse relaxation time distribution, in order to compute clay-bound water, capillary-bound water and free fluid volumes ($T_2$ cutoffs of 3 ms and 33 ms, typical for sands / sandstones, were used to separate the three types of pore fluids), as well as a CMR-derived water saturation.
6) Linear regression analysis on the good quality formation pressure data, in order to derive the pressure gradients ($\Delta p/\Delta H$, where $p$ – measured formations pressure, $H$ – true vertical depth), the formation fluids in-situ densities and to identify the probable gas-water contacts (GWC).

7) Establishing petrophysical cutoffs, in order to delineate net reservoir and net pay intervals. The cutoffs were obtained by analyzing the relationships between the permeability ($k$) measured on core plugs at appropriate confining pressure and the log-derived $\phi$ and $S_w$. For $k = 0.1$ mD as a minimum acceptable limit appropriate for gas, we obtained $\phi$ cutoff $\approx 8 - 11\%$ and $S_w$ cutoff $\approx 70\%$, for both Field A and Field B.

## 4 Results and discussion

Figure 5 illustrates an example of petrophysical interpretation results for the wireline logs recorded in the exploration well from Field A (Fig. 2): track 1 – measured depth (MD) and borehole's temperature; track 2 – true vertical depth subsea (TVDSS); track 3 – zonation of the reservoir intervals; track 4 – total gamma ray, caliper and bit size; track 5 – water saturations in the uninvaded and flushed zones ($S_w$, $S_{xo}$); track 6 – bulk volumes of formation water, movable and residual (immobile) gas; track 7 – lithology analysis (volume fractions of clay, silt, sand and effective porosity); track 8 – oriented borehole electrical image (light tones: high resistivity, dark tones: low resistivity); track 9 – processed nuclear magnetic resonance (CMR) data showing the volumes of clay-bound water, capillary-bound water and free fluids; track 10 – computed pressure gradients and formation fluids densities; tracks 11 and 12 – delineation of net reservoir and net pay intervals using the cutoffs $\phi = 10\%$ and $S_w = 70\%$, corresponding to $k = 0.1$ mD.

Two distinct gas-bearing reservoirs resulted from interpretation in the "Sand" and "Silt" units. By applying the $\phi$ and $S_w$ cutoffs, these reservoirs are located on the 1391.6 – 1425.5 m MD (1119.8 – 1143.6 m TVDSS) and, respectively, 1458.6 – 1466.7 m MD (1166.7 – 1172.3 m TVDSS) intervals. The main, 23.8 m thick, upper reservoir has an average $\phi$ of 15.6%, locally reaching 31.9%, the minimum computed $S_w$ being 4.4% (a maximum gas saturation of 95.6%). The 5.6 m thick secondary reservoir has an average $\phi$ of 14.8%, locally reaching 24.2%, and a minimum computed $S_w$ of 49.7%.

As observed in Fig. 2 and Fig. 5, the maximum recorded resistivities and the corresponding maximum computed gas saturation and gas volume do not occur at the top of the main "Sand" reservoir, but 13 to 15 m lower. The resistivity suppression in the upper part of the main reservoir, especially on the 1392 – 1404 m MD interval, is due to a large amount of capillary-bound water (up to 14 – 15% volume) at the reservoir's top, as shown by the CMR results. This abnormal amount of water, which is irreducible (immobile), can be explained by the presence of a siltstone interval at the top of the generally coarser-grained "Sand" reservoir. Such siltstones typically have low resistivities but can produce clean (water-free) hydrocarbons.

Another significant cause of resistivity suppression comes from the thinly laminated nature of some reservoirs, especially in Field B (Fig. 3), as indicated by the drill cores photographs and descriptions. The interbedded thin and resistive sands/silts and conductive mudstones, with millimeter to decimeter bed thicknesses, cannot be properly resolved by the electrical logging tools and their resistivities are averaged. This leads to a severe underestimation of the thin reservoir beds true resistivities (maximum recorded resistivities in the main reservoir intervals of Field B are one order of magnitude smaller than the ones measured in Field A) and to computing unrealistically high water saturations and low gas saturations. In this regard, the minimum $S_w$ values obtained for the "Sand" interval in the wells of Field B range from 9.1% to 34.9%, whereas in Field A they range from 4.4% to 8.2%. One way of mitigating this problem would be to investigate the ability of modern tensor / 3D or azimuthal resistivity tools (such as Schlumberger's *Rt Scanner* or the ARI – *Azimuthal Resistivity Imager*) to resolve the thinly laminated sand beds and determine more realistic true resistivities for water saturation determination. Another way would be a deconvolution processing of resistivity logs, in order to improve their vertical resolution.

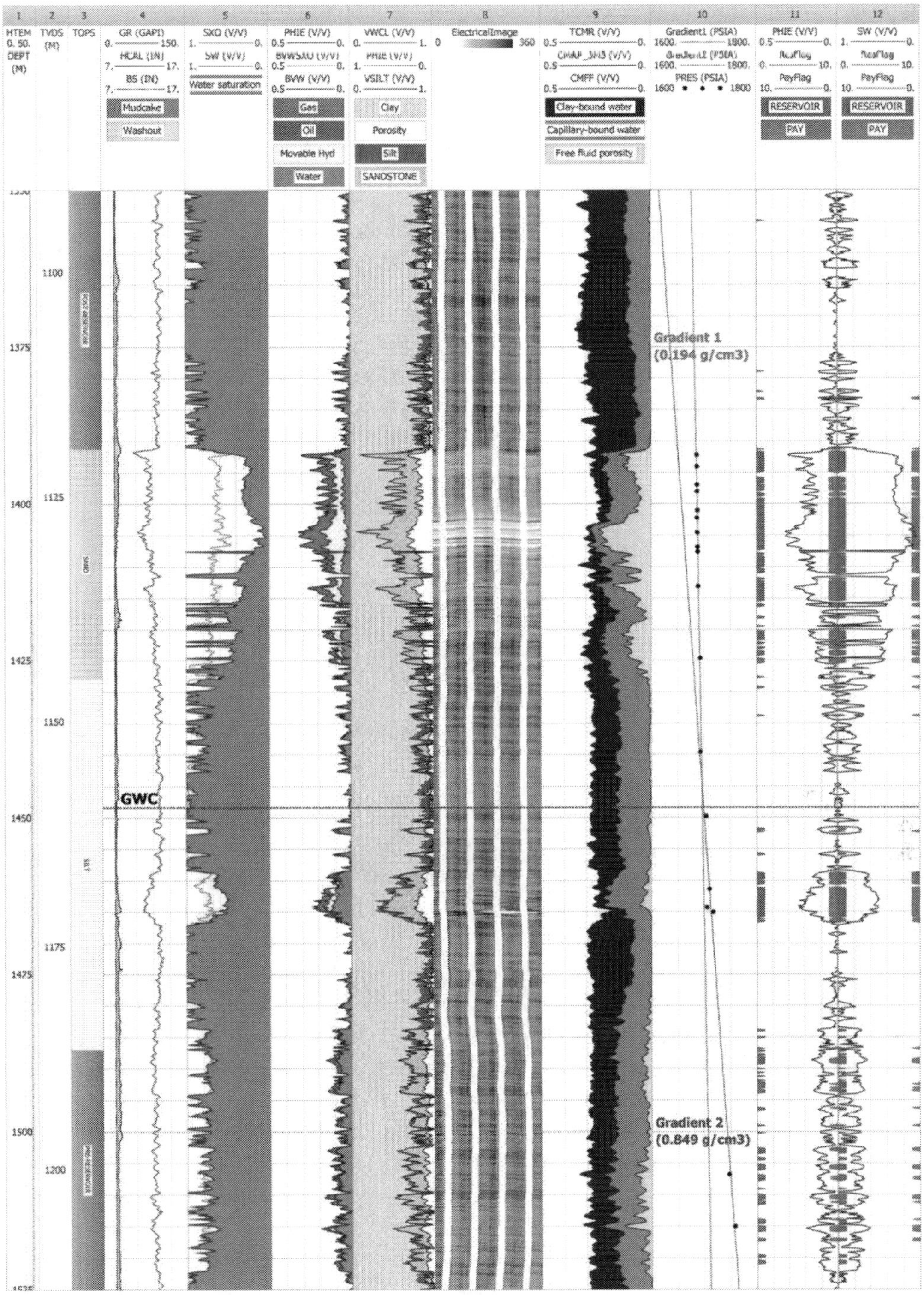


**Fig. 5.** Petrophysical interpretation of the wireline well logs recorded in an exploration well from Field A. The GWC level marks the gas-water contact suggested by the intersection of the pressure gradients at 1448.4 m MD (1159.6 m TVDSS). This apparent GWC is located between two gas-bearing reservoirs resulted from the interpretation, at a depth where the processed nuclear magnetic resonance data show no indication of free fluids, but only bound (immobile) water.

The pressure gradients analysis in Fig. 5 indicates the presence of two fluids and corresponds to the hydrocarbon pressure gradient and the hydrostatic pressure gradient. The densities of these fluids, computed from the pressure gradients, are $\delta_{f1}$ = 0.194 g/cm$^3$ (gas) and $\delta_{f2}$ = 0.849 g/cm$^3$ (water), higher and, respectively, lower than the true densities of gas and formation water in the studied fields. For only one out of six analyzed wells a correct gas density (0.083 g/cm$^3$) was obtained from the pressure gradients, in the other cases the resulted gas densities were too high (> 0.1 g/cm$^3$). Formation water estimated densities were too low (< 1 g/cm$^3$) for the wells in Field A, but realistic for the wells in Field B, ranging from 1.014 to 1.038 g/cm$^3$. A possible cause for the differences between true and estimated fluid densities may consist in the formation testers not reading pressures corresponding to pure fluids, but to mixtures of gas and water or water and gas. In this case, the gas-water contact depth inferred solely from the pressure gradients intersection may be significantly inaccurate, leading to errors in the volumetric estimation of gas resources. This GWC uncertainty, which cannot be ruled out, could also manifest itself in the form of pressure gradients intersection being located at a depth level with no free fluids indications on the results of nuclear magnetic resonance log, i.e. within an impermeable formation.

## 5 Conclusions

We analyzed complex borehole data (wireline geophysical well logs, formation pressures and fluid samples, petrophysical measurements on cores and X-ray diffraction measurements) available for several exploration and appraisal wells drilled on the Romanian continental shelf of the Black Sea, which targeted Early Pliocene (Dacian) gas-bearing reservoirs. The analysis revealed some particular challenges related to the petrophysical evaluation of these formations, such as the low *GR* contrast of the reservoirs, the frequent suppression of resistivity logs readings and the uncertainty of gas-water contacts identification.

## References

1. G. Ionescu, M. Sisman, R. Cataraiani, *Source and reservoir rocks and trapping mechanism on the Romanian Black Sea shelf*. In: C. Dinu and V. Mocanu (eds.), *Geology and Tectonics of the Romanian Black Sea Shelf and its Hydrocarbon Potential*, Bucharest Geoscience Forum, Special Volume **2**, 67–83 (2002)
2. C. Dinu, H.K. Wong, D. Ţambrea, L. Maţenco, *Stratigraphic and structural characteristics of the Romanian Black Sea shelf*, Tectonophysics **410**, 417– 435 (2005)
3. I. Moroşanu, *Romanian Continental Plateau of the Black Sea: Tectonic-Sedimentary Evolution and Hydrocarbon Potential*, Oscar Print Publishing House, Bucharest, Romania, 176 pp., ISBN 978-973-668-167-7 (2007)
4. I. Moroşanu, *The hydrocarbon potential of the Romanian Black Sea continental plateau*, Romanian Journal of Earth Sciences **86**, 2, 91–109 (2012)
5. G. Georgiev, *Geology and Hydrocarbon Systems in the Western Black Sea*, Turkish Journal of Earth Sciences **21**, 723–754 (2012)
6. G.R. Pickett, *Pattern recognition as a means of formation evaluation*, The Log Analyst **14**, 4, 3–11 (1973)
7. A. Poupon, J. Leveaux, *Evaluation of water saturation in shaly formations*, The Log Analyst **12**, 4, 3–8 (1971)
8. G.E. Archie, *The electrical resistivity log as an aid in determining some reservoir characteristics*, Transactions of the AIME **146**, 1, 54–62 (1942)

# Structural characterization of the metamorphic basement from North Dobrogean Promontory (Romania) using geophysical well logging

*Constantin-Laurian* Ciupercă[1,*], *Bogdan Mihai* Niculescu[2] , and *Bogdan Marian* Popina[3]

[1] WEATHERFORD International
[2] University of Bucharest, Faculty of Geology and Geophysics, Department of Geophysics, Bucharest, Romania
[3] OMV PETROM S.A., Bucharest, Romania

**Abstract.** Metamorphic processes, leading to mineralogical and structural changes of the rocks in response to physical (pressure, temperature) and chemical conditions, can be associated with the development of sin-metamorphic or post-metamorphic fractures. The post-metamorphic ones are directly related to tectonic stress. In addition, tectonic stress may cause the reorientation of minerals on a direction perpendicular to the stress direction, generating foliations (schistosity).
This paper presents a structural analysis of the pre-Alpine metamorphic basement and its relations with the sedimentary cover by using geophysical data recorded in an exploration well located in the North Dobrogean Promontory (Romania). The analysis was based on quad-combo wireline logs, spectral gamma ray, sonic cross-dipole and borehole electrical imaging data. The imaging analysis allowed the identification and characterization of metamorphic foliations, sedimentary bedding, natural fractures and the determination of tectonic stress orientation. A brittleness index was computed by means of elastic parameters derived from density and sonic compressional and shear logs. Also, a fracture intensity characterization by using fracture area per volume of rock was conducted. The integration of geophysical well loging with mud logging and drilling data allowed us to carry out a detailed analysis of the metamorphic basement in the studied area.

**Keywords:** fractures, metamorphic basement, North Dobrogean Promontory, structural characterization, well logging

## 1 Introduction

Metamorphic rocks, resulted from transformation of pre-existing sedimentary or magmatic rocks in response to physical (pressure, temperature) and chemical conditions, may represent reservoirs for hydrocarbons only if they are affected by fracture systems that allow the

* Corresponding author: ciupercaconstantin@yahoo.com

hydrocarbons accumulation and migration. Many systems of natural fractures can coexist in the metamorphic reservoirs, having various origins and being formed in different stages of metamorphism or pre-metamorphism. Epimetamorphic rocks may preserve the natural fracture system of the pre-existing rocks. Other type of natural fractures can occur after a metamorphic basement exondation, being subject to subsequent alteration and dissagregation. Generally, the post-metamorphic fractures are directly related to tectonic stress. In addition, tectonic stress may cause the reorientation of minerals in a direction perpendicular to the stress direction, generating foliations (schistosity). The erosion of metamorphic rocks after a basement exondation may lead afterwards to buried relief structures, the sedimentary cover rocks showing differential compaction processes. These can be associated with the creation of domes over the crystalline buried relief.

In this study we present a structural analysis of the pre-Alpine metamorphic basement and its relations with the sedimentary cover, by using geophysical data recorded in an exploration well located in the North Dobrogean Promontory (Romania). The analysis was based on quad-combo openhole wireline logs (total gamma ray, electrical resistivity curves with investigation radii ranging from very shallow to deep, bulk density, thermal neutron porosity), spectral gamma ray, sonic cross-dipole and borehole electrical imaging data (CMI – *Compact MicroImager*). The borehole electrical image analysis allowed the identification and characterization of metamorphic foliations, sedimentary bedding, natural fractures and the determination of tectonic stress orientation. A brittleness index, as proposed by Rickman et al. [1] and Gray et al. [2], was computed by means of elastic parameters (Young's modulus and Poisson's ratio) derived from density and sonic compressional and shear velocity logs. Also, a fracture intensity characterization by using the fractures area per unit volume of rock, as defined by Dershowitz and Herda [3], was conducted. The integration of geophysical well logging with mud logging and drilling data allowed us to carry out a detailed analysis of the metamorphic basement in the study area.

## 2 Geological and tectonic background

The study area is located within the North Dobrogean Promontory [4, 5], a unit delimited by the regional Peceneaga-Camena Fault and the Central Dobrogea unit towards S and SW, and by the regional Sfântu Gheorghe Fault and the Pre-Dobrogean Depression towards N and NE (Fig. 1 and Fig. 2). The North Dobrogean Promontory represents a prolongation towards NW, underneath Neogene deposits, of the intracratonic North Dobrogean Orogen - a fold-thrust zone of highly deformed Proterozoic, Paleozoic and Mesozoic metamorphic, igneous and sedimentary rocks, which are outcropping E and S of Danube river, in North Dobrogea unit [5]. The North Dobrogean Orogen is the westernmost part of a Cimmerian folded belt extending through southern Crimea to the Asian Cimmerides. It is a nappe complex, consisting of three thrust sheets with northern vergence (Fig. 2): the Niculiţel nappe with oceanic affinities, and the Măcin and Tulcea nappes with a Late Proterozoic – Cambrian metamorphic basement [8].

The North Dobrogean Promontory resulted from the sinking of the NW part of North Dobrogean Orogen starting with the Paleogene and consists of a basement (*1*: the Orliga, Megina and Boclugea meso- and epimetamorphic groups [5, 9, 10, 11], comprising amphibolites, gneisses, micaschists, meta-tuffs, quartzites, phyllites and chlorite-sericite schists; *2*: acidic and basic magmatic rocks such as granites, pyroclastites, basalts and gabbros; *3*: Paleozoic sedimentary formations, with relatively limited development) and a post-tectonic Neogene sedimentary cover with Badenian (Middle Miocene) – Romanian (Late Pliocene) and Quaternary deposits.

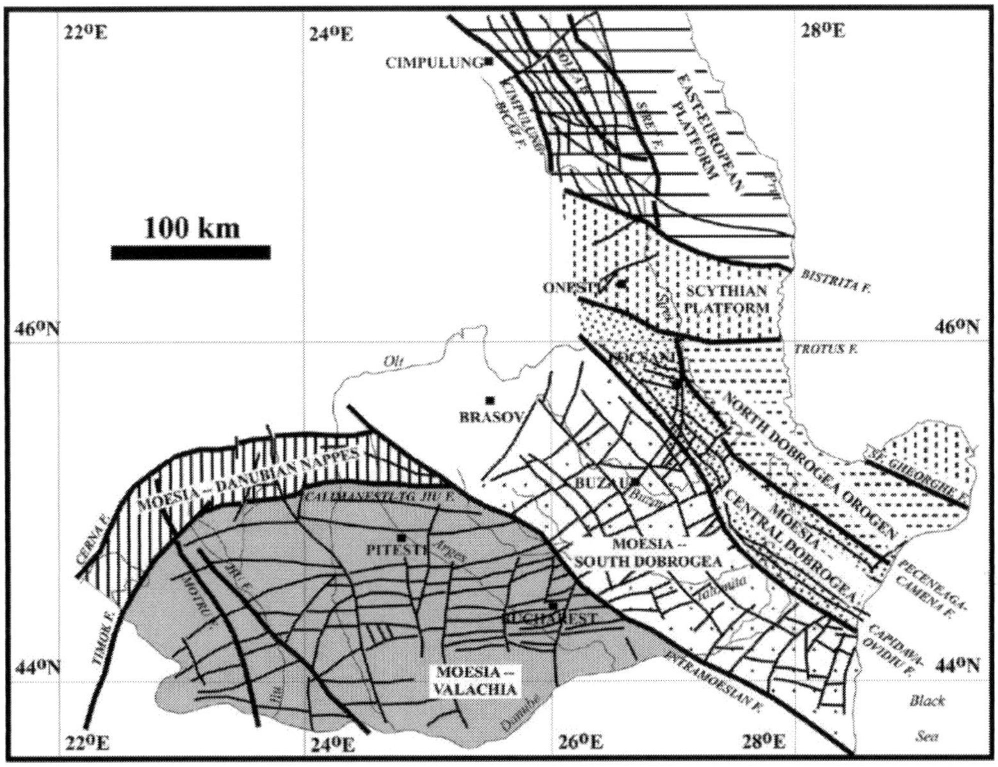


**Fig. 1.** Simplified tectonic map showing the main units of the South and East Carpathians foreland (East-European Platform, Scythian Platform and Moesian Platform) and the North Dobrogean Orogen and Promontory (based on [6]).

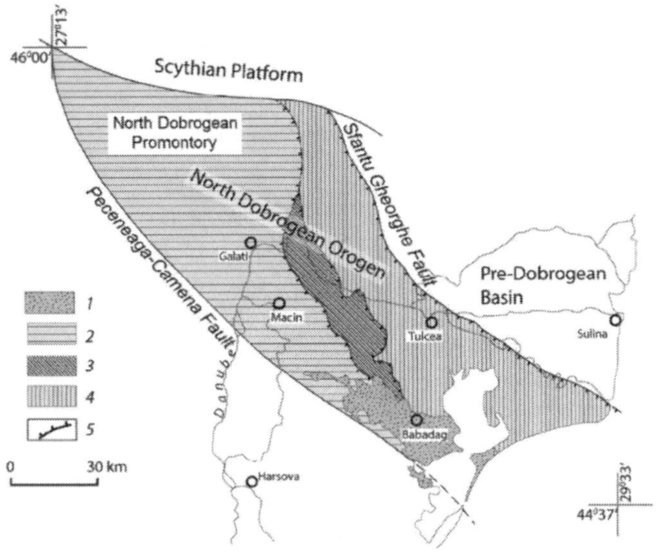


**Fig. 2.** Location of North Dobrogean Promontory and structure of the North Dobrogean Orogen, as known from outcrops and drill-holes. *1* – Babadag late-Cretaceous cover (post-Peceneaga-Camena fault); *2* – Măcin nappe; *3* – Consul and Niculițel nappes; *4* – Tulcea nappe; *5* – Sfântu Gheorghe fault (modified from [5] and [7]).

Regarding the crystalline basement of North Dobrogean Orogen (and North Dobrogean Promontory), some preferential orientations can be observed. For the Orliga medium-grade metamorphic group (considered the oldest metamorphic sequence that outcrops in North Dobrogea area), the identified structures have a E-W orientation, being refolded afterwards on a NW-SE direction. The Megina and the Boclugea medium-grade and, respectively, low-grade metamorphic groups, are characterized by structures with predominantly NW-SE orientations, subsequently refolded by younger deformation phases [5].

In the area of the analyzed exploration well, the Neogene cover of the North Dobrogean Promontory consists mainly of Pliocene deposits, overlying transgressively in a discordant manner the predominantly crystalline basement. Due to the irregular topography of the basement, the overlaying Pliocene formations frequently appear as domes separated by depressionary areas or by faults. From a lithological standpoint, the Pliocene comprises alternations of fine-grained unconsolidated sandstones, marly sandstones, marls, sandy marls and shales, as revealed by the mud logging records and cores taken from exploration wells drilled in this area.

## 3 Geophysical Well Logging Data: Sedimentary Cover – Crystalline Basement Relationship

The composite plot from Fig. 3 shows the conventional openhole logs and the CMI borehole electrical image - with static normalization of the recorded microconductivity data - for an exploration well drilled in the study area, along with some significant data processing results such as the computed brittleness index $B$ resulting from the elastic properties of the rocks [1, 2] and the volumetric fracture intensity $P_{32}$ [3].

The $P_{32}$ parameter, with $m^{-1}$ units, is defined as the total area of fractures per unit volume, i.e.

$$P_{32} = A_f / V \tag{1}$$

where $A_f$ is the total area of the fractures and $V$ is the volume of the region being measured. $P_{32}$ is direction- and scale-independent and it can be used to classify a given rock mass, being the preferred measure of fracture intensity [12].

The brittleness index $B$, expressed as a percentage, was computed using the expressions

$$B_E = 100 \cdot (E - E_{min}) / (E_{max} - E_{min}) \tag{2}$$

$$B_\nu = 100 \cdot (\nu - \nu_{max}) / (\nu_{min} - \nu_{max}) \tag{3}$$

$$B = (B_E + B_\nu) / 2 \tag{4}$$

where $E$ is Young's modulus, $\nu$ is Poisson's ratio, $B_E$ is the brittleness estimated from Young's modulus, $B_\nu$ is the brittleness estimated from Poisson's ratio and the subscript *min* and *max* denote minimum and maximum values [2].

The boundary between the overlying Pliocene sedimentary rocks and the metamorphic crystalline basement is represented by a nonconformity at approximately 774.6 m depth (Fig. 3 and Fig. 4). This is clearly marked both on the conventional well logs (increased resistivities and bulk density below the contact, decreased neutron porosity as well as decreased compressional and shear sonic transit times below the contact, indicating compact rocks) and on the color-coded borehole electrical image (rapid change from conductive rocks – dark tones to highly-resistive rocks – light tones). Also, the Pliocene – crystalline basement nonconformity is evidenced by a sharp increase of the brittleness index $B$, whose variations in the basement are generally correlated with the fracture intensity $P_{32}$. Usually, a nonconformity suggests a depositional hiatus, in which a long period of erosion occurred prior to deposition of sediments.

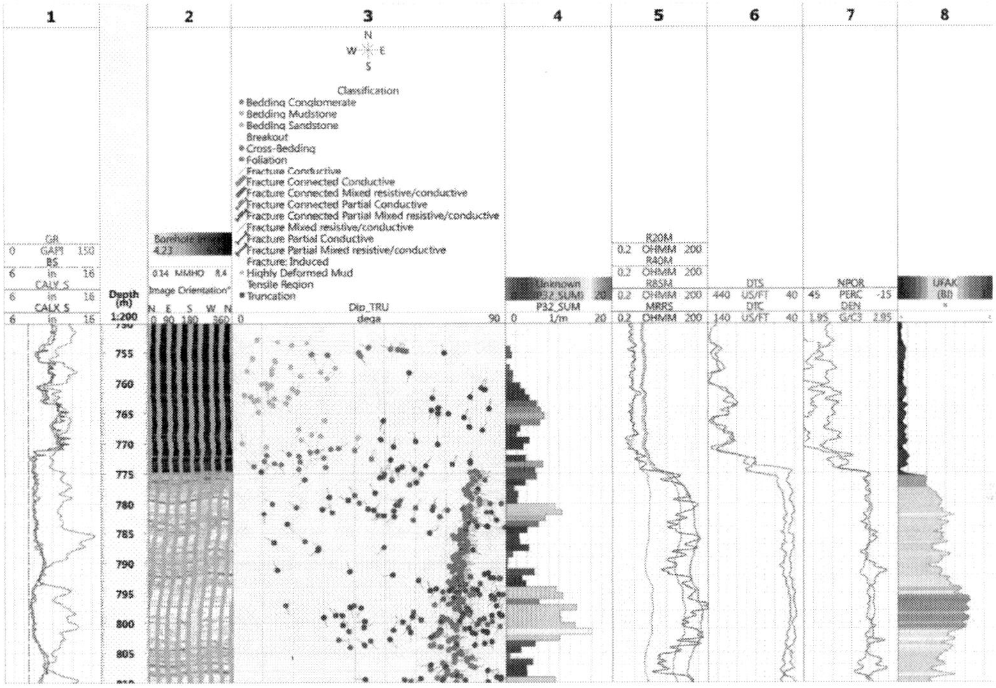

**Fig. 3.** Composite plot showing the basement – sedimentary cover nonconformity. Tracks: 1 – auxiliary curves; 2 – CMI statically normalized image; 3 – dip / azimuth tadpoles plot; 4 – total fractures ($P_{32}$); 5 – resistivity curves; 6 – sonic curves; 7 – nuclear curves; 8 – brittleness index ($B$).

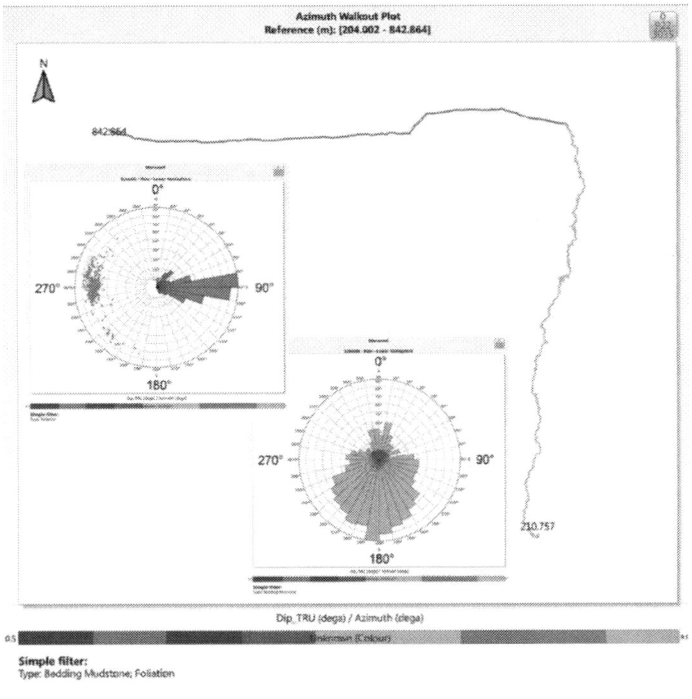


**Fig. 4.** Azimuth Walkout Plot showing the basement – sedimentary cover nonconformity.

Just above the crystalline basement rocks, a conglomerate layer of about 4 m thickness is encountered, as revealed by the borehole microresistivity image texture. This ortho-conglomerate formation can be attributed to the dewatering and alteration of the basement rocks. The fracture intensity parameter is increased in the conglomerate area and another 10 m above. Fracturing processes in this area could be associated with the creation of domes over the crystalline buried relief.

The azimuth vector plot ("walkout plot") in Fig. 4 was constructed by cumulatively stacking unit vectors up through the investigated well. The azimuth of each vector is derived from the dip data ("tadpoles") resulted from the borehole electrical image analysis. The two inset plots / stereonets in Fig. 4 illustrate the distribution of dip angles and dip azimuths (using azimuth frequency plots with 10° bins) of the planar features identified as shale bedding in the overlying Pliocene deposits and, respectively, interpreted as metamorphic foliations or schistosity in the crystalline basement. The basement is characterized by high-angle dips (frequently 60° to 80°), with predominantly E (≈ 90°) dip azimuth trend. This eastward dip azimuth implies an approximate N-S strike of the foliations / schistosity. The shale bedding in the sedimentary formations shows very low-angle dips and a general dip azimuth trend towards S to SSW (with a much larger variability of the azimuth values). On the azimuth walkout plot, the Pliocene cover – crystalline basement contact / nonconformity is evidenced by the sudden change in the dip azimuths of the planar features picked on the borehole electrical image.

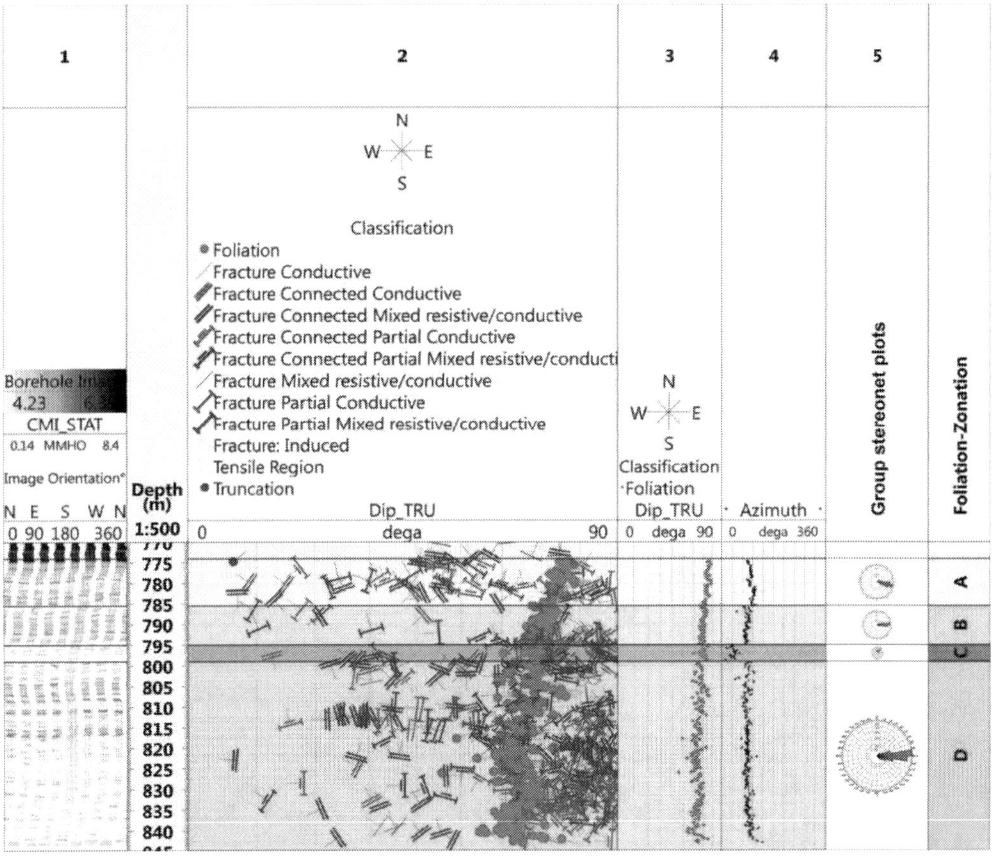


**Fig. 5.** Borehole electrical image of the crystalline basement, with manually picked dips and azimuths of the various planar features identified on the image. Tracks: 1 – CMI statically normalized image; 2 – dip / azimuth tadpoles plot; 3 – foliation dip; 4 – foliation azimuth; 5 foliation Rose plot.

## 4 Geophysical Well Logging Data: Crystalline Basement Analysis

Figure 5 presents the crystalline basement interval investigated using borehole electrical imaging (statically normalized image in track 1; lighter tones – low conductivity / high resistivity, darker tones – high conductivity / low resistivity), with the planar features picked from the image (the tadpoles in track 2). These features correspond to metamorphic foliation / schistosity; other picked features correspond to natural or induced fractures within the basement. Tracks 3 and 4 in Fig. 5 show the dips and dip azimuths of the features classified as foliations / schistosity.

A sudden change in azimuth is observed in zone C (795 to 799 m depth interval), from 90° (E) to 45° (NE), with respect to adjacent zones. Possible explanations for such an event may include the occurrence of a local metamorphic differentiation or the presence of another protolith. In Fig. 3, this zone corresponds to the depth interval with the highest values of the fracture intensity $P_{32}$ (track 4) and of the brittleness index $B$ (track 8).

Azimuth vector plots such as the one in Fig. 6 are useful in identifying even subtler azimuth changes throughout the crystalline basement interval, as the one at 785.5 m depth, used to delineate the upper boundary of zone B / lower boundary of zone A from Fig. 5. Such zonations can be carried out based on the dip, azimuth, fracture intensity and fracture type variability.

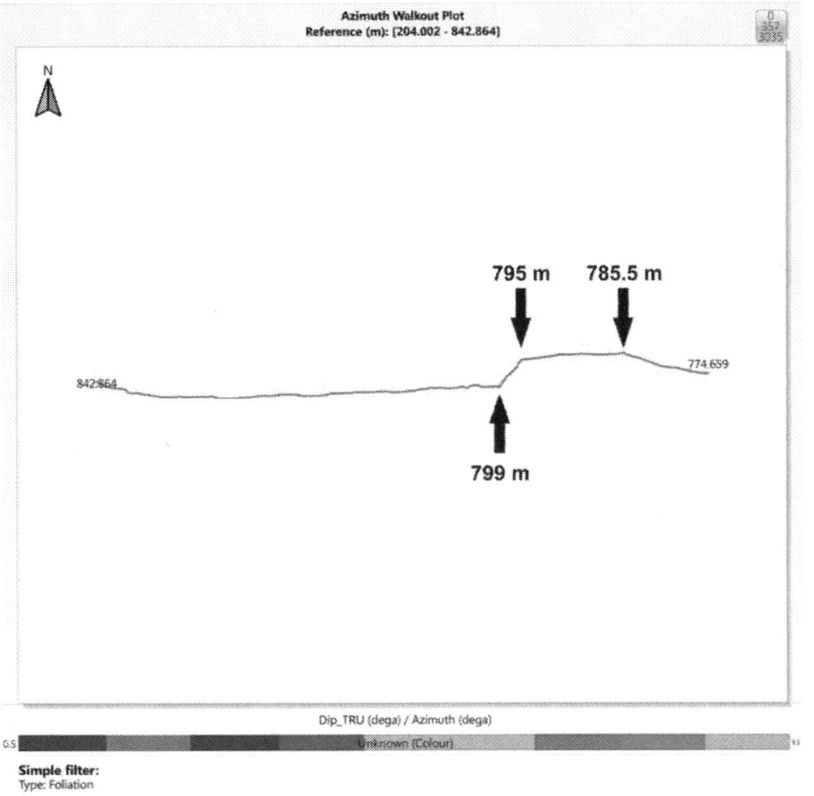


**Fig. 6.** Azimuth Walkout plot corresponding to the metamorphic foliation or schistosity within the crystalline basement. The marked depths correspond to the boundaries of zones B and C from Fig. 5.

## 5 Geophysical Well Logging Data: Tectonic Stress Analysis

Borehole breakouts and drilling-induced fractures are important indicators of horizontal stress orientation. The breakouts are stress-induced enlargements of the wellbore cross-section and occur when the stresses around the borehole exceed those required to cause compressive failure of the borehole wall [13]. The enlargement of the wellbore is caused by the development of shear fractures sub-parallel to the borehole wall, causing pieces of the wall to spall off and creating an ovalized borehole. The long axes of borehole breakouts are oriented approximately perpendicular to the maximum horizontal compressive stress ($S_H$) orientation, i.e. in the minimum horizontal stress ($S_h$) direction [14]. Drilling-induced fractures are created when the stresses concentrated around a borehole exceed those required to cause tensile failure of the wellbore wall [15]. They usually appear as narrow sharply defined features that are sub-parallel or slightly inclined to the borehole axis in the vertical wells and are not associated with significant borehole enlargement in the fractures direction. Such fractures develop approximately parallel to the $S_H$ orientation [16].

Figure 7 shows the elements which allow the identification of the present-day tectonic stress direction. The oriented 8-arm caliper data of the CMI tool were used to locate borehole ovalizations corresponding to the breakouts orientation $S_h$ (track 6 – "Potato Plot" representation). The long axes of the ovalizations are oriented N-S to NNE-SSW, suggesting an approximately E-W or WNW-ESE direction of $S_H$. The drilling-induced fractures and the tensile zones were identified too on the electrical image (Fig. 8 – tracks 2, 3, 4) and indicate an approximate WNW-ESE direction of $S_H$, confirming the stress direction obtained from borehole breakouts.

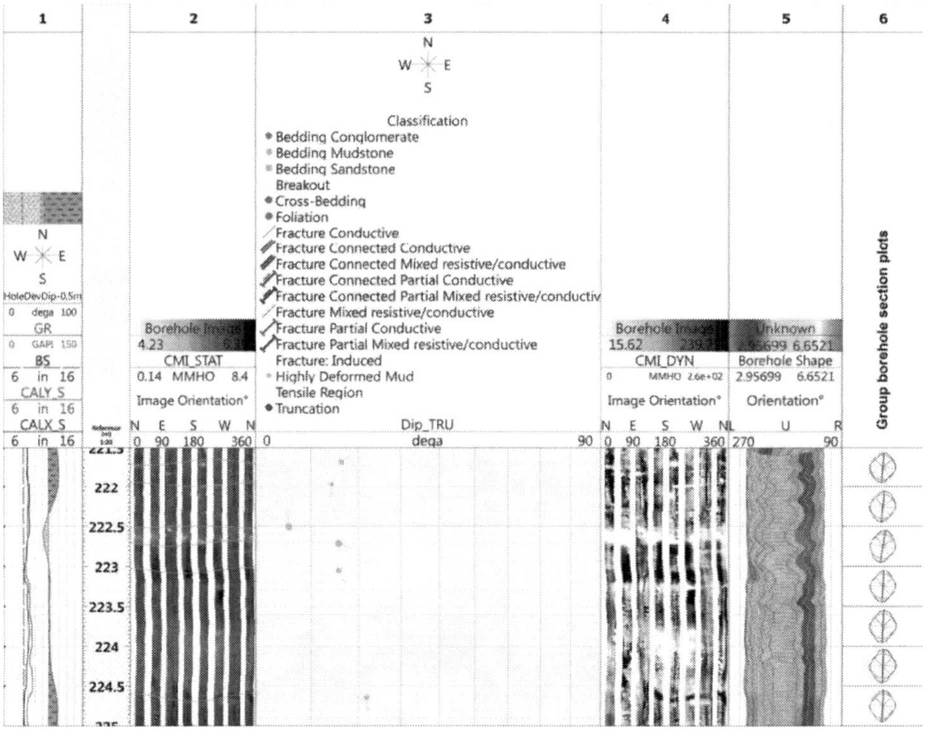


**Fig. 7.** Present-day in situ stress direction estimated from borehole breakouts identified on 8-arm caliper data recorded by the CMI tool, together with the electrical microconductivity data. Tracks: 1 – auxiliary curves; 2 – CMI statically normalized image; 3 – dip / azimuth tadpoles plot; 4 – CMI dynamically normalized image; 5 – borehole wireframe model; 6 – "Potato Plot".

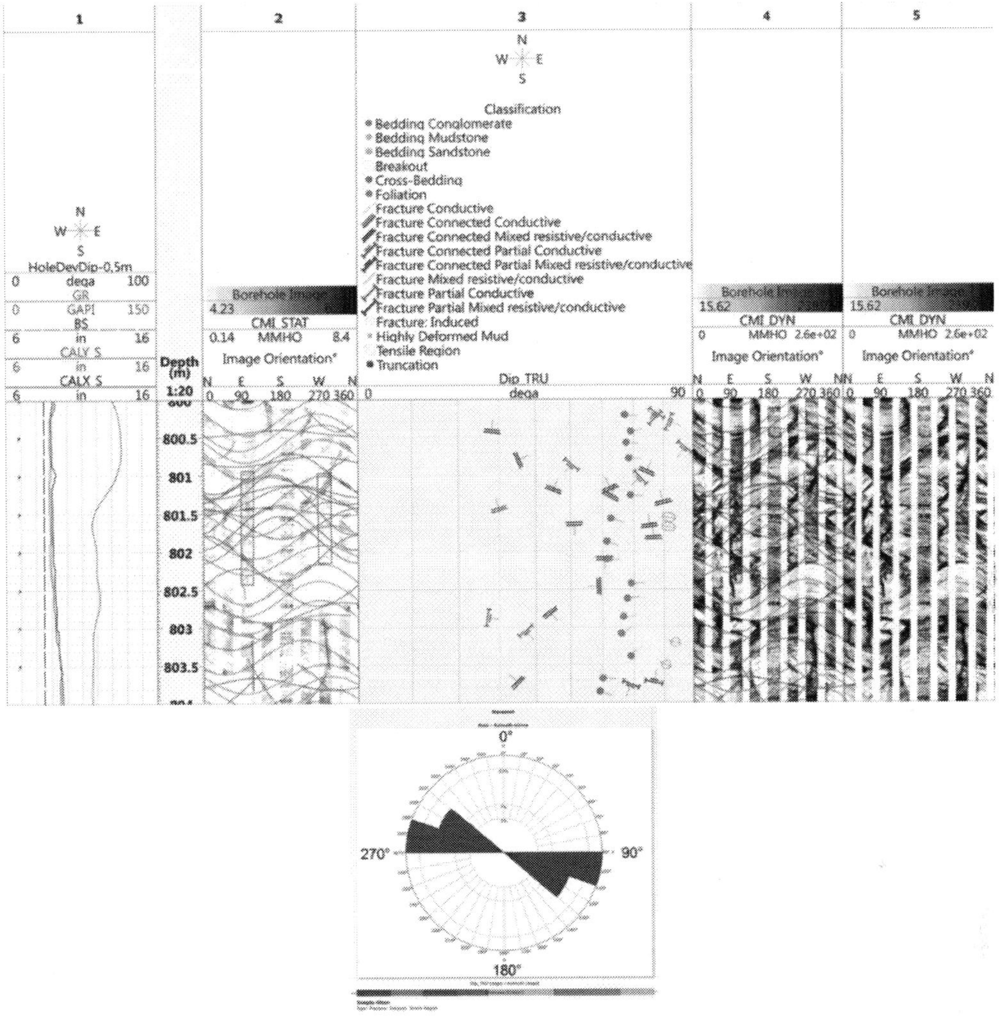


**Fig. 8.** Present-day in situ stress direction estimated from drilling-induced fractures and tensile zones. Tracks: 1 – auxiliary curves; 2 – CMI statically normalized image; 3 – dip / azimuth tadpoles plot; 4 and 5 – CMI dynamically normalized image. The azimuth frequency plot (Rose diagram) shows the predominant orientation of fractures.

## 6 Conclusions

Geophysical well logging investigation methods provide relevant and meaningful insights into the structure of the crystalline basement and its relationships with the sedimentary cover. We have analyzed and interpreted conventional and high-resolution (borehole electrical imaging) wireline logging data recorded in an exploration well drilled in North Dobrogean Promontory (Romania). This study demonstrates the applicability and the potential of these methods in order to identify and characterize the metamorphic foliations/schistosity and, possibly, metamorphic differentiations or mineralogical changes within the basement. Volumetric fracture intensity and formation brittleness characterization, as well as the identification of tectonic stress orientation via breakouts or drilling-induced fractures, are also illustrated in this research.

## References

1. R. Rickman, M.J. Mullen, J.E. Petre, W.V. Grieser, D. Kundert, *A Practical Use of Shale Petrophysics for Stimulation Design Optimization: All Shale Plays Are Not Clones of the Barnett Shale,* SPE Annual Technical Conference and Exhibition, paper SPE-115258-MS, 840–850 (2008)
2. D. Gray, P. Anderson, J. Logel, F. Delbecq, D. Schmidt, R. Schmid, *Estimation of stress and geomechanical properties using 3D seismic data*, First Break **30**, 3, 59–68 (2012)
3. W.S. Dershowitz, H.H. Herda, *Interpretation of fracture spacing and intensity*. In: J.R. Tillerson and W.R. Wawersik (eds.), Proceedings of the 33rd U.S. Symposium on Rock Mechanics, A.A. Balkema, Rotterdam, 757–766 (1992)
4. Gavăt, I. Cornea, A. Gheorghe, R. Gaspar, L. Tomescu, *La structure du soubassement moesien entre le Danube et les rivières du Ialomiţa, Buzău et Siret,* Revue Roumaine de Géologie, Géophysique et Géographie, Serie Géophysique **11**, 1, 23–32 (1967)
5. M. Săndulescu, *Geotectonics of Romania* (in Romanian), Technical Publishing House, Bucharest, 336 pp. (1984)
6. M. Tărăpoancă, *Architecture, 3D geometry and tectonic evolution of the Carpathians foreland basin*, Ph.D. Thesis, Vrije Universiteit, Amsterdam, ISBN 90-9017847-3 (2004)
7. I. Balintoni, C. Balica, *Peri-Amazonian provenance of the Euxinic Craton components in Dobrogea and of the North Dobrogean Orogen components (Romania): A detrital zircon study*, Precambrian Research **278**, 34–51 (2016)
8. J.-C. Hippolyte, *Geodynamics of Dobrogea (Romania): new constraints on the evolution of the* Tornquist–*Teisseyre Line, the Black Sea and the Carpathians*, Tectonophysics **357**, 33–53 (2002)
9. O. Mirăuţă, E. Mirăuţă, *The Paleozoic from the southern part of the Măcin Mountains (Cerna-Hamcearca region)* (in Romanian). Dări de Seamă ale Şedinţelor Comitetului Geologic (1958–1959), **XLVI**, 129–142 (1962)
10. O. Mirăuţă, *Contributions to the knowledge of the Paleozoic Formations from the southern part of the Măcin Mountains* (in Romanian). Studii şi Cercetări de Geologie, Geografie şi Geofizică, Seria Geol. **11**, 2, 497–512 (1966)
11. D. Patrulius, E. Mirăuţă, M. Mureşan, M. Iordan, *Stratigraphic and structural sinthesis of North Dobrogea. I. Paleozoic formations* (in Romanian), Unpublished report, Archives of the Geological Institute of Romania, Bucharest (1973)
12. C.-C. Lee, C.-H. Lee, H.-F. Yeh, H.-I. Lin, *Modeling spatial fracture intensity as a control on flow in fractured rock*, Environmental Earth Sciences **63**, 6, 1199–1211 (2010)
13. M.D. Zoback, D. Moos, L.G. Mastin, R.N. Anderson, *Well bore breakouts and in situ stress*, Journal of Geophysical Research **90** (B7), 5523–5530 (1985)
14. R.A. Plumb, S.H. Hickman, *Stress-induced borehole elongation: A comparison between the Four-Arm Dipmeter and the Borehole Televiewer in the Auburn Geothermal Well*, Journal of Geophysical Research **90** (B7), 5513–5521 (1985)
15. B.S. Aadnoy, *Inversion technique to determine the in-situ stress field from fracturing data*, Journal of Petroleum Science and Engineering **4**, 127–141 (1990)
16. B.S. Aadnoy, J.S. Bell, *Classification of drilling-induced fractures and their relationship to in-situ stress directions*, The Log Analyst **39**, 6, 27–42 (1998)

E3S Web of Conferences **66**, 01006 (2018) https://doi.org/10.1051/e3sconf/20186601006
*AG 2018 – 4th International Conference on Applied Geophysics*

# Linear discontinuous deformations in the light of investigations performed with ERT method

*Piotr* Strzałkowski, *Roman* Ścigała, *Katarzyna* Szafulera*, *Marek* Kruczkowski

Silesian University of Technology, Faculty of Mining and Geology, Gliwice, Poland

**Abstract.** The article presents an example of using the electrical resistivity tomography method to assess the condition of shallow rock mass layers in the area of linear discontinuous deformations created in the past due to underground mining activity. The research concerned the mining area of one of the Upper Silesian Coal Basin mines, where intensive mining operations have been conducted for several decades. In the considered area, linear discontinuous deformations were created in the form of ground steps. Their location is related to characteristic layout of deposit accessing roadways and extraction fronts in several coal seams. The article analyzes the geological structure of the deposit and the state of deformation of the rock mass caused by mining operations. In order to evaluate the hitherto impacts, appropriate calculations of the extraction influence were performed, assuming different views on the summation of horizontal strain in long time intervals. The calculations were carried out using the theory of W.Budryk - S. Knothe. Theoretical considerations were supplemented with geophysical surveys performed with using electrical resistivity tomography. Obtained results of the near-surface layers ERT imaging in the form of resistivity profiles confirm the existence of rock loosening zones in the areas of discontinuous deformations occurrence.

**Keywords:** post-mining discontinuous deformations, linear deformations, ERT method

## 1 Introduction

Underground mining extraction causes two types of surface deformations : continuous - where no visible breaks in near surface layers is visible and discontinuous - where discontinuity on the surface is observed and some relative movement on both sides of discontinuity occurred in vertical and/or horizontal direction. Continuous deformations manifest on the surface in the shape of so-called subsidence troughs. They always accompany underground mining - with deeper extraction they tend to be more smooth but occupy wider area on the surface [2, 7, 10]. A number of prediction models in this field was developed in Poland, e.g. [6, 8, 12].

---

* Corresponding autor: katarzyna.szafulera@polsl.pl

Discontinuous deformations, according to their shape, may be divided into two types : surface-type and linear-type. They not always take place - some favorable additional factors must exist to trigger them [3, 4, 10, 14].

Typical example of surface-type discontinuous deformation is a sinkhole [3]. Presently, in Upper Silesian Basin they are not as common as it was in the past, because of greater depth of extraction nowadays and connected with it different mechanism of their creation. It is usually tied with instability of old shallow workings caused by influences of presently performed deep extraction.

Linear discontinuous deformations are significant problem arising on the surface as the effect of underground mining operations. The main causes for their formation include:

- mining extraction in the vicinity of tectonic faults (in particular on one side of dislocation)
- asymmetric extraction of several deposit layers to given (same) border line - usually the limit of different pillars (protective, safety etc.) - see Fig. 2.

Linear deformations commonly arise on the land surface in the shape of ground steps (Fig. 1), trenches, cracks and fissures. Characteristic is, that they usually spread along one direction, sometimes for several hundreds of meters, while perpendicular dimensions stay relatively small. Very often, parallel system of such deformations is created - in Fig. 1 two ground steps are visible, crossing the road.

**Fig. 1.** Typical ground step created as the effect of mining of several coal seams to the border of protective pillar (photo by R. Ścigała).

The results of some research show, that significant influence on the formation of linear discontinuous deformations have also high values of horizontal tensile strain, which may occur in case of very intensive one-sided extraction or mutual influence of extraction edges located at the distance not greater than the influence range - Fig. 3. In such case, horizontal strain resulting from every extraction edge is going to be summed, which finally may double its value. According to [9], the critical value of horizontal strain necessary to trigger ground steps amounts approximately 12-15 ‰, while the activation of already existing (created earlier) deformations may occur with strain greater than 2 ‰.

Regardless of the cause of deformation, they pose a serious threat to the surface. In order to be able to recognize such threat, it is necessary to conduct detailed analysis of geological and mining conditions, as well as to assess the condition of shallow rock mass layers in the considered location.

One of the methods of determining the anomaly zones in the geological structure of the near-surface layers of the rock mass in a non-destructive way is the use of geophysical methods.

In the framework of this paper, practical example has been shown of ERT investigation in the area of linear discontinuous deformations triggered due to mining activity. The aim of the study was to confirm the usability of ERT imaging for recognition of the threat of discontinuous deformation arising in the areas of intensive underground mining operations.

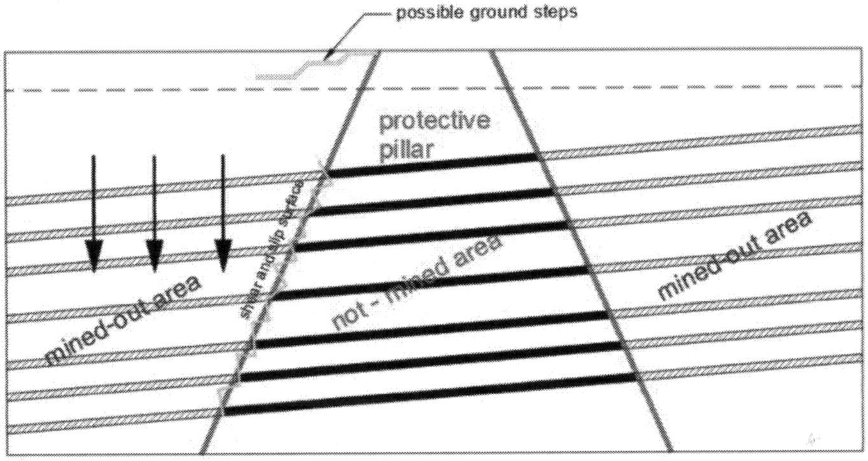


**Fig. 2.** Extraction scarps in the border area of protective pillar as the cause of ground steps creation.

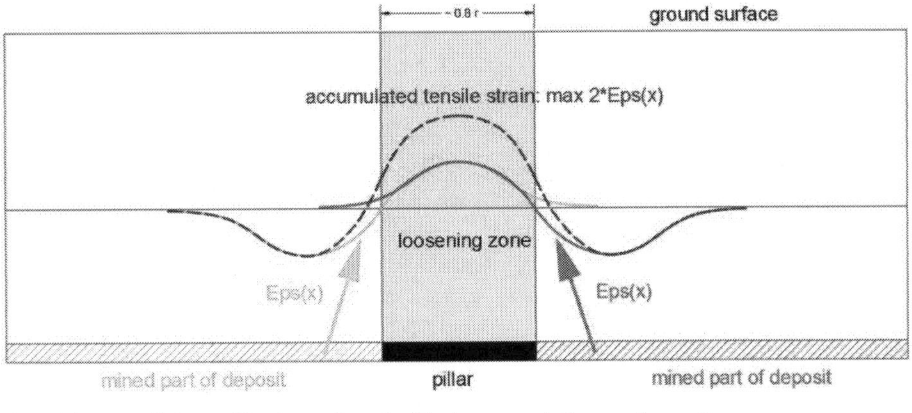


**Fig. 3.** Mining conditions of horizontal strain (Eps) accumulation on the surface.

## 2 Mining and geological conditions in the investigation area

**The description of discontinuous deformations**

As the effect of long term underground mining to the same border of safety pillar, system of ground steps was created on the surface, crossing grassland - Fig. 4-6. The throw of every step amounts from 0.5 m to 1.0 m. Because they arose in arable area, the discontinuity was not clearly visible - steps formed a kind of fissure. The length of deformation exceeded 200m, while in perpendicular direction the average distance between every step was about 50m.

**Geological conditions**
In the area of interest, rock mass is built of the following layers: Quaternary, Tertiary, Triassic and Carbon.

- Quaternary layers cover the whole considered area. They are built mainly of sand, gravel and loam. Maximum thickness of Quaternary layers reaches 32m.
- Tertiary part of the rock mass of thickness about 248m consists of loam with limestone inserts and blue-gray shale with gypsum.
- Triassic layers with thickness of 40m are built mainly of shell limestone.
- Carbon up to the depth 900m consists of "orzeskie" and "rudzkie" layers. They are formed mainly by claystones, mudstones, hard coal layers and coal shales.
- In the considered area, there is lack of important tectonic faults.

**Underground mining extraction conducted in the area of interest**
"Orzeskie" layers were extracted in the period 1985 – 1997, with thickness from 1.3 m to 3.5 m, at the depth from 380 m to 580 m. Extraction was led by using longwall system with caving.

Coal seams from "rudzkie" layers were extracted from the year 1962 up to 2002, with thickness from 1.3 m to 3.4 m, at the depth from 380 m to 820 m. Extraction was led by using longwall system with caving.

In Fig. 4 the map of total thickness of extracted deposit is shown - different colors mark different intensity of extraction. White area marks unextracted part - the protective pillar, designed for protection of main deposit accessing roadways.

As it can be seen from this picture, there is high diversity of extracted deposit - from 0 m at the pillar location up to 20 m towards south and 15 m towards north. Such situation created conditions for ground steps creation - on both sides of the pillar a large amount of deposit has been extracted, which favors the possibility of rock mass vertical shear-slip along the plane of multiple extraction edges. Additionally, above the pillar high tensile horizontal strain occurred, which undoubtedly was an additional factor favorable for rock mass vertical slip.

**Performed calculations**
For calculation of maximum strain, which probably took place above the protective pillar, Budryk - Knothe theory [6] was used, implemented into dedicated software [15]. Because of long history of extraction in this area, different strain summation methods were used :

- simple summation which usually leads to exaggerated (abnormally high) strain values in case of considering long time periods,
- summation in 10-years independent extraction periods [7],
- summation with the assumption of strain relaxation [1, 11].

For calculation purposes, the following parameters of Budryk-Knothe theory where taken : coefficient of roof control $a=0.84$, parameter describing the influence dispersion $tg\beta=2.2$, parameter responsible for description of horizontal component of deformations: $B=0.32r$ (where r is the radius of main influence range). Parameters were determined on the basis of surveys led in this area in the past.

Computer aided simulation of extraction was then performed in order to assess possible maximum values of strain during different positions of extraction front. Obtained maximum values of tensile horizontal strain are presented in table 1.

As it comes from performed calculations, the value of strain greater than 12 [‰] took place in the end of 70s of the last century. Unfortunately no information is available about ground steps creation in those years. First their findings come from the year 2002, where they probably were activated by extraction led at that moment, with smaller strain values - at the level of 2[‰].

**Fig. 4.** The map of total thickness of extracted deposit in the vicinity of created linear discontinuous deformations.

**Table 1.** Total horizontal strain $\varepsilon_{max}$ calculated over long time periods with different summation methods

| Horizontal strain summation method | Maximum strain value |
|---|---|
| 1. Simple summation of horizontal strain | **+28.9 [‰]** |
| 2. Horizontal strain calculated over 10-years extraction periods | **+12.9 [‰]** |
| 3. With using the assumption of relaxation of horizontal strain | **+13.6 [‰]** |

## 3 ERT methodology, results and its discussion

### 3.1 ERT methodology

In September 2017, a study was carried out to assess the structure of surface rock mass layers using the ERT method. The terrain morphology was not complicated, surface is generally horizontal - planar, with average height of 250 m above sea level. Most of the study area is occupied by arable fields and grassland (Fig. 5). For measurements, the GF-Instruments Ares II unit was used [5]. The profile was located transversely in relation to the course of 4 linear discontinuous deformations (Fig. 6). The discontinuous deformation zones crosses the ERT profile at the positions of electrodes: No 15 - 17, No 27 - 29 and 40 - 42.

The ERT profile line consisted of 48 electrodes arranged along a straight line stretched from NE (electrode No. 1) towards SW (electrode No. 48), at the intervals of 4 m. The total length of the profile was 188 m, hence the maximum penetration depth, which may be estimated at about 1/5 of the maximum spacing of electrodes [13] was about 38 m.

**Fig. 5.** General view of survey site.

ERT survey was performed with using three arrays: "Wenner α" (360 measurements), "Wenner-Schlumberger" (529 measurements), "Dipol - Dipol HD" (1288 measurements). The measurement results were processed using the Res2DInv software [13]. Fig. 7 presents the obtained ERT profiles in the form of maps of resistance distribution obtained with classical inversion procedure implemented in Res2DInv.

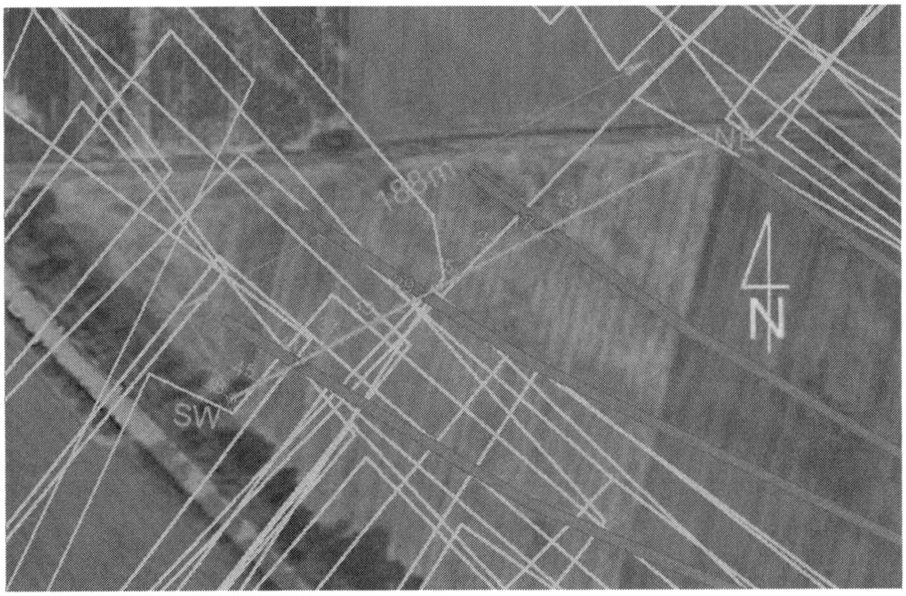


**Fig. 6.** Location of ERT profile (blue) on the background extracted fields edges (gray) and the course of linear deformations (red).

### 3.2 Results and its discussion

In Fig. 7, on the background of resistivity distribution, location of ground steps is shown, marked with black rectangle. Analyzing this image, it can be seen, that in the area of ground steps, the zone of higher resistivity is visible in subsurface layers. So it may be stated, that there is evident change in rock mass structure, especially its loosening due to high values of tensile horizontal strain. The second important factor influencing the anomalies visible in ERT image is a shear-slip of fractured rocks in the vicinity of system of extraction edges, which finally revealed on the surface in the shape of linear discontinuous deformations.

It should also be mentioned, that all used arrays gave similar image of resistivity changes, although Dipole-Dipole HD presents more detailed distribution of resistivity near discontinuity zones.

Of course presented interpretation of ERT investigation must be treated with some attention - another recognition methods should be used (geophysical or mining) to confirm in-situ deformation state in near-surface layers.

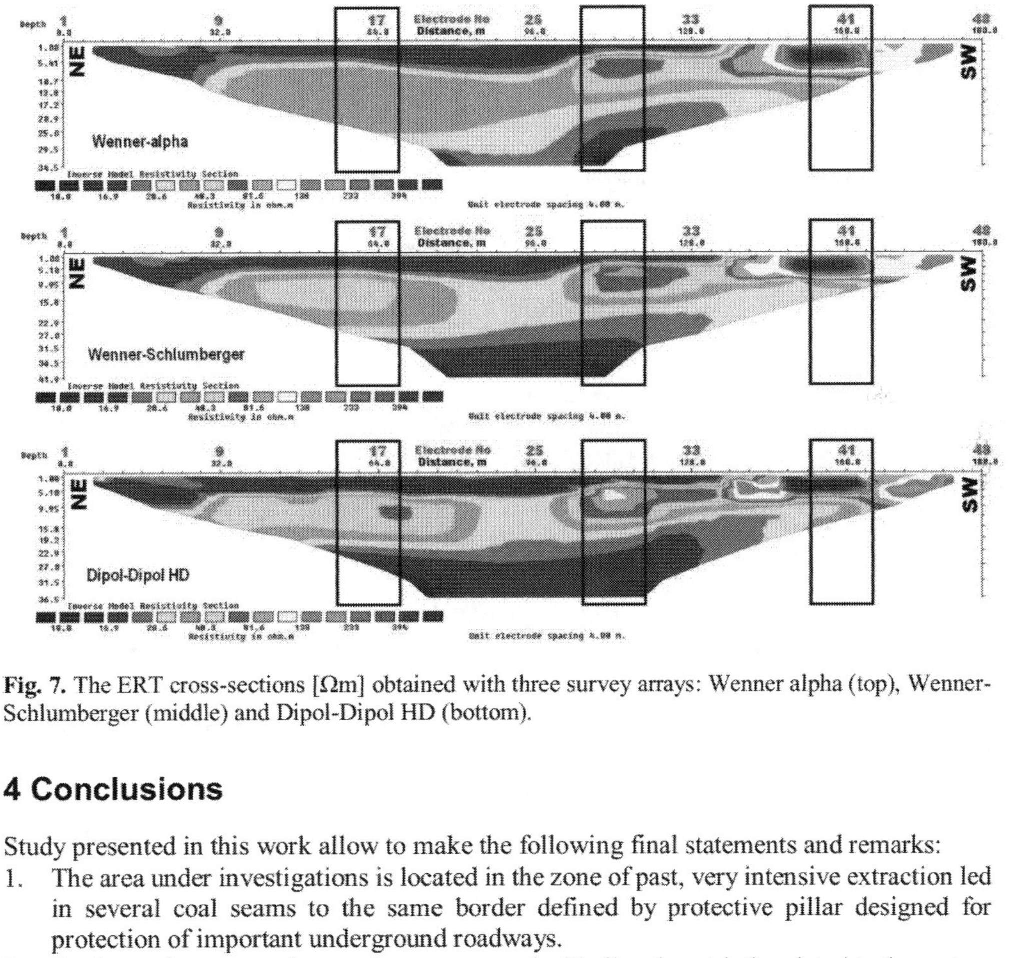


**Fig. 7.** The ERT cross-sections [Ωm] obtained with three survey arrays: Wenner alpha (top), Wenner-Schlumberger (middle) and Dipol-Dipol HD (bottom).

## 4 Conclusions

Study presented in this work allow to make the following final statements and remarks:

1. The area under investigations is located in the zone of past, very intensive extraction led in several coal seams to the same border defined by protective pillar designed for protection of important underground roadways.
2. On the surface, several ground steps occurred with direction strictly related to the system of finished extraction edges, outside extracted area. Due to this fact, additional adverse element occurred - high values of tensile horizontal strain, which undoubtedly was an element favourable for rock mass loosening.

3. Presented ERT images show existence of several high resistivity anomalies in the areas, where discontinuous deformation were created on the land surface. So it may be stated, that there is the evident change in rock mass structure (loosening) due to high values of tensile horizontal strain and shear-slip of fractured rocks in the vicinity of extraction edges.
4. Due to the fact, that there was no possibility of practical confirmation of mentioned above findings - for example by drilling, presented interpretation must be treated with care. It is necessary to remember, that quality of ERT surveys depends on many factors, which cannot always all be taken into account.
5. ERT surveys performed in the area of intensive underground extraction conducted in the past, may help in finding zones of rock mass loosening, which in turn may enable more precise determination of areas exposed to the threat of linear-type discontinuous deformation creation.

## References

1. J. Białek, *The description of transient phase of mining area subsidence with taking into account the asymmetry of final influences*, Research Journal of Silesian University of Technology, Mining series **194**, Gliwice (1991) (in Polish)
2. M. Chudek, *Geomechanics and basics of management of environmental protection in mining and post-mining areas*, Publishing House of the Silesian University of Technology, Gliwice (2010) (in Polish)
3. M. Chudek, W. Janusz, J. Zych, *A study concerning recognition of creation and prediction of discontinuous deformations due to underground extraction of deposits*. Research Journal of Silesian University of Technology, Mining series **141**, Gliwice (1988) (in Polish)
4. Z. Fajklewicz, W. Piwowarski, J. Radomiński, E. Stewarski, A. Tajduś, *The investigation of rock mass deformation state aiming at recovering the post-mining land development capability*, Publishing House of the AGH University of Science and Technology, Kraków (2004) (in Polish)
5. GF-Instruments, *ARES II, Automatic resistivity & IP system, User guide*, Czech Republic, Brno (2014)
6. S. Knothe, *The equation of finally formed profile of subsidence trough*, Archive of Mining and Metallurgy **1**, 1 (1953) (in Polish)
7. S. Knothe, *Forecasting the impact of mining exploitation*, „Śląsk" Publishing House. Katowice (1984) (in Polish)
8. T. Kochmański, *Integral Theorie der Bodenbergungen unter Einnirkung des Abbaus*, Freiberger Forschungshefte, A.l/18, Freiberg (1959) (in German)
9. A. Kowalski, *Recognition and possibilities of prediction of linear discontinuous deformations*, Proceedings of the Conference: 8th Days of Mining Surveys and Protection of Mining Areas, Ustroń, Poland (2005) (in Polish)
10. H. Kratzsch, *Mining Subsidence Engineering*, Springer-Verlag, Berlin, Heidelberg, New York (1983)
11. J. Kwiatek, *On rheological aspects of the threat of building constructions located in mining areas*, Scientific Works of Central Mining Institute **827**, Katowice (1997) (in Polish)
12. J. Litwiniszyn, *Differential equation of rock mass displacement*, Archive of Mining and Metallurgy **1**, 1 (1953) (in Polish)
13. M. H. Loke, *2D & 3D electrical imaging surveys*, Geotomo software: www.geotomosoft.com, Malaysia (1996-2018)

14. P. Strzałkowski, *The outline of the protection of mining areas,* Publishing House of the Silesian University of Technology, Gliwice (2010) (in Polish)
15. R. Ścigała, *Computer aided prediction of rock mass and land surfaces deformations caused by underground mining extraction*, Publishing House of the Silesian University of Technology, Gliwice (2008) (in Polish)

# ERT method in the study of chemical pollution of the hydrogeological environment – numerical analysis of 2D and 3D models

*Grzegorz* Bania*

AGH University of Science and Technology, Faculty of Geology, Geophysics and Environment Protection, Kraków, Poland

**Abstract.** Electrical resistivity tomography (ERT) method is often used to solve problems related to chemical pollution of the hydrogeological environment. The source of such contamination can be, e.g. industrial heaps, tailings ponds and municipal landfills. The contaminants spreading often takes the form of a (3D) pollution tongue. In this case, the standard interpretation of 2D ERT surveys may be difficult. Numerical modeling was carried out in order to show the specificity of pollutants propagation. Both 2D and 3D models were analysed. Those refer to the real geological situation – vicinity of Nowa Huta (Krakow, Poland) – where heaps and tailings ponds of the metallurgical plant are present. An appropriate geoelectrical model referring to the above situation was prepared. It was assumed that highly mineralized waters in the form of a contamination tongue are spreading within the aquifer. Resulting interpreted resistivity distributions for 2D and 3D variant of ERT method were analysed. It has been shown that the method is promising in terms of the ability to detect and assess the nature of the transitional zone between clean and contaminated waters. Additionally, it has been shown that 3D modeling can be a useful, complementary element in interpreting the results of DC-resistivity methods.

**Keywords:** DC-resistivity method, ERT, electrical resistivity tomography, modeling, groundwater pollution

## 1 Introduction

The DC-resistivity method, including the electrical resistivity tomography (ERT) method, is used particularly often in studies related to groundwater contamination caused by the impact of different types of landfills [e.g. 1-9]. The aim of the study was to show the possibility of using the ERT method for solving problems related to groundwater pollution of wastes from metallurgical origin (and not only) through numerical modeling. Modeling is an additional tool that allows to check how a given situation, e.g. spatial hydrogeological structure described by physical parameters (geophysical model), will be reflected in simulated measurements results and inversion. Moreover it allows to analyse various scenarios and to

* Corresponding author: bania@agh.edu.pl

consider potential limitations of the DC-resistivity method in order to select the appropriate measurement methodology.

The work concerns two-dimensional (2D) and three-dimensional (3D) models of the hydrogeological center in relation to the real terrain situation – vicinity of Nowa Huta (Krakow, Poland) [9-10]. These models simulate a hypothetical pollution tongue/cloud that penetrate into the groundwater. The source of contamination may be, e.g. a heap or tailings pond of the metallurgical plant. This pollution cloud indicates a targeted displacement of low resistivity contaminants within the assumed aquifer. During the construction of the models, the results and conclusions from the research concerning groundwater pollution in vicinity of the Zelazny Most tailings pond (derived from the copper ore enrichment in KGHM) [4, 11] were taken into account. In that case, the use of geoelectrical methods proved to be very effective.

## 2 Creation of geoelectrical models

Referring to the real geological situation from the vicinity of Nowa Huta (Krakow, Poland), a typical lithological profile was created (Fig. 1). There is 3 m of overburden (soil, sandy loam). Deeper, there are layers of sand and sandy gravel which constitute aquifer with a thickness of 7 m. The deepest layer (10 m under the surface) is made of clay which is impermeable and prevents penetration of contaminants into the deeper parts of the center. On the basis of the lithological profile, two resistivity models were created - A, which assumes unpolluted water within the aquifer and B, which simulates contaminated medium (Fig. 1). Due to the similar resistivity characteristics of sands and gravels, they are treated as one layer. For simplification, the appropriate rock model was used [12]. In this model, one component is the rock skeleton (spherical, homogeneous grains) and the second one, mineralized water that fills its pores.

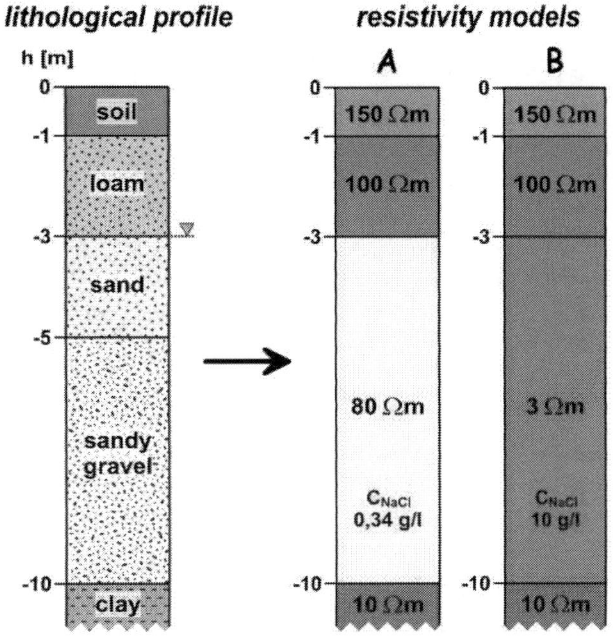


**Fig. 1.** Typical lithological profile from the vicinity of Nowa Huta and corresponding resistivity models.

In the model (Fig. 1) it was assumed that the aquifer had a temperature of 10 °C and water filling the pores was saturated with NaCl with a concentration of 0.34 g/l. Based on the dependence of resistivity to the mineralization of the NaCl solution [13], the water resistivity was determined to be 22.7 Ωm. Additionally, it was assumed that the sand grains (quartz) are practically non-conductive and taking the porosity coefficient K = 0.37 the resultant resistivity of the aquifer was calculated – 80 Ωm. In addition, in the considered geoelectrical models, overburden resistivities of 150 and 100 Ωm were assumed. The resistivity of the clay layer was assumed to be 10 Ωm. The chosen value is characteristic for clays that can be found in the vicinity of Nowa Huta, i.e. approx. 10-20 Ωm [10].

For the needs of 2D and 3D modeling it was assumed that the layers of the geoelectrical models (Fig. 1) are homogeneous and isotropic and a layer with a resistivity of 80 Ωm (model A) will be the background for the pollution cloud spreading within it. It was assumed that the mineralization of contaminated waters is 10 g/l. The calculated resistivity of contaminated water is 0.87 Ωm. The resultant resistivity value for the contaminated layer was determined - 3 Ωm (model B).

## 2.1 2D models

2D models were created in the Res2Dmod software [14]. Two situations were analysed - M1 and M2 (Fig. 2) which simulate two different types of the front of moving pollution tongue. In both models 2 m electrode spacing and two mesh blocks between adjacent electrodes were used. Vertically, up to a depth of 10 m, the mesh blocks have a fixed height of 0.5 m, and from 10 m in depth their height is gradually increasing. The visibility of the models and inversion results was limited to about 15 m in depth and it was assumed that the last layer continues deep into infinity. It should be emphasized that the 2D model is being considered, i.e. it does not change its properties in the direction perpendicular to the measurement line. In fact, the medium is heterogeneous and anisotropic and as a result of inversion this leads to the interpretation of artifacts related to the 3D effect (side effect) [e.g. 15].

The M1 model (Fig. 2) simulates a situation in which the front of pollution cloud is fuzzy, i.e. changes in the resistivity proceed in a gradient manner creating a transitional zone between the contaminated (3 Ωm) and clean (80 Ωm) waters. Such situation may be suitable for a relatively slow moving pollution tongue or in the case where the tongue will stop moving and contaminated waters will gradually mix with the sweet ones. Model M2 (Fig. 2) considers the situation of a rapid movement of pollution cloud which, in the resistivity model can be presented with sharp boundary between the resistivity layers of 3 and 80 Ωm. In addition, for both models, the border between contaminated and clean centers is not vertical but tilted.

Simulated apparent resistivity pseudosections were calculated for dipole-dipole array with current and potential dipoles $a$ = 1, 2, 3, 4, 5, 6 Δx and distances between these dipoles, respectively $n$ = 1, 2, 3, 4, 5, 6 $a$. Inversion of the data was carried out in the Res2Dinv software [16]. The inversion process based on the L1-norm (robust) inversion method have been applied. The obtained inverse model resistivity sections have been compiled under the models in Figure 2. For better comparison of modeled situations with inversion results, the real horizontal boundaries between the model layers were marked in the sections. In addition, the external borders of the transitional zone for the M1 model and the border between contaminated and clean centers in the case of the M2 model were marked.

 https://doi.org/10.1051/e3sconf/20186601007

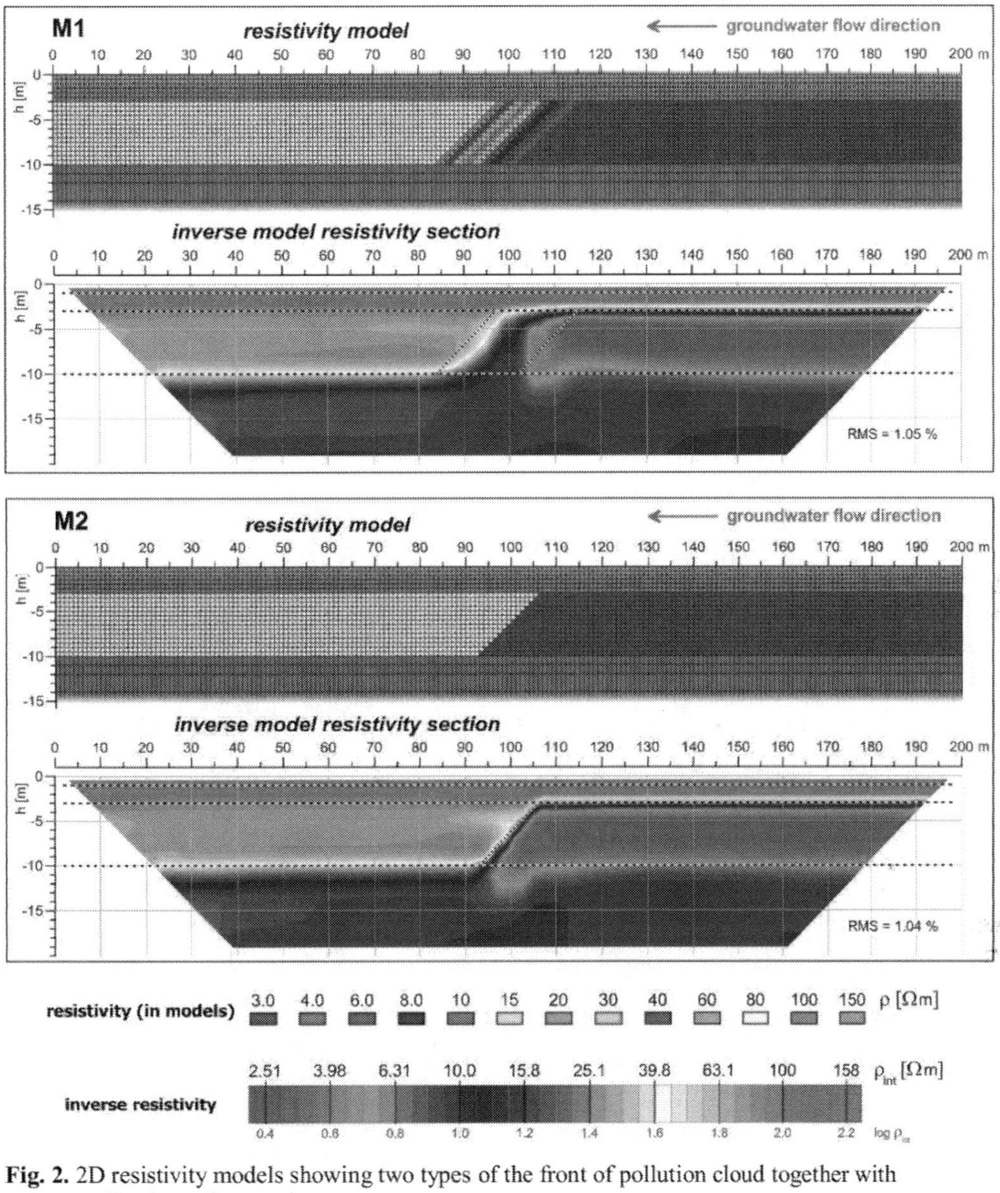


**Fig. 2.** 2D resistivity models showing two types of the front of pollution cloud together with corresponding inversion results.

## 2.2 3D model

The 3D model of the pollution tongue was created in the Res3Dmodx64 software [17]. Creating this model (Fig. 3a), subsequent layers / depth levels were reflected exactly as in the assumed geoelectrical model (Fig. 1). Similarly to the 2D models (Fig. 2), the heights of individual blocks to a depth of 10 m are 0.5 m which gives 20 depth levels. Below this depth there are additional layers with gradually increasing height of blocks, which is partially visible in the section in Figure 3c. Horizontal grid dimensions (x, y) in all model depth levels are 100 x 100 (Fig. 3b). Two nodes between adjacent electrodes (with 2 m spacing) in the x and y directions were applied. In total, this gives the number of electrodes 51 x 51 = 2601. Note that using such a large number of electrodes in field measurements would be very troublesome. For this reason, the proposed 3D model (Fig. 3) should be considered only hypothetically. Figure 3b shows the selected depth level of the model - from -5.0 to -5.5 m. It can be noticed that the shape of the tongue is not symmetrical. This is to refer to real conditions and causes an additional "difficulty" for the inversion process, which task is to recreate the original model. The hypothetical source of pollution is placed outside the model, over 30-50 m on the x axis. It was assumed that the dominant direction of groundwater flow is consistent with the A-A' section line marked in Figure 3. The front of the pollution cloud, well visible on the section A-A' (Fig. 3c), was designed in a similar way to the 2D M2 model (Fig. 2). In the direction corresponding to the x axis (Fig. 3b), there is also the spread of the contaminated waters. It is consistent with the visible gradient zone proposed in the 2D M1 model (Fig. 2).

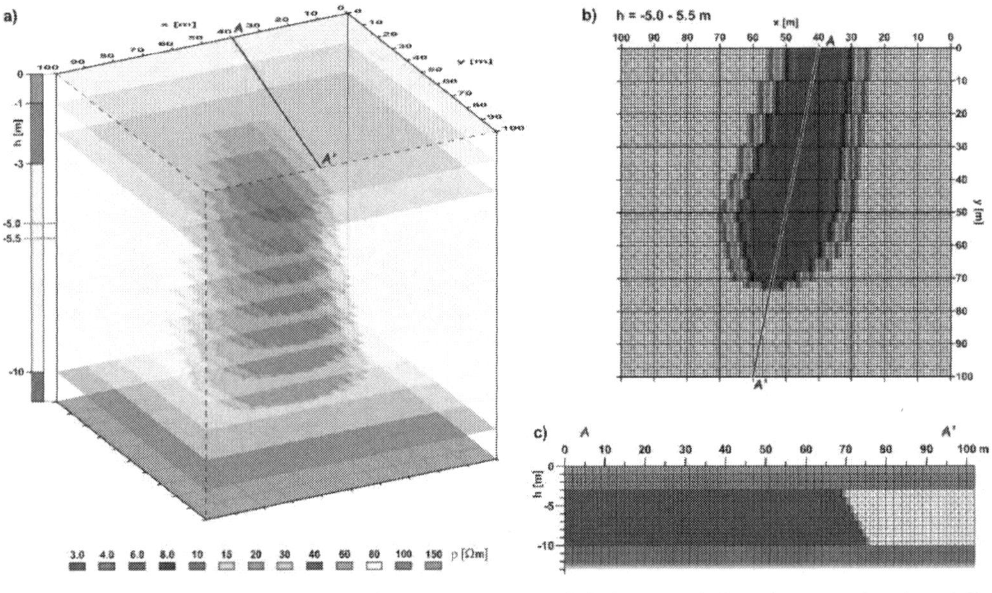


**Fig. 3.** 3D resistivity model of a pollution tongue: a) simplified, general view; b) example of model's depth level (bird's eye view); c) A-A' section.

The spatial distribution of apparent resistivity for the 3D model was calculated for the dipole-dipole array with the same parameters as for the 2D models. Inversion of the obtained data was performed in the Res3Dinvx64 software [18] using similar parameters as for the inversion of 2D data. Due to the different nature of the 3D inversion, it is not possible to use identical parameters that can be used in 2D inversion. The results of 3D inversion are presented in two ways in Figure 4. In the first case, from the interpreted resistivity block, the overburden layer and unpolluted aquifer were removed, thanks to which the interpreted pollution tongue which lay on the low resistivity substrate was exposed (Fig. 4a,b). The second way consisted in showing selected 2D sections being "cuts" made by the interpreted block (Fig. 4c,d,e).

## 3 Results and analysis

In the case of both 2D models (Fig. 2), the horizontal boundaries have been interpreted well. Interpreted resistivity values of individual layers are almost identical to those assumed. The sharp boundary for the M2 model has also been interpreted well. The only difference to the modelled situation is the appearance of an undervalued resistivity zone of the clay layer under the mentioned boundary. In the resultant section it looks like pollution tongue has penetrated into the clay layer. The gradient boundary zone for the M1 model has been interpreted in a generalized way. From the right side it takes a vertical character, while from the left it shows the correct tilt. This is probably related to the subtle resistivity changes within the transitional zone assumed in the model. Such changes, due to the resolution of the DC-resistivity method, are not large enough to be interpreted. Nevertheless, the general shape and location of the discussed zone have been reflected satisfactorily.

Comparing the 3D modeled situation (Fig. 3) with the inversion results (Fig. 4), it can be concluded that its interpretation is very good. The shape of the pollution tongue is similar to the modelled one. Also, the interpreted resistivity values are almost identical comparing to modelled ones. Within the pollution cloud, the interpreted resistivity value is about 3 Ωm. In the case of clay layer, the interpreted values are generally the same as the assumed value of 10 Ωm.

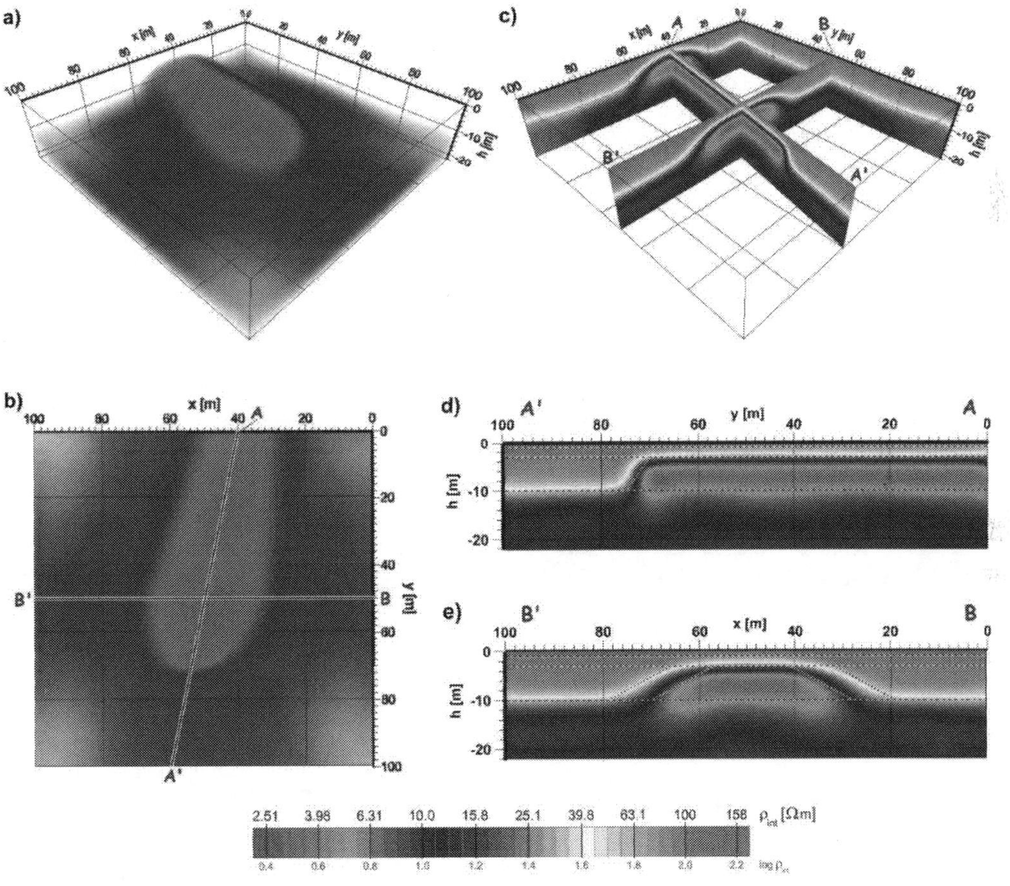


**Fig. 4.** Inversion results for the 3D pollution tongue model. Exposed substrate layer together with the tongue: a) general view; b) bird's eye view. 2D sections obtained from the 3D inversion results: c) general view; d) section A-A'; e) section B-B'.

In the A-A' section (Fig. 4d) the front of pollution tongue is not as sharp as for the result for a 2D M2 model (Fig. 2). Both situations, although they simulate a similar type of transitional zone between the clean and contaminated centre, may reflect it in different ways. It seems that better results were obtained for the inversion of 2D data. The difference is probably related to the fact that the 2D model assumes the constancy of the medium in the direction perpendicular to the section plane, while in the case of the 3D model it is more complicated. In the direction corresponding to the x-axis there are changes in model structures and the 3D inversion algorithm takes them into account. However, in both situations, the phenomenon of the pollution tongue penetrating the low resistivity clay layer is visible. It has a much stronger character for 2D data (Fig. 2 - M2). It follows that the 3D inversion more accurately reflects the assumed model.

The case of the B-B' section (Fig. 4e) raises the problem of interpreting the gradient zone between the contaminated and clean centers. As can be seen it is interpreted very well. Moreover, the phenomenon of pollution cloud penetrating the bottom layer is observed, similarly as in 2D M1 model (Fig. 2). Due to the interpretation of the transitional zone fragment mentioned for the 2D case, it can be concluded that the situation has been better represented for 3D inversion results, where we do not deal with such a phenomenon.

## 4 Conclusions

On the basis of the analysis of the presented geoelectrical models and the obtained inversion results, the following conclusions were formulated:

- The 2D and 3D numerical modeling shows that the ERT method in both (2D, 3D) variants is suitable for solving the problem of groundwater contamination with soluble chemical substances;
- The ERT method appears to be promising in terms of the ability to detect and assess the nature of the transitional zone between clean and contaminated waters. However, for a real geological structure, especially within Quaternary formations, this task may be difficult and ambiguous. It seems simpler to effectively use electrical resistivity tomography method to study the behaviour of the pollution tongue in the aquifer in a monitoring system, especially using the 2D ERT;
- 3D ERT, due to the technical requirements (e.g. the use of a large number of electrodes to cover the research area), can be difficult to apply in the real field situation. Nevertheless, it may be very useful in the form of modeling for directing 1D and 2D DC-resistivity measurements;
- Analysis of 3D models can help to reduce to a certain extent the equivalence associated with the impact of spatial center structure on the 2D ERT measurements results.

The research presented in this paper was supported by the Dean's Grant no. 15.11.140.221.

## References

1. T. Dahlin, First Break **14**, 1137 (1996) doi: 10.3997/1365-2397.1996014
2. C. Bernstone, T. Dahlin, T. Ohlsson, H. Hogland, Environ. Geol. **39**, 360–371 (2000) doi: 10.1007/s002540050015
3. L.N. Meads, L.R. Bentley, C.A. Mendoza, Can. Geotech. J. **40**, 551–558 (2003) doi: 10.1139/t03-017
4. S. Lasocki, J. Antoniuk, W.J. Mościcki, *Environmental Protection Problems in the Vicinity of the Zelazny Most Flotation Wastes Depository in Poland*, J. Environ. Sci. Heal. A **38**, 1435–1443 (2003)
5. J.E. Chambers, O. Kuras, P.I. Meldrum, R.D. Ogilvy, J. Hollands, Geophysics **71**, B231-B239 (2006) doi: 10.1190/1.2360184

6. P. Martinez-Pagan, A.F. Cano, E. Aracil, J.M. Arocena, *Electrical Resistivity Imaging Revealed the Spatial Properties of Mine Tailing Ponds in the Sierra Minera of Southeast Spain*, J. Environ. Eng. Geoph. **14**, 63–76 (2009)
7. R. Clement, M. Descloitres, T. Gunther, L. Oxarango, C. Morra, J.P. Laurent, J.P. Gourc, Waste Manage. **30**, 452–464 (2010) doi: 10.1016/j.wasman.2009.10.002
8. G. Bania, M. Ćwiklik, 76th EAGE Conference and Exhibition 2014, Amsterdam, Netherlands, EAGE Publications BVNetherlands (2014) doi: 10.3997/2214-4609.20140966
9. G. Bania, M. Ćwiklik, 77th EAGE Conference and Exhibition 2015, Madrid, Spain, EAGE Publications BVNetherlands (2015) doi: 10.3997/2214-4609.201412539
10. W.J. Mościcki, G. Bania, M. Ćwiklik, M. Florek-Odrazil, *Budowa utworów przypowierzchniowych na terenie zespołu dworsko-parkowego w Branicach koło Krakowa – wyniki badań elektrooporowych (The subsurface structure of manor-park complex in Branice near Krakow - the geoelectrical research)*, Nawarstwienia historyczne miast Europy Środkowej, Wydawnictwa AGH, Krakow, 407–423 (2016) (in Polish)
11. J. Antoniuk, W.J. Mościcki, *Investigation of polluted underground water with electrical and electromagnetic methods. Case of the „Zelazny Most" settling reservoir*, 2nd meeting "Environmental & Engineering Geophysics", Nantes, France, 339–342 (1996)
12. J.J. Fried, *Groundwater pollution. Theory, methodology, modelling, and practical rules*, Elsevier Scientific Pub. Co, Amsterdam, New York. Developments in water science 4 (1975)
13. N.B. Dortman, *Petrophysics*, Reference Book, Book one, Rocks and mineral deposits. Nedra, Moscow (1992)
14. M.H. Loke, *Rapid 2D resistivity and I.P. forward modeling using the finite-difference and finite-element methods*, Geotomo Software, Manual (2016)
15. G. Bania, M. Ćwiklik, Geology, Geophysics & Environment **39**, 331 (2013) doi: 10.7494/geol.2013.39.4.331
16. M.H. Loke, *Rapid 2D Resistivity & IP Inversion using least-squares method*, Geotomo Software, Manual (2010)
17. M.H. Loke, *3-D resistivity & IP forward modeling using the finite-difference and finite-element methods*, Geotomo Software, Manual (2014)
18. M.H. Loke, *Rapid 3-D Resistivity & IP inversion using the least-squares method*, Geotomo Software, Manual (2016)

# Seismicity induced by hard coal mining in the vicinity of faults

*Zbigniew* Burtan*, *Jerzy* Cieślik and *Dariusz* Chlebowski

AGH University of Science and Technology, Faculty of Mining and Geoengineering, Cracow, Poland

**Abstract.** An inherent feature of Polish collieries within the Upper Silesia Coal Basin is the high level of mining induced seismicity, resulting in elevated rockburst hazard levels. One of the major causes of high-energy seismic events is that mining operations are continued in the vicinity of major faulting zones. The study summarises the results of geo-mechanical and statistical analysis of mining-induced seismic activity in the region of major faults, in a selected section within a colliery. Seismic activity assessment involves the categorisation of seismic events due to tectonic movements in the context of various face development systems with respect to the faulting zone: perpendicular (advancing towards the faulting zone or retreating) or parallel (along the faulting zone). Registered seismic activity was analysed in the context of epicenter locations and variations of seismic activity in relation to the developing face operations in the function of time and energy ratings (Gutenberg-Richter formulas). Results have demonstrated that increased levels of seismic activity in the strata can be attributable to mining operations in the vicinity of major faulting zones.

**Keywords:** hard coal mining, faults, induced seismicity

## 1 Introduction

The geomechanical and statistical analysis of mining-induced seismicity performed in preparing this paper assumes the form of a case study and covers the geological and mining conditions of one of the collieries of the former Katowicki Holding Węglowy S.A. company. The coal beds in the mining area of the selected colliery are found in three major tectonic blocks separating the respective mining sections (J, K and L). The boundaries of these sections constitute fault zones V, VI, VII, VIII, Środkowy and Kłodnicki (Fig. 1) with various azimuths and throw amplitudes (KHW S.A. 2000-2014).

In mining to date, the highest rockburst hazard level was associated with the extraction of coal beds in parts K and L in the area of the Kłodnicki Fault and in part J in areas affected by mining events. Of the 19 rockbursts recorded since 1980, the majority (14) occurred near faults, where exploitation was conducted in the direct surroundings of extensive zones of mainly fault VI (Fig. 2) and the Kłodnicki Fault. In analysing the location of seismic foci related to mining within the area of tectonic disturbances, it is difficult to clearly determine

* Corresponding author: burtan@agh.edu.pl

whether a specific seismic event is actually the effect of fault activation and its influence, or the collapse of rigid roof strata. The results of the seismic moment analysis might provide some insight, as the mechanisms occurring alongside faults are mainly connected with shearing [1, 2]. Unfortunately, there is a shortage of such data and they are not available for every seismic event. Seismic events are usually considered as located within faults, related to fault activation, or to a certain extent connected with the presence of faults when the hypocentral distance of their focus from the edge of the fault is no longer than 150÷200m (Fig. 2), although such distances are assumed only on the basis of experiences and intuitions regarding the specific (local) conditions of mining. In regional scale, the relations were established between major faults locations and occurring of strong seismic events [3].

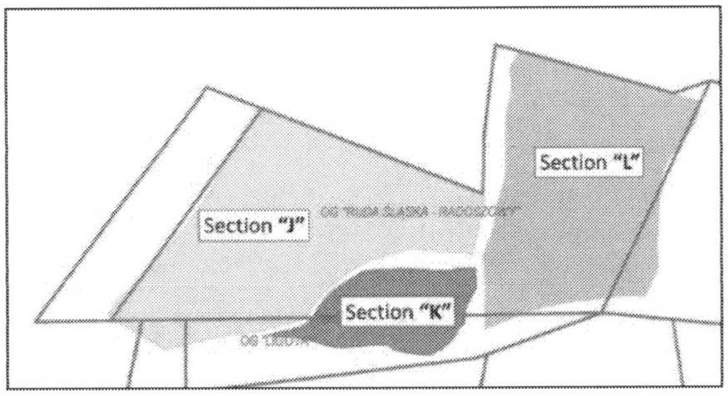


**Fig. 1.** The colliery's mining area by mining parts J, K, L.

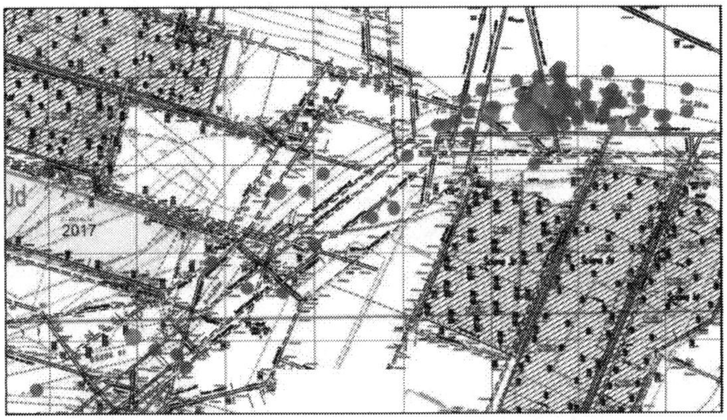


**Fig. 2.** Location of seismic foci ($\geq 10^5$J) resulting from the activation of fault VI.

In the context of rockburst prevention, it would be valuable, for specific geological and mining conditions, to determine the distance between the longwall face and the rupture surface of the fault which increases the probability of seismic events connected with the activation of the latter. However, due to the complexity of this problem, it might prove impossible to find a solution based on analytical and numerical calculations. An attempt to determine such distances was made in the course of analysing data from the selected colliery's mining geophysics station on seismic activity observed during the mining of four longwalls of coal bed 502/1 in part K. The study was based on the authors' previous experience regarding the locations of seismic foci during mining operations [5], where the vast majority of seismic events occurred in front of the mining face. Such regularities were also confirmed in other studies [6–9].

The analysis demonstrated that for every specific mining face a group of seismic events can be found which are directly connected with the influence of faults through the anomalous location of their foci in front or behind the wall front. Using this methodology, it is also possible to determine the extent and the critical distance between the mining face and fault edge at which its activation might occur. An analysis of the mining data demonstrates that the anomalies of seismic foci location is observed both when the face is retreating from and advancing towards the fault or moving along the faulting zone. The results of anomalous location analyses were verified using Gutenberg-Richter (GR) distribution [10], which confirmed the different nature of seismic activity in the designated ranges of fault zones.

## 2 Seismic activity analysis during the mining of coal bed 502/1 in part K

The analysis was based on the data on the scale of seismic activity during the mining of bed 502/1 in part K, which was delimited on three sides by the Kłodnicki, VI and VII faults with throws ranging from 75 to 220m (Fig. 3). Coal bed 502/1 with the thickness of 8m was mined on two strata, from top to bottom with a cave-in and the presented data concern mining operations in the first stratum, which was associated with considerably greater seismic activity than mining in the second stratum. The roof of the coal bed was composed of shale, arenaceous shale and fine-grained sandstone. The floor was largely composed of fine-grained sandstone and shale formations. The mining operations in the analysed area were conducted on four exploitation front – EF1, EF2, EF3 and EF4 (Fig. 3). EF1 was mined along fault VII, retreating from the Kłodnicki Fault and advancing towards fault VI. EF2, EF3 and EF4 were exploited from the Kłodnicki Fault towards fault VI. Table 1 presents the profile of seismic activity during mining operations on coal bed 502/1 in part K with regard to the consecutively selected longwalls EF1÷EF4.

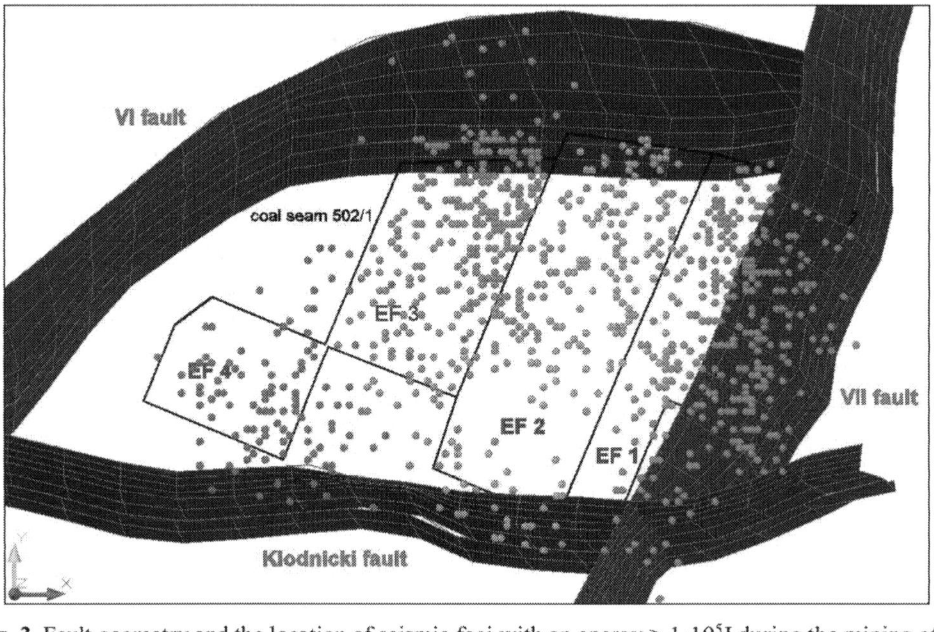


**Fig. 3.** Fault geometry and the location of seismic foci with an energy > $1\cdot10^5$J during the mining of longwalls EF1, EF2, EF3 and EF4 in part K of bed 502/1.

Perpendicular distances from the seismic foci to the mining face and to the fault edge were calculated on the basis of the coordinates of foci locations during the exploitation of individual longwalls and the current mining face location in relation to the fault line. Vector calculus was used in the calculations [11].

**Table 1.** Seismic activity during the exploitation of coal bed 502/1 in part K.

| Longwall | Year | Number of seismic events in energy intervals [J] | | | | | |
|---|---|---|---|---|---|---|---|
| | | $10^3$ | $10^4$ | $10^5$ | $10^6$ | $10^7$ | Total |
| EF1 | 1997 - 2000 | 982 | 270 | 36 | 3 | 0 | 1496 |
| EF2 | 1997 - 2000 | 795 | 147 | 18 | 1 | 0 | 1263 |
| EF3 | 2002 - 2003 | 1399 | 266 | 61 | 7 | 0 | 2105 |
| EF4 | 2005 - 2006 | 573 | 97 | 18 | 3 | 0 | 826 |

## 2.1 An analysis of seismic events locations relative to the fault

Fig. 4 and Fig 5. illustrate the correlations between the distances of seismic foci for seismic events recorded during the operation of wall EF1 from the line of the Kłodnicki Fault and from the line of fault VI, respectively, based on the runway of the mining face. The positive values of seismic event distances (on the y axis) mean that the seismic event was located outside the fault zone and negative values mean that the seismic focus was located on the fault (at a maximum distance of 200m from its line). EF1 began its run 15m from the line of the Kłodnicki Fault and ended 70m from the line of fault VI with a runway of 655m. In accordance with the adopted 200m criterion, the ranges marked with the red vertical line were identified with the fault's range of influence.

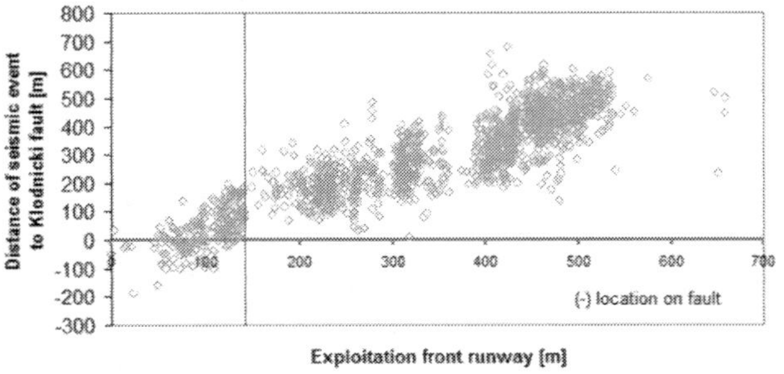


**Fig. 4.** Distance of seismic event to the line of the Kłodnicki Fault based on runway length.

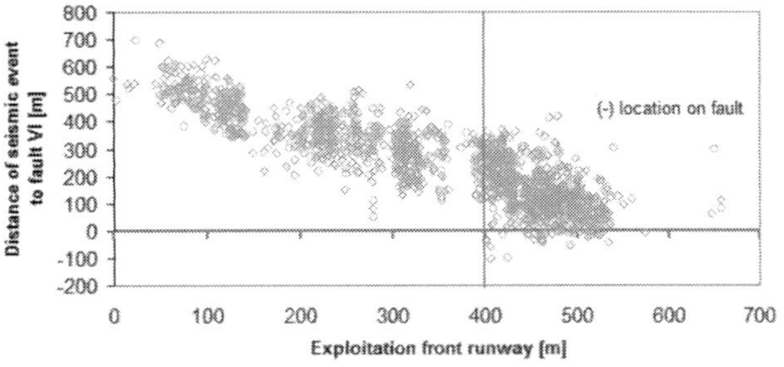


**Fig. 5.** Distance of seismic event to the line of fault VI based on runway length.

In accordance with the methodology adopted for EF1, the range of influence can be estimated as ~140m for the Kłodnicki Fault and ~300m for fault VI.

### 2.2 An analysis of seismic events locations relative to the mining face

The next stage involved an analysis of the seismic event locations relative to the EF1 mining face to determine whether the focus was in front (positive distance value) or behind the mining face (negative distance value) (Fig. 6). These analyses were performed for all seismic energy intervals. As observed [5], ca. 75% of seismic events during mining operation are located in front of the mining face.

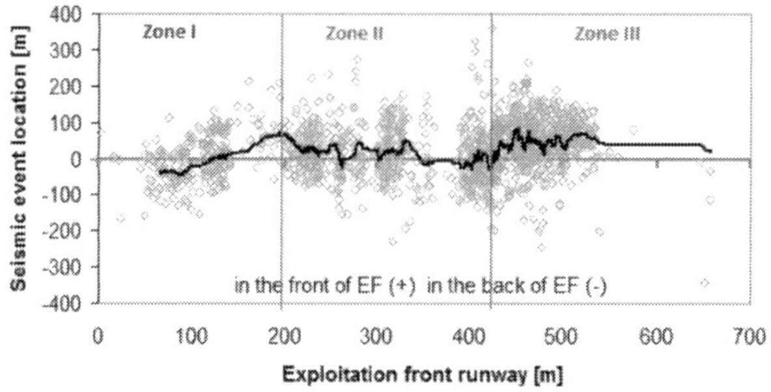


**Fig. 6.** The variation of seismic event locations in front of and behind the face during the operation of EF1.

The analysis demonstrated that at the initial run of EF1, most seismic foci were located behind the mining face. As it developed (and the runway became longer), the percentage of events located in front of the face was increasing; for a runway of 150m the moving average with 30 seismic events (Fig. 6 - black line) assumed only positive values. The distance for which the rising trend in this correlation changes is marked with a red vertical red line (Zone I), which is assumed as an estimated boundary of the range of influence of the Kłodnicki fault on the seismic activity during the mining operations in EF1. In Zone II a similar number of seismic events was recorded in front of and behind the mining face or at a small distance from it. For higher runway lengths during advance towards fault VI, another anomaly occurs, i.e. the number of seismic events with foci located in front of the mining face increases. The beginning of this trend, marked with a second red vertical line, is assumed as the boundary of the range of influence of fault VI on the seismic activity values during the mining operation of EF1 (Zone III). Following the adopted methodology, the range of influence of the Kłodnicki Fault for the mining operation of EF1 can be estimated as ~185m, and ~320m for fault VI.

On the basis of the analysed data regarding the seismic activity recorded during the mining operation of EF1, an attempt was made to present the correlations between the seismic energy of individual seismic events and the distance of the seismic focus to the Kłodnicki Fault and fault VI (Fig. 7 and Fig. 8). The presented distributions do not indicate a correlation between seismic event energy and the distance of the focus to the line of the fault, which means that the seismic events related to fault activity have a similar energy as other seismic events.

Qualitatively similar trends can also be observed for data obtained during the mining of other longwalls – EF2, EF3 and EF4 (Fig. 9÷11).

Following the adopted methodology, for the mining operation of EF2 both the range of influence of the Kłodnicki Fault and fault VI can be determined as ~150m, for EF3 the range of influence of fault VI was ~230m, and for EF4 the range of influence of the Kłodnicki Fault was ~150m. The combined ranges of influence obtained for individual mining faces are listed in Table 2.

The highest values obtained for the range of influence of faults during the mining operation of EF1 could result from mining along fault VII, and in the case of EF3 the influence of the presence goaves in the previously extracted EF2 wall.

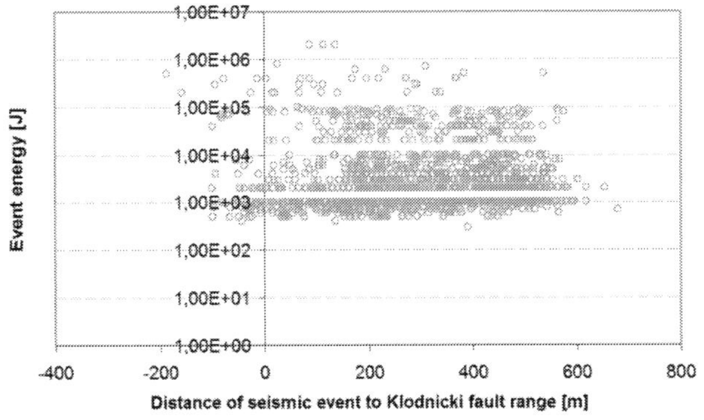


**Fig. 7.** The energy distribution of individual seismic events along with their distance to the Kłodnicki Fault.

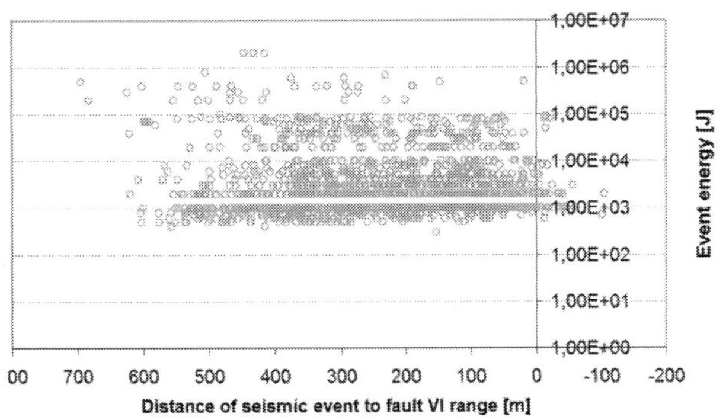


**Fig. 8.** The energy distribution of individual seismic events along with their distance to fault VI.

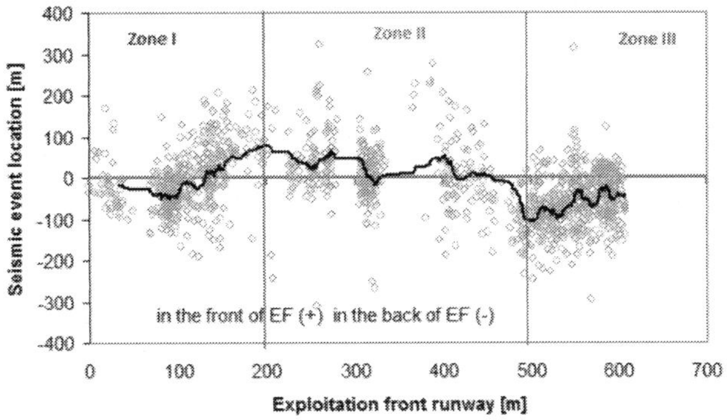


**Fig. 9.** The variation of seismic event locations in front of and behind the wall during the operation of EF2.

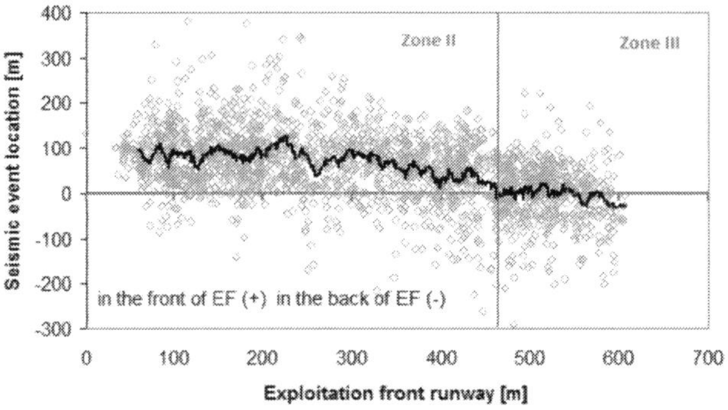


**Fig. 10.** The variation of seismic event locations in front of and behind the wall during the operation of EF3.

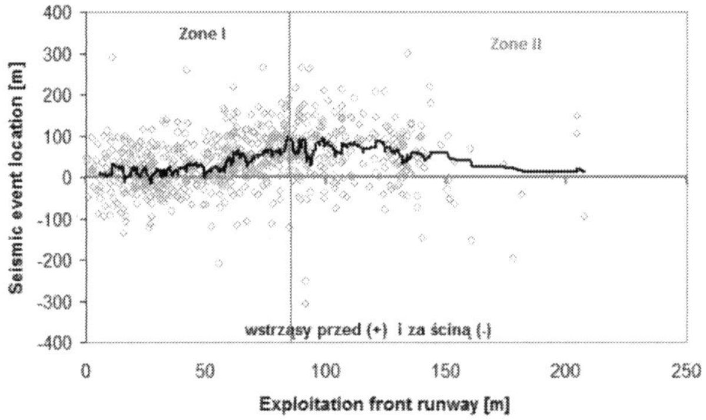


**Fig. 11.** The variation of seismic event locations in front of and behind the wall during the operation of EF4.

**Table 2.** The range of influence of faults on seismic activity for individual mining faces.

| Longwall | Kłodnicki fault range [m] | Fault VI range [m] |
|---|---|---|
| EF1 | 140 | 300 |
| EF2 | 150 | 150 |
| EF3 | - | 230 |
| EF4 | 150 | - |

## 5 Seismic activity analysis using Gutenberg-Richter distribution

Important data on the magnitude of seismic hazard is provided by the Gutenberg-Richter distribution (the distribution of seismic energies) [10] used in seismology, which describes the correlation between the number of seismic events recorded in respective energy classes and the value of the seismic energy emitted.

$$LogN = a - b\ LogA_s \qquad (1)$$

where $N$ means the number of seismic events with an energy $A_s$ and $a$ and $b$ are distribution parameters $a$ is a measure of seismic activity level and $b$ describes the relative correlation between the number of strong seismic events and the number of weak seismic events. The lower value of $b$ (slope), and therefore lower line steepness ($b=tg\alpha$) described by Gutenberg-Richter distribution, indicating a greater percentage of strong seismic events, is a sign of greater seismic hazard levels, and, in consequence, the dynamic events threat affecting underground workings. Using the Gutenberg-Richter distribution to compare seismic activity from various periods and mining regions facilitates the evaluation of variations in seismic hazard levels, including activity changes with regard to the range of occurrence of high-magnitude events. Figs. 12÷15 present Gutenberg-Richter distributions and the equations describing them for seismic events with seismic energies $A_s \geq 10^3$J recorded during the mining operation of EF1, EF2, EF3 and EF4 in coal bed 502/1, part K, for designated extraction zones relative to faults, and the parameters of those distributions are listed in Table 3.

**Table 3.** GR parameters distribution for longwalls EF1, EF2, EF3 i EF4 in coal bed 501/ part K.

| Exploitation front | Zone number | Gutenberg-Richter parameters | |
|---|---|---|---|
| | | a [-] | b [-] |
| Exploitation front EF1 | Zone I | 3.63 | 0.47 |
| | Zone II | 5.15 | 0.73 |
| | Zone III | 5.40 | 0.76 |
| Exploitation front EF2 | Zone I | 4.03 | 0.58 |
| | Zone II | 5.13 | 0.78 |
| | Zone III | 5.58 | 0.85 |
| Exploitation front EF3 | Zone II | 5.13 | 0.70 |
| | Zone III | 5.59 | 0.73 |
| Exploitation front EF4 | Zone I | 6.64 | 1.18 |
| | Zone II | 4.47 | 0.59 |

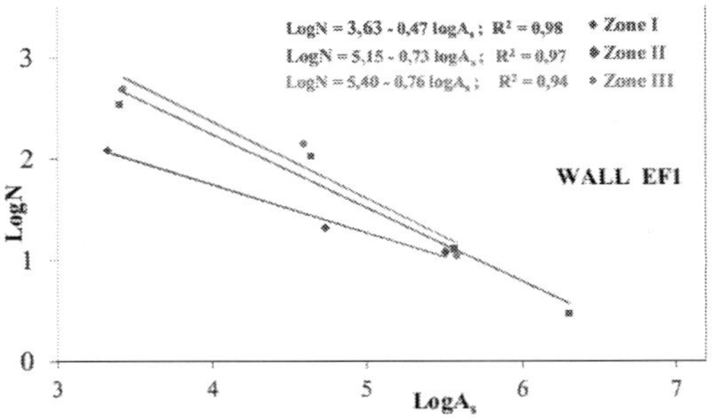


**Fig. 12.** GR parameters for longwall EF1 in coal bed 502/1 part K.

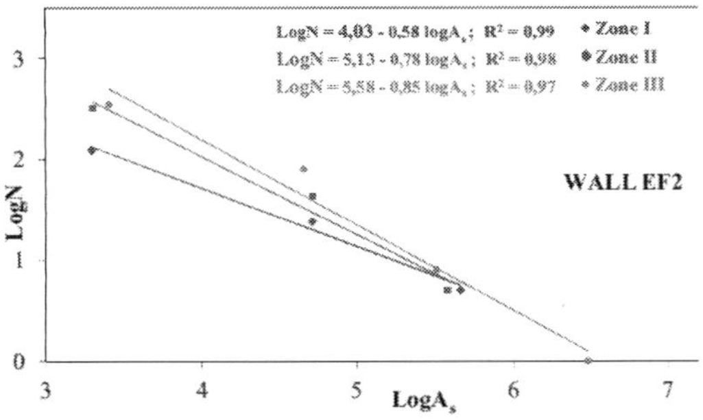


**Fig. 13.** GR parameters for longwall EF2 in coalbed 502/1 part K.

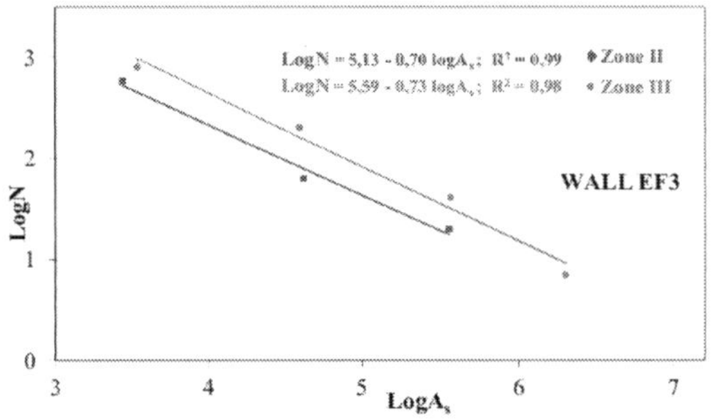


**Fig. 14.** GR parameters for longwall EF3 in coalbed 502/1 part K.

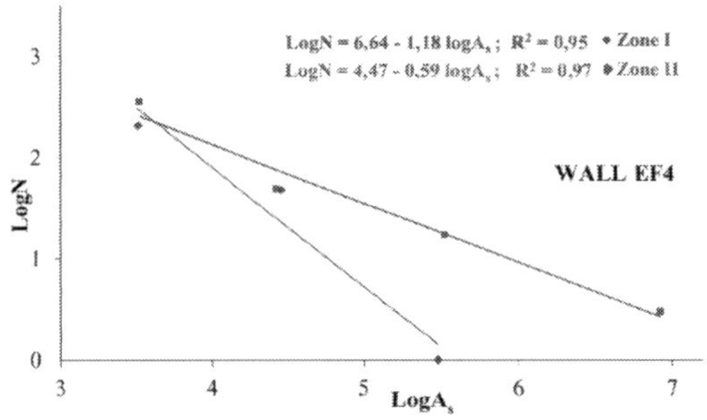


**Fig. 15.** GR parameters for longwall EF4 in coalbed 502/1 part K.

On the basis of the presented parameter distributions and values for respective fronts selected in coal bed 502/1, part K, it can be concluded that there are significant variations in Gutenberg-Richter distributions in relation to specific zones designated in previous analyses. It can also be observed that:

- for EF1, EF2 and EF3 correlations are present at individual mining stages; at stage I the seismic hazard level measured by the value of the ***a*** parameter is the lowest, and at stage III it is the highest,
- for EF1 and EF2 at stage I and for EF1, EF2 and EF3 at stages II and III there is a similar percentage of strong and weak seismic events measured by the value of the ***b*** parameter (lower at stage I and higher at stages II and III), which might indicate that there are different mechanisms behind most seismic events occurring at stage I and those occurring at stages II and III,
- for EF4 there are inverse correlations at stages I and II relative to EF1, EF2 and EF3.

## 5 Summary

The presented correlations of the influence of selected discrete tectonic faults on seismic activity during mining operations within coal bed 502/1, part K (using the example of a selected mine of the former KHW S.A.) determine the range, i.e. the distance of the exploitation front from the line of the Kłodnicki Fault and fault VI, at which the influence of those tectonic dislocations on the intensity of seismic events can be recorded. The most similar distances found in the course of range of influence analysis for specific fronts were obtained for the Kłodnicki Fault (~140÷150m). In the case of fault VI, it had a greater range of influence (~150÷300m), which, for EF1, might have resulted from mining along fault VII, and for EF3, from the presence of goaves in the extracted EF2. The obtained relationships and range values might prove useful in designing rockburst prevention measures used for performing mining operations in fault zones.

Nevertheless, the adopted methodology for determining the range of influence of a fault on seismic activity based on seismic event locations in front of/behind the exploitation front requires further study and verification on a larger number of cases of mining operations under restricted tectonic and mining conditions.

*The paper was prepared within AGH-UST statutory research, No 11.11.100.005*

## References

1. Ł. Rudziński, G. Lizurek, *Mechanizm zjawiska sejsmicznego oraz tąpnięcia w O/ZG Rudna w Polkowicach z 19.03.2013 r. z wykorzystaniem lokalnych i regionalnych sieci Cuprum (Seismic event mechanism and rockbursts in O/ZG Rudna in Polkowice on 19 March 2013 using local and regional Cuprum networks)*, Czasopismo Naukowo-Techniczne Górnictwa Rud **3**, 61–72 (2015) (In Polish)
2. G. Kwiatek, Acta Geophys. **52**, 2, 155–171 (2004)
3. E. Pilecka, Z. Pilecki, *Analysis of relation between induced seismic activity and satelite data.* 19th Symposium on Application of Geophysics to Engineering and Environmental Problems SAGEEP, EEGS, 2-6 April 2006, Seattle, USA, 346-355. (2006)
4. E. Pilecka, Z. Pilecki, *The changes in lineaments locations in the zones of seismicity induced by mining.* Proc. 13th European Meeting of Environmental and Engineering Geophysics "Near Surface 2007", EAGE, 3-5 Sept. 2007, Istanbul, Turkey, (2007) (Electronic document)
5. J. Cieslik, Z. Burtan, D. Chlebowski, A. Zorychta, Journal of Sustainable Mining **16**, 3, 94–103 (2017) doi: 10.1016/j.jsm.2017.10.002
6. M. Kozłowska, *Ilościowa analiza odległości epicentralnych wstrząsów względem frontu eksploatacji w O/ZG Rudna (LGOM) (Quantitative analysis of epicentral tremor distance in relation to exploitation front in O/ZG Rudna LGOM),* Przegląd Górniczy **68** (12) 75–85 (2012) (In Polish)
7. M. Kozłowska, Acta Geophys. **61**, 5, 1156–1169 (2013) doi: 10.2478/s11600-013-0137-x
8. J. Dubiński, Z. Pilecki, W. Zuberek, *Badania geofizyczne w kopalniach – przeszłość, teraźniejszość, i zamierzenia na przyszłość (Geophysical research in mines - past, present and future plans)* Wyd. IGSMiE PAN, Kraków (2001) (in Polish)
9. Z. Pilecki, *Zastosowanie rejonowej obserwacji sejsmoakustycznej do kontroli stanu zagrożenia tąpaniami (The use of zonal seismoacustic observations for rock burst hazard monitoring).* Publs. Inst. Geophys. Pol. Acad. Sc. Series M, M-16, 245, 203-216 (In Polish)
10. B. Gutenberg, C. F. Richter, *Seismicity of the Earth and Associated Phenomena*, 2nd ed. Princeton University Press., New York (1954)
11. M. Paluch, *Matematyka bezstresowa w życiu codziennym i technice (Non-stress mathematics in everyday life and technology)*, Agencja wydawniczo-poligraficzna Art-Tekst, Kraków (2015) (In Polish)
12. Report, *Kompleksowe projekty eksploatacji pokładów zagrożonych tąpaniami (Comprehensive mining projects in coal beds threaten by rockbursts),* KHW S.A. KWK „WS", Katowice, (unpublished material) (2014) (In Polish)

# Rockburst hazard assessment based on seismic tomography and analytical modelling – a case study

*Dariusz* Chlebowski* and *Zbigniew* Burtan

AGH University of Science and Technology, Faculty of Mining and Geoengineering, Krakow, Poland

**Abstract.** In accordance with the formal regulations currently in force in Poland, the criteria for rockburst hazard assessment with respect to coal deposits include the records of seismic events and their impacts, de-stressing of the entire seam or its parts and expert opinions of mine operation engineers. Effectiveness of the de-stressing can be verified by geophysical test data whilst the expert opinions are mostly based on mathematical modelling using specialist software or dedicated simulation algorithms. This study collates and synthesises the results of tests carried out in a seam section in a colliery within the Upper Silesia Coal Basin, obtained by the outlined methods. Geophysical survey results are interpreted basing on seismic tomography procedures utilising the geo-tomography techniques for velocity field reconstruction in data processing. For comparison, the stress state modelling data are provided, based on conventional engineering solutions applicable to mechanics of deformable media. The actual assessment of the rockburst threat level is based on observations of distributions of the longitudinal wave velocity in relation to the actual value of the vertical component of stress concentration within the coal seam.

**Keywords:** rockburst hazard, strata de-stressing, analytical modelling, seismic tomography

## 1 Introduction

Mining operation in the conditions of rockburst hazard require effective prevention measures, including long-term and short-term control strategies developed in the mining practice [1]. Broadly-understood rockburst control and forecasting of the rockburst threat levels rely mostly on geophysical methods: seismic and seismoacoustic observations, seismic measurements involving seismic profiling in coal seams and seismic tomography [2–9]. In recent years analytical and numerical methods were also developed and effectively used to determine the range and extent of stress concentration zones, de-stressed regions in the proximity of longwall/shortwall faces and near the development entries [10–11].

De-stressing strategies, understood as technical and organisational measures, are put in place to reduce the stress levels within the deposits and, consequently, to minimise elastic

* Corresponding author: chlebo@agh.edu.pl

strains within the rock strata, besides two types of de-stressing effects can be produced: stress relief or strata destruction. The stress relief effect consists in reduction of the stress tensor components within the seam being mined, whilst the destructive effect gives rise to formation of the cracking and fracturing zones within the deposit such that it is no longer able to accumulate elastic energy. According to a widely adopted stress relief strategy, the seams that are not threatened or where the rockburst threat levels are low should be mined first. As a result, in areas adjacent to the mined-out seams most desirable stress changes occur, leading to reduction of respective stress components.

The conditions prevailing in a plot in the coal deposit (shortwall panel AD/510) to be mined in one of the collieries within the former Katowice Mining Holding S.A. are investigated and used in the comparative study of the rockburst threat levels derived by geophysical methods [12] and analytical simulations [13]. Geophysical methods include seismic tomography with the technique of P-wave velocity field reconstruction whilst analytical modelling relies on conventional studies of the state of stress in rock strata affected by the presence of old exploited mine workings, recalling the principles of mechanics of elastic media [14–16].

## 2 Geological and mining conditions

The plot being mined is the shortwall (open-end) panel AD located in the III stratum (roof substrata) in the seam 510, at the depth between 570 to 600m. The working method involves the cross-end mining with the roof caving whilst the shortwall has its beginning at the cross-entry AD and stretches in the north-east direction. The heading limiting the plot on the western side is the upwards heading no 139, and from the eastern end- the upwards heading no 11/III-115 (Fig. 1).

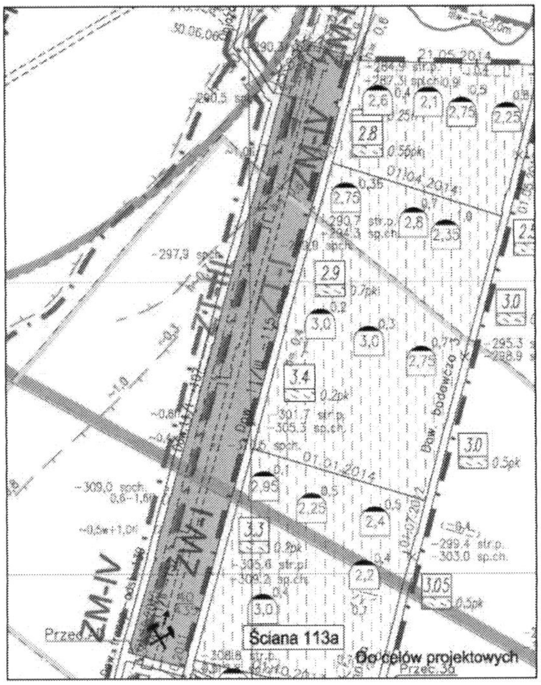


**Fig. 1.** Map of mine workings in the coalbed 510/III in the region of the shortwall AD.

Geometric parameters of the shortwall section are: length 40m, the range 435m and height 3.0m. Initially, the total thickness of the seam 510 in the investigated region ranged from 9.1÷10.0m; the compressive strength of rock strata being 16.3÷29.0MPa (22.2MPa on the average). From the eastern side, the panel has the mined-out sections: I (roof strata) and II (intermediate strata) where hydraulic backfilling was employed. The thickness of the panel (roof strata III) to be mined within the seam 510 ranges from 2.7 to 3.7m, the seam being inclined by about 4° in the south-west direction [12–13].

Features making up the roof rock and floor strata developed in the form of illite shales and sand-containing illite shales whose uniaxial compressive strength falls in the range 20.5÷38.5MPa (32.7MPa on the average). Previous mining operations revealed the presence of a major fault SW-NE, having the displacement of 0.1÷0.6m in the NW direction. Furthermore, in the northern section of the shortwall panel AD there is a faulting zone SW-NE, with the displacement 0.9m in the SE direction. In this region other disturbances due to sedimentation can potentially occur: thinner seam sections, faulting, and brittle, fallible rocks in the immediate roof, hence the coal layer 1.0m thick has to be left in the shortwall roof. As a result of previous mining activities, the mined-out voids in the layer II of the coalbed 510 are filled with sand left over after backfilling operations, that is why a coal shelf 0.5m in thickness was left to prop and stabilise the powered supports, which is a requisite for the effective face advance.

At the distance of about 25m from the roof strata in the coalbed 510 there are residues of the backfilling operations and in the final sections of the face range the shortwall mining is continued beyond the edge and in the strata which have not been yet overcut. Beneath the coalbed 510, at the distance of 134m, there is the coal seam 620, worked by the longwall mining. Mining operations in this coalbed will be continued beyond the working edges [12–13].

## 3 State of stress and rockburst hazard assessment in the context of seismic data

Geophysical methods relying on seismic tomography were applied to evaluate the conditions of the rock strata in the shortwall region, limited by the upwards heading no 139 and 11/III-115 and by the cross-cuts AD and 49 in the coalbed 510 [12]. The main objective was to obtain good-quality seismic images and to derive the times of seismic wave propagation within the coalbed and in its proximity, whilst the actual configuration of geophones and source locations was controlled by availability for testing of the existing mine workings (Fig. 2). The recording time was 500ms whilst the length of seismic rays ranged from 34÷304m, enabling the P-waves and S-waves to be effectively registered within the coalbed and in its vicinity. The sampling time in the recording procedure (1000 samples/channel) was 0,5ms, hence the registered frequency had to be restricted to the threshold value of 1000Hz. Following the filtering and phase correction, registered seismograms yield the first P-wave arrival times in rock strata surrounding the coalbed being mined. In the further step the P-wave and S-wave velocities along the ray paths were obtained and the adequacy of the computation procedure was verified via the statistical treatment of sets of ray lengths and wave velocities along the ray paths. Correlation analysis and linear regression of the paths and arrival times allow the systematic errors involved in individual path measurements to be eliminated through introduction of the constant correction term, which is of key importance when registering the first arrival times of refracting waves. After incorporation of a correction terms, the observation procedures can be extended beyond the edge of the mined coal bed and to the surrounding rock strata. The analysis of correlation and linear regression in the ray path-velocity system proves most useful in examination of seismic records in the context of refracted wave propagation in the gradient medium, either continuous or discrete, and allows for elimination of error due to the first refracted wave arrival being mistaken for the refracting wave [12].

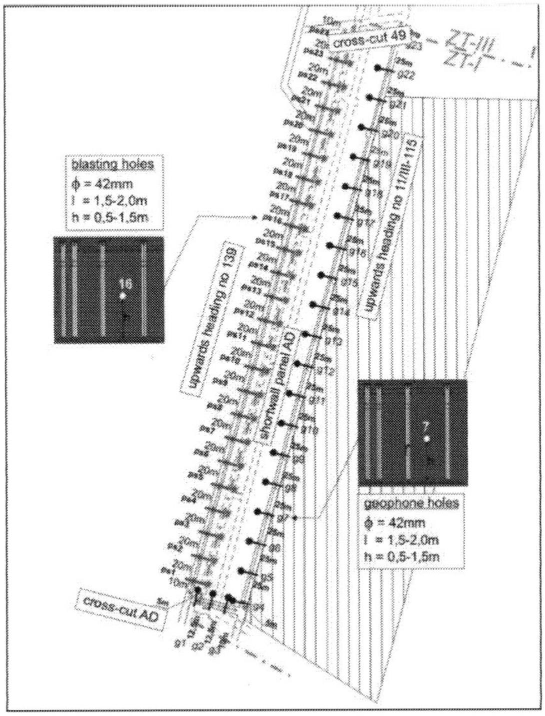

**Fig. 2.** Layout for seismic tomography survey [12].

In the next step of the procedure the velocity fields of respective waves were reconstructed by the seismic tomography utilizing the SIRT (Simultaneous Iterative Reconstruction Technique). The iterative procedure was repeated until the error involved in reconstruction was effectively reduced to the noise level or levelled off. Accordingly, the measure of the reconstruction error was the standard deviation of the set of differences between the measured and computed times, derived basing on the reconstructed field of wave velocities along the ray paths.

To provide a reliable comparison with simulation data, geophysical measurement data are limited to selected parameters of the velocity field of seismic waves registered within the coalbed whilst the respective wave velocity fields in surrounding rock strata are neglected. The results are collated as contour maps of P-wave velocity (Fig. 3) and the plots of seismic anomaly parameters (Fig. 4). Thus obtained distribution patterns indicate that P-wave velocity ($V_{PW}$) along the ray path when propagating within the coalbed is variable in the range 1400÷1900m/s. Reconstructed velocity fields fall in the similar range and the mean radial velocity and mean velocity of the reconstructed field becomes 1725m/s. The anomaly parameter ($A_S$) is calculated in relation to the P-wave velocity determined in the natural conditions of the coal deposit (1950m/s) on the basis of previous seismic measurements in the coalbed 510.

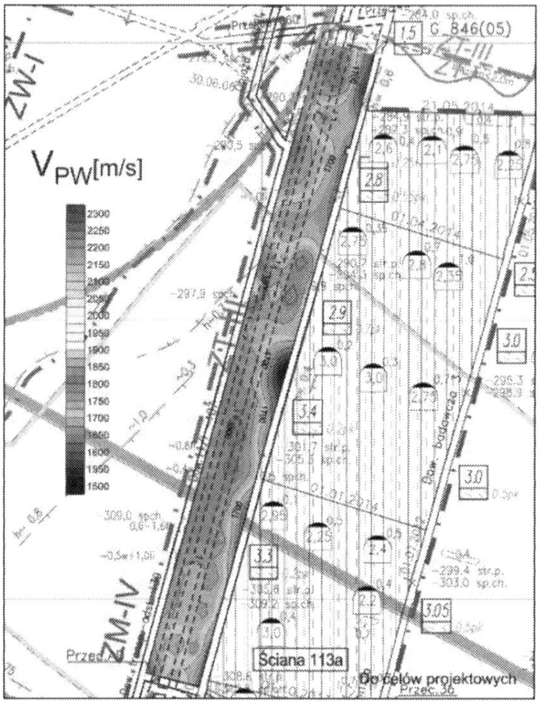


**Fig. 3.** P-wave velocity field within the coalbed (shortwall panel AD) [12].

Actually, the distribution of the seismic anomaly parameter is associated with redistribution of stresses within the coalbed, hence the positive value of the anomaly parameter in excess of 5% is indicative of stress increase whilst the negative values correspond to the weaker zones within the coalbed, including the de-stressing zones. Within the shortwall panel the averaged value of the seismic anomaly parameter equals –11% (Fig. 4), indicating that the mean vertical stresses are below the levels registered in the lithostatic conditions (associated with depth of deposition). For a single de-stressing operation involved in the mining system with hydraulic backfilling, the computed time required for effective de-stressing based on an averaged value of the seismic anomaly parameter will exceed one year; in the case of repeated de-stressing operations (involving the backfilling system) this time period equals three years from the date of the survey. Alongside the upwards heading 139 (within the belt 5m wide) the average value of the seismic anomaly parameter becomes –9%, varying from –4 to –14%, whilst in the proximity of the upwards heading 11/III-115 its varies from –10 to –25% (16% on the average). Actually, the seismic anomaly parameter distribution suggests that the stress concentration zone in the vicinity of the upwards heading 139 is connected to the fracturing zone due to the coalbed 510 being mined at the edge of its layer II, and to the zone caused by mining operations in the top layers (from the eastern end). Generally, the value of seismic anomaly parameter tends to decrease in the direction from the upwards heading 139 towards the heading 11/III-115. As the values of the anomaly parameter in excess of –7.5% are associated with the limited de-stressing efficiency zone, the de-stressing efficiency along the upwards heading 11/III-115 and its direct vicinity is on the average level, due to one or two layers in the coalbed 510 being mined-out. Within the belt of the undisturbed coal body extends up to 5m from the upwards heading 139, the fracturing zone emerging around the mine working and along the edge in the coalbed 510 gives rise to the weak de-stressing effect. Locally, over short segments, the negative value of the anomaly parameter corresponding to the de-stressing action tends to increase beyond –7.5%; however, there have been no reports of its value exceeding zero, which would be indicative of stress increase [12].

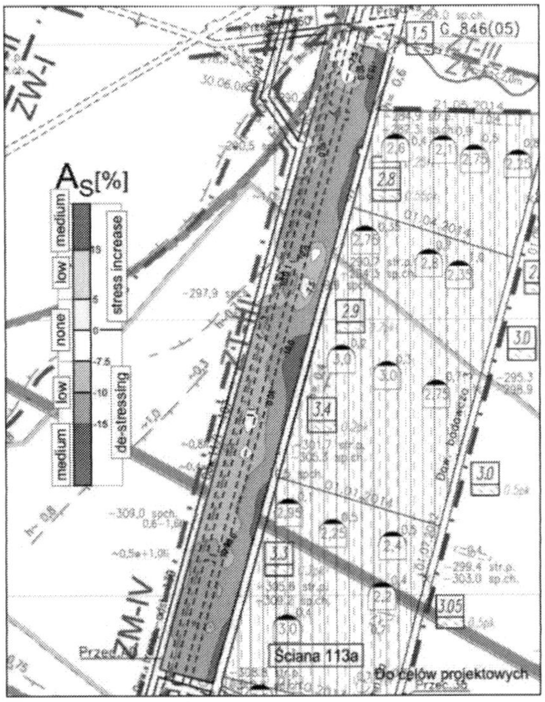


**Fig. 4.** Distribution of the anomaly parameter in the coalbed (shortwall panel AD) [12].

In the light of these considerations and the fact that the areas of lower P wave velocity (with respect to the natural conditions) identified in seismic surveys and characterised by the negative value of the seismic anomaly parameter are indicative of the stress de-stressing zones (in relation to the lithostatic conditions), it is reasonable to expect low and stabilised rockburst threat levels in the shortwall plot AD.

## 4 Analysis of the state of stress and rockburst hazard in the light of analytical modelling

From the geomechanical point of view, the decisive factors controlling the rockburst hazard include:

- the primary state of stress, associated with the mining depth, the impacts of the previous mining operations and tectonic disturbances,
- geomechanical properties of the rock media in the regions of mining activities,
- induced seismicity of the rock strata, due to potential activation of shock-prone strata.

At the stage of design of new mining operations or in the absence of reliable seismicity data, the rockburst hazard assessment has to rely in the first place on prediction of stress levels (by analytical or numerical methods) related to the resistance parameters of the rock medium. Actually, the factor controlling the rockburst occurrence is the presence of critical stress concentration zones in the proximity of mine excavations. Consequently, this evaluation of the rockburst threat in the shortwall panel AD relies on analytical simulation data (in the form of contour maps) [13] defining the vertical components of the primary stress tensor (Fig. 5) and the coefficient of stress concentration (Fig. 6).

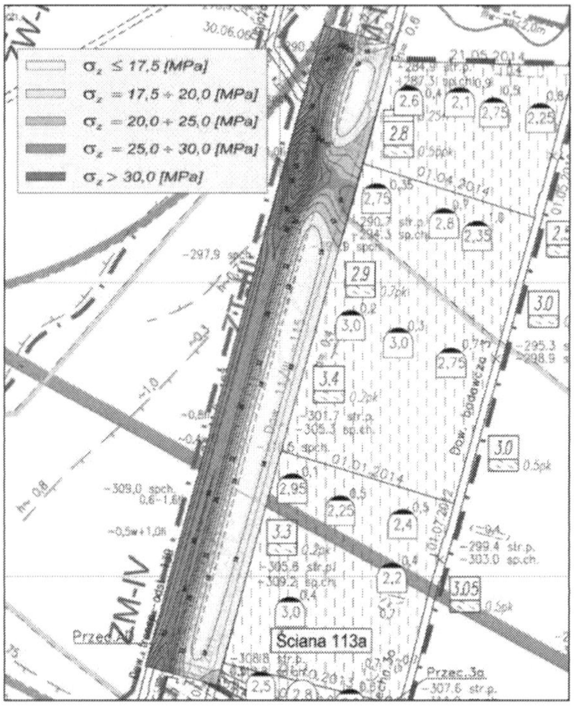


**Fig. 5.** Vertical stresses in the region of the shortwall panel AD [13].

Calculations show that the state of stress in the investigated shortwall plot (between the upwards headings 139 and 11/III-115, the cross-cut AD and the final segment of the face range 113a) is non-uniform, which can be attributed to the impacts of previous mining operations. Simulation data indicate that the value of the vertical stress component falls in the range 14.9÷43.3MPa (Fig. 5), so locally they exceed the uniaxial compressive strength of the coal, giving rise to the stress concentration zones within the coalbed. On the other hand, variability of the stress concentration factor expressed as the product of the primary and gravity-induced stresses varies from 1.03 to 2.98 (Fig. 6), which suggests that the de-stressing (stress relief) effect does not occur within the shortwall panel whilst the undercutting can lead to destructive de-stressing, of primary importance in the context of rockburst threat control. Stress concentration zones revealed on the contour maps are attributable to superimposed interactions of edges of the face end 111 (layer I within the coalbed 510) and 112 (layer II); of exploited excavations adjacent to the longwall face 113a (layer III) and the impacts of previous mining activities in the coalbed 510 distant by about 20m, along the final segment of the shortwall section. In its major part, the shortwall plot is affected by previous undercutting of two layers of the coalbed 510 (top roof and middle layer) so the destructive de-stressing effect is manifested through changes of geo-mechanical properties of the top layer within the investigated coalbed. The existing edge 113 in the seam 620 does not significantly contribute to the state of stress in the shortwall panel AD because the vertical distance between the respective seams is sufficiently large (140m).

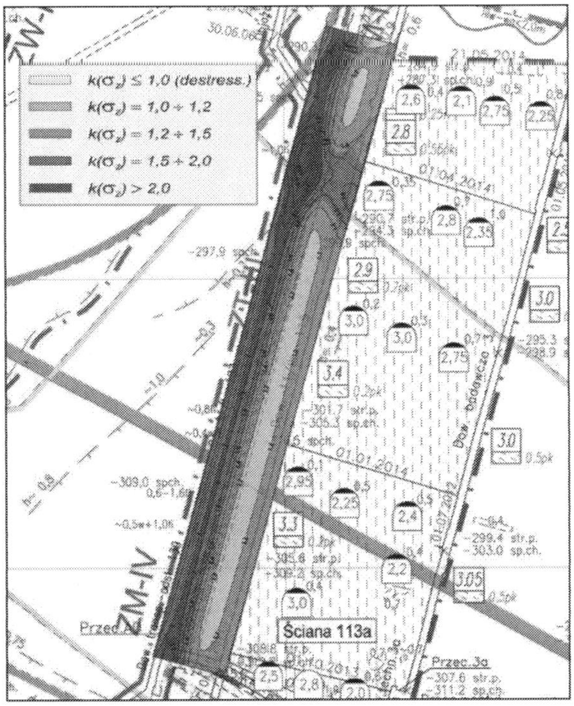


**Fig. 6.** Map of vertical stress concentration factor in the shortwall panel AD [13].

When evaluating the rockburst threat, two distinct phases of mining operations have to be considered when the involved threat levels will be different. In the first phase, when mining activities are continued in the section underneath the mined-out seam 510 the rockburst threat level will be relatively low because the major portion of seismic energy of potential quakes will be dissipated in old exploited excavations. In the second phase, when the mined section of the face range extends beyond the exploited workings within the seam 510, the rockburst threat levels will be elevated because the physical effect of seismic energy dissipation will not occur. The most threatened site (in the context of rockburst threat forecast and the presence of coal in the floor strata) will be the upwards heading 139 and the belt of undisturbed body of the coal deposit abutting to it.

## 5 Conclusions

With reference to the geological and mining conditions in the area, the study synthesises the key aspects involved in the development of the state of stress and the de-stressing effects experienced during mining operations in one of the collieries in the Upper Silesia Coal Basin. Seismic tomography data and the results of analytical simulations are interpreted and compared. Of particular interest was the belt-like section of undisturbed coal panel (40x35m) within the shortwall panel AD in the top layer of the seam 510, at the depth of 600m. The main objective of the investigations was to compare the geophysical survey data with analytical results obtained by the methods of strata mechanics in the context of rockburst hazard assessment prior to projected mining operations.

The comparative analysis leads us to the following conclusions:

- Even though the results obtained by the methods are not fully correlated, the state of stress in the shortwall panel AD is actually regarded as non-uniform, which can be attributed to previous mining of the coal seams 510 and 501.

- Analytical results suggest the occurrence of stress concentration zones due to the superimposed impacts of the edges of old exploitation, which is not corroborated by the geophysical survey data.
- Seismic survey data indicate a well-identified though non-uniform de-stressing effect in the shortwall plot due to previous mining of the two layers in the coal seam 510; its effectiveness being categorised as low or medium.
- According to the geophysical survey data, the coal seam 510 within the shortwall plot AD is generally not rockburst threatened even though its structure still retained to accumulate elastic energy. Analytical data suggest that rockburst threat in the course of mining operations will vary, depending on the actual position of the face front in relation to energy-dissipating gob areas within the coal seam 501.

## References

1. J. Dubiński, W. Konopko, *Tąpania - ocena, prognoza, zwalczanie (Rock bursts – assesment, prognosis, defeating),* Wyd. GIG Katowice (2000) (In Polish)
2. J. Dubiński, Z. Pilecki, W. Zuberek, *Badania geofizyczne w kopalniach – przeszłość, teraźniejszość, i zamierzenia na przyszłość (Geophysical research in mines - past, present and future plans)* Wyd. IGSMiE PAN, Kraków (2001) (in Polish)
3. J. Dubiński, *Sejsmiczna metoda wyprzedzającej oceny zagrożenia wstrząsami górniczymi w kopalniach węgla kamiennego (Seismic method of shock threat assessment in hard coal mines)* Central Mining Institute, Katowice (1989) (in Polish)
4. Czarny R., Marcak H., Nakata N., Pilecki Z., and Z. Isakow, Monitoring Velocity Changes Caused By Underground Coal Mining Using Seismic Noise. *Pure and Applied Geophysics* 173 (6), 1907–1916 (2016) doi: 10.1007/s00024-015-1234-3
5. Z. Pilecki, *Zastosowanie rejonowej obserwacji sejsmoakustycznej do kontroli stanu zagrozenia tapaniami (The use of zonal seismoacoustic observations for rockburst hazard monitoring).* Publ. – Institute of Geophysics, Polish Academy of Sciences, Series M, M-16 (245) 203-216 (In Polish)
6. Z. Szreder, Z. Pilecki, J. Kłosiński, *Comparison of Profiling Results of Attenuation and Velocity of Refracted P-wave in Coal-seam,* Proc. 14th EAGE European Meeting of Environmental and Engineering Geophysics (2008) doi: 10.3997/2214-4609.20146265
7. E. Glowacka, Z. Pilecki, Seismo-acoustic anomalies and evaluation of seismic hazard at the 'Marcel' Coal Mine, *Acta Geophysica* **37**, 1, 47–59 (1991)
8. Z. Pilecki, An Example of Rock Burst Hazard State Control Using a Zonal Seismoacoustic Observation. *Proc. Fifth Conf. on Acoustic Emission/Microseismic Activity,* Trans. Tech. Publications, Clausthal-Zellerfeld, 313–332 (1995)
9. A. Jakubów, Z. Szreder, Z. Pilecki, E. Pilecka, *Zabezpieczenie robót górniczych poprzez obserwacje sejsmoakustyczne na przykładzie KWK Jastrzębie (Application of seismoacustic observations for safety measures of mining works: example from Jastrzebie coal mine).* Publ. Inst. Geophys. Pol. Acad. Sc. M-19 (281) 209–215 (1995) (In Polish)
10. M. Kwaśniewski, J. Wang, *Analiza numeryczna deformacji górotworu wywołanych eksploatacją górniczą (Numerical analysis of mining-induced deformations of the rock mass),* Proc. of the School of Underground Mining, 153–187 (1996) (in Polish)
11. Z. Pilecki, *Dynamic analysis of mining tremor impact on excavation,* Proc. Int. FLAC Symp. on Num. Model. in Geom., Detournay & Hart, 397–400 (1999)

12. Report, *Ocena stanu zagrożenia tąpaniami oraz określenie stopnia odprężenia III warstwy pokładu 510 w rejonie planowanej eksploatacji ubierką na wybiegu 450m na podstawie pomiarów geofizycznych (Assessment of rockburst threat and de-stressing effect in the layer III within the coal seam 510 in the region of projected shortwall mining activities on the face range section 450 m, basing on geophysical survey data)*, Agos-Gemes Sp. z o.o., Katowice (2016) (in Polish, unpublished materials)
13. Report, *Ocena i analiza naprężeń w górotworze skłonnym do tąpań w aspekcie eksploatacji w latach 2015-2017 w zakresie projektowanej eksploatacji ubierką AD w partii wschodniej pokładu 510 (Assessment of state of stress in burst-prone rock strata due to mining activities over the years 2015-2017 in the context of projected shortwall mining operations in the eastern section of the coal seam 510, with the main focus on rockburst threat assessment procedures)*, Research and Technical Bureau Geotech II, Krakow (2016) (in Polish, unpublished materials)
14. A. Sałustowicz, *Zarys mechaniki górotworu (Fundamentals of strata mechanics)*, Slask Publishers (1968) (in Polish)
15. M. Borecki, M. Chudek, *Mechanika górotworu (Strata mechanics)*, Slask Publishers (1972) (in Polish)
16. Zorychta, *Kryterium powstawania tąpań przy eksploatacji pokładu węgla kamiennego (Criteria of rockburst occurrence during coal mining)*, AGH Journal of Mining and Geoengineering **125**, (1984) (in Polish)

# Effectiveness of reducing seismic hazard by means of group winning blasting - case study from a copper ore mine in Poland

*Anna Barbara* Gogolewska[1,*], and *Natalia* Czajkowska[2]

[1]Wrocław University of Science and Technology, Wrocław, Poland
[2]Graduate from Wrocław University of Science and Technology, Wrocław, Poland

**Abstract.** The copper ore deposit situated in the south-west of Poland is mined by three underground mines owned by KGHM Polish Copper JSC. Exploitation has been accompanied by rock burst hazard since the beginning. Thus, numerous different preventing measures have been developed such as temporary, organizational and long-term ones. However, no one has been able to predict the time, place and energy of a seismic event. The group winning blasting, with maximum number of blasted faces, is the most effective operation to reduce seismic threat. The more faces are blasted the more seismic energy should be reduced. The study aims at assessing the blasting effectiveness in inducing rock bursts and tremors. For this purpose, the seismic activity induced by mining and blasting were investigated. The number of blasting works and blasted faces as well as length of time between subsequent blasting works were analysed and related to provocation effectiveness. The linear correlation and different regressions were calculated to determine these relations. Moreover, the seismic energy reduction in the rock mass was evaluated by means of SRMS Index, which is a factor measured directly before and after blasting. The analyses covered one mine panel in the Polkowice-Sieroszowice copper mine over four-year period.

**Keywords:** copper ore deposit, underground mining, induced tremors, rock burst, group winning blasting

## 1 Introduction

Underground copper ore mining in Poland is associated with the Legnica-Głogów Copper District (LGCD) located in the south-western part of the country, where in 1957 the copper ore deposit was discovered and KGHM Polish Copper JSC has been operating since 1961. The company consists of three mines Rudna, Lubin, Polkowice-Sieroszowice and three copper smelters Cedynia, Głogów, Legnica [1].

Factors such as exploitation under constrained conditions at great depth in a single seam deposit, sandstone zones without copper mineralization, the use of room-and-pillar mining

* Corresponding author: anna.gogolewska@pwr.edu.pl

systems not rarely leaving remnants, cause a rock burst hazard, which is the most dangerous threat in the mines. This hazard also results from the geological structure and tectonics of the deposit [2].

In the LGCD mines, rock burst prevention must be used as an inseparable element of mining technology. Long-term, temporary and organizational preventive methods have been developed. The temporary methods include, but are not limited to, group excavation and face blasting, which consists in firing simultaneously several or a dozen or more faces (possible maximum number) in order to provoke tremors or rock bursts while waiting after blasting and thus induce elastic energy reduction in rock mass. The key preventive measure is monitoring the condition of the rock mass and the effectiveness of rock burst prevention by means of convergence measurements, continuous recording of seismic activity (mining seismology), passive seismic tomography as well as geotechnical and geological observations. In addition, observations of seismoacoustic activity induced by group blasting are carried out with the use of the Microprocessor Meter of Crackles (MMC) before and after blasting works, whose results are presented as the Synthetic Index of Rock Mass Stability (SRMS Index) [3].

To-date, the group blasting has been the most important active method for reducing rock burst hazard, hence the aim of the study was to assess its effectiveness in inducing seismic events and to determine its relationship with rock mass seismicity and blasting parameters. The necessity of evaluating the condition of the rock mass is governed by state legal regulations as well as internal mine ordinances. The analyses covered the D-IE block, which belongs to the D field in the G-54 division in the Polkowice-Sieroszowice mining plant. Four years (2009-2012) of high seismic activity were selected. For the study purpose the group blasting parameters, seismic activity and rock burst hazard were characterised and investigated. The relationship between the SRMS Index and seismic energy was analysed but only for two years (2009 and 2010), in which the majority of blasting operations were done at regular intervals and seismic hazard was higher compared to 2011-2012 years. Similarly, the correlation between the number of blasted faces, the number of group blasting works and the time length between these operations and the provocation effectiveness related to the number and energy of tremors was analysed for those two years due to the same reasons.

## 2 Research area

The first operational works in the Polkowice mine were conducted in 1968. The Sieroszowice mine was established in 1978 as a result of drilling new excavations from the Rudna and Polkowice mines and crossing the boundaries of the Sieroszowice mining area with these excavations. The new mine did not have its own shaft, therefore the Rudna mine's shaft was used to transport the ore to the surface [1, 4].

The existing Polkowice-Sieroszowice mine plant, was created in 1996 by merging the Polkowice mine with the Sieroszowice one [1, 5]. The mine is mining in the following mining areas: Sieroszowice (in 85% of this area, because its 15% is exploited by the Rudna mine), Polkowice, Radwanice-East (Radwanice-Wschód), and in the part of the Deep Industrial Głogów (Głogów Głęboki-Przemysłowy-GGP) (in 50% of this area, because 50% of the area will be mined by the Rudna mine) (Fig. 1).

The mine operates in 20 mining fields at a depth of 600 to 1200 m, and the GGP plant will exploit the deposit at a depth of over 1200 m. The copper ore exploitation conditions are good in most of the area occupied by the mine. Difficulties may occur only in the southern part of the deposit because of strongly developed tectonics [1].

The mine excavates two lithological types of copper ore i.e. sandstone and carbonate-slate ores. The thickness of the deposit is relatively small and depends on the part being exploited; in the Polkowice area it is in the range of 0.6 to 4.5 m, with the average thickness of 2.5 m; in the area of Sieroszowice it varies from 0.7 to 4.5 m, and the average thickness reaches 2.0 m. Due to the thin layers the mine does not use hydraulic backfill but roof self-deflection to protect and liquidate the empty post mining space. The bolting system is used to support the roof and walls of excavations [3 ,4, 5].

**Fig.1**. Mining areas of the Polkowice-Sieroszowice copper ore mine (marked in yellow) (Sieroszowice, Polkowice, Radwanice-Wschód (Radwanice-East) and Głogów Głęboki-Przemysłowy (Deep Industrial Głogów) (based on [5]).

The exploitation is carried out with various types of a one-stage room-and-pillar system with roof self-deflection, which are adapted to the geological and mining conditions. The deposit is excavated with the use of a blasting technique. Mining and transport are carried out with machines adjusted to a small deposit thickness. The one-stage room-and-pillar system with roof self-deflection (J-UG-PS- the acronym in Polish) is used in four mining divisions of the Polkowice-Sieroszowice mine including the G-54 one [3, 4, 5].

The G-54 division belongs to the mining area of Sieroszowice. the D field lies at a depth of 925 to 1050 m and consists of the D1E floor, D2E floor, D3E floor and D-IE block. The D field is bounded on the northern side with inclines, from the south and east it borders with the Rudna mine, while on the western side it is limited by a drift and a zone without mineralization located within the D1E floor (Fig. 1).

## 3 Results and discussion

Analysis of the provocation effectiveness of blasting works covered the D-IE block in the G-54 division in 2009-2012 years. To determine the appropriate mine field and years for the research, seismic activity was calculated. The effectiveness of seismic energy reduction in rock mass with the use of the group blasting was calculated and assessed taking account of tremors and their effects in form of rock bursts. The parameters of the blasting works were determined, including their number, number of faces to be blasted and the time between successive blasting works. The variability of the SRMS Index in relation to seismic activity was observed. The correlation between blasting work parameters and the effectiveness of the inducement of tremors and rock bursts was calculated.

### 3.1 Seismic activity

Seismic activity in the G-54 division and in the D-IE block in years 2009-2012 was high. It was described by the number and energy of tremors and rock bursts. It has been assumed that tremors are seismic phenomena with $E \geq 1 \cdot 10^3$J energy while high-energy tremors have $E \geq 1 \cdot 10^5$J energy. The seismic activity is shown in Table 1 and Figure 2. In 2009-2012 years, in the G-54 division 948 tremors with total energy of $5.53 \cdot 10^8$J ($5.84 \cdot 10^5$J/1tremor) occurred including 129 high-energy ones with total energy of $5.42 \cdot 10^8$J ($4.20 \cdot 10^6$J/1tremor).

In the D-IE block there were 644 tremors with total energy of $4.42 \cdot 10^8$J ($6.86 \cdot 10^5$J/1tremor), including 88 high-energy ones with total energy $4.34 \cdot 10^8$J ($4.93 \cdot 10^6$J/1 tremor). The majority of tremors (250) occurred in 2011, in 2009 the least (95) were registered. In 2010 there were 147 seismic events and in 2012, 152 ones. In 2010, the energy of tremors was the highest and amounted to $2.55 \cdot 10^8$J and in 2009 it reached $8.15 \cdot 10^7$J. The total energy of tremors was the smallest ($1.02 \cdot 10^7$J) in 2012. The largest amount of energy per 1 tremor was $1.74 \cdot 10^6$J/1tremor in 2010 and the lowest in 2012 ($6.74 \cdot 10^4$J/1tremor).

In the D-IE block, the majority (29) of high-energy tremors occurred in 2009, in 2010 only one less tremor was recorded (28) (Figure 2). In 2010, the total energy of high-energy tremors was the highest and amounted to $2.54 \cdot 10^8$J and in 2009 it reached $8.04 \cdot 10^7$J. In 2011 there were 18 high-energy tremors with total energy of $9.19 \cdot 10^7$J. The smallest number (13) and energy ($7.98 \cdot 10^6$J) of tremors appeared in 2012. This year, there was no tremors of energy equal to or larger than $1 \cdot 10^7$J. The largest energy per one high-energy tremor ($9.07 \cdot 10^6$J) was in 2010.

It can be stated that the energy of tremors decreased in subsequent years of 2009-2012 period, which means that the seismic risk was reduced. The seismic activity in the G-54 division was determined by that of the D-IE block, so the analyses were focused on this block. The years 2009-2011 were more active compared to 2012 year, which justified the selection of 2009-2011 period to be assessed in further investigations.

At the beginning of October 2012, the progress of all the faces of the rooms directed to the block border was stopped by the stony zone. Thus, the development of the exploitation front of the D-IE block was completed; only liquidation of workings was carried out. In October, November and December no tremors were provoked.

#### *3.1.1 Rock bursts in G-54 mining division over 2009-2012 period*

In the Polkowice-Sieroszowice mine within the years 2009-2012, there were 22 rock bursts with a total energy of $1.11 \cdot 10^9$J; 11 of them with a total energy of $4.21 \cdot 10^8$J occurred in the G-54 division, including 8 ones with energy of $3.43 \cdot 10^8$J occurred in the D-IE block. In 2009, there were 9 rock bursts in the entire mine ($1.7 \cdot 10^8$J), in the G-54 division 3 rock bursts ($6.41 \cdot 10^7$J) and 2 ($4.8 \cdot 10^7$J) occurred in the D-IE block. In 2010, there were 9 rock bursts in the entire mine ($8.09 \cdot 10^8$J), in the G-54 division there were 5 events ($2.44 \cdot 10^8$J) and 4 ones occurred in the D-IE block ($2.13 \cdot 10^8$J). In 2011, the number and energy of rock bursts dropped significantly. There were 4 rock bursts in the whole mine ($1.35 \cdot 10^8$J), in the G-54 division 3 events ($1.13 \cdot 10^8$J) appeared, 2 in the D-IE block ($0.82 \cdot 10^8$J). In 2012, there were no rock bursts in the mine, which could be related to the completion of the cutting in the D-IE block. It can be noticed that the number and energy of rock bursts increased in 2010 compared to 2009 and then considerably decreased during the period 2011 to 2012. The rock burst hazard in the entire mine was primarily affected by that of D-IE block.

**Table 1**. Seismic activity in G-54 mining division and D-IE block related to energy classes in years 2009-2012.

| Year | Energy Class [J] | | | | | Number of Tremors | Total Energy [J] | Energy per one Tremor [J] |
|---|---|---|---|---|---|---|---|---|
| | $10^3$ | $10^4$ | $10^5$ | $10^6$ | $10^7$ | | | |
| **G-54 mining division** | | | | | | | | |
| **2009** | 46 | 36 | 27 | 3 | 3 | 115 | $9.87 \cdot 10^7$ | $8.58 \cdot 10^5$ |
| **2010** | 104 | 59 | 26 | 7 | 6 | 202 | $2.93 \cdot 10^8$ | $1.45 \cdot 10^6$ |
| **2011** | 209 | 96 | 17 | 6 | 3 | 331 | $1.33 \cdot 10^8$ | $4.02 \cdot 10^5$ |
| **2012** | 174 | 95 | 25 | 6 | 0 | 300 | $2.84 \cdot 10^7$ | $9.46 \cdot 10^4$ |
| **Number of Tremors** | 533 | 286 | 95 | 22 | 12 | 948 | $5.53 \cdot 10^8$ | $5.84 \cdot 10^5$ |
| **D-IE block** | | | | | | | | |
| **2009** | 41 | 25 | 24 | 3 | 2 | 95 | $8.15 \cdot 10^7$ | $8.58 \cdot 10^5$ |
| **2010** | 81 | 38 | 18 | 5 | 5 | 147 | $2.55 \cdot 10^8$ | $1.74 \cdot 10^6$ |
| **2011** | 156 | 76 | 13 | 3 | 2 | 250 | $9.50 \cdot 10^7$ | $3.80 \cdot 10^5$ |
| **2012** | 78 | 61 | 11 | 2 | 0 | 152 | $1.02 \cdot 10^7$ | $6.74 \cdot 10^4$ |
| **Number of Tremors** | 356 | 200 | 66 | 13 | 9 | 644 | $4.42 \cdot 10^8$ | $6.86 \cdot 10^5$ |

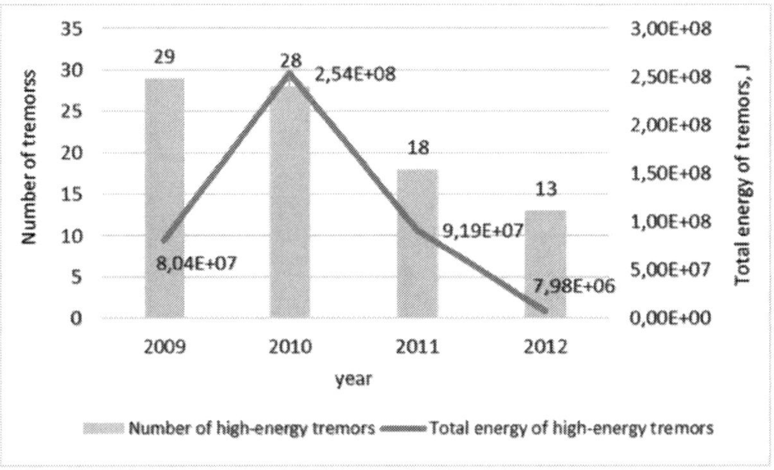


**Fig.2.** Number and total energy of high-energy tremors in D-IE block in years 2009-2012.

## 3.2 Group winning blasting parameters

In rock bursting prevention active methods are widely used to reduce threats triggered by rock bursts. They are based on special blasting works carried out not only in the solid face, but sometimes also in the floor and roof. Blasting works are performed in the excavations to mine the ore; in roof layers to trigger a roof to collapse and to eliminate the mined-out space; at the floor of workings and walls of the pillars, to reduce the rock mass seismicity. Provoking the rock mass to vibrate with group blasting works proved to be a very effective method reducing the rock burst threat used in Polish copper ore mines [3, 4, 6]. The winning group blasting works were characterized by their frequency, which was defined and calculated by the time length between subsequent blasting operation (in hours), the number of blasting works within one month, the number of faces blasted simultaneously, the number of faces blasted in a month in the D-IE block. The group blasting parameters were analysed over 2009-2012 years. During the first three years of the period (2009-2011) a relatively significant number of regular blasting works were performed compared to 2012 year. Therefore, majority of further analyses were referred to 2009-2011 years.

The number of blasting works during each year is presented in Table 2. In 2009 the number of blasting works was 218, in 2010 it was 217, in 2011 decreased to 137 and in 2012 decreased to 77. During 2009-2012 period 649 blasting works were done. In 2009, the monthly average number of blasting works was 18.2, in 2010 it slightly decreased to 18.1, in 2011 it decreased to 11.4 and in 2012 it was only 8.6.

**Table 2**. Monthly average number of group blasting works in D-IE block in years 2009-2012.

| Year | Number of Blasting Works | Monthly Number of Blasting Works per year |
|---|---|---|
| 2009 | 218 | 18.2 |
| 2010 | 217 | 18.1 |
| 2011 | 137 | 11.4 |
| 2012 | 77 | 8.6 |

Table 3 includes the number of faces blasted in particular months in each year. In year 2012 from October to December no blasting works were done, which was connected with decreasing area of exploitation and difficult mining and geological conditions. So, nine months (January to September) of 2012 year should be considered in the further calculations. In 2009, 3,124 faces were blasted, and in 2010, 3,450 ones. The monthly number of faces per one blasting in 2009 was 14.3 and in 2010 it reached 15.9. Monthly number of faces varied from 89 to 397 and was relatively stable. In 2011, 2,666 faces were blasted, and in 2012, 1,369 ones. The monthly number of faces per one blasting in 2011 increased to 19.5 and in 2012 to 17.8 compared to 2009-2010 years. Monthly number of faces varied from 10 to 397 and was relatively unstable.

During 2009-2012 period the monthly average time between blasting works varied from 22 h to 624 h and was relatively stable excluding that in 2012 year (Table 4.) The average monthly time interval between subsequent blasting works in year 2009 was 32.1 h, in 2010 it slightly decreased to 30.2 h, in 2011 it significantly increased to 61.3 h and in 2012 it enlarged to 150.8 h. Therefore, in 2009 the faces of headings were blasted every 32.1 hours, in 2010 they were blasted slightly more frequently i.e. every 30.2 hours, in 2011 the blasting operations were performed every 61.3 hours and in 2012 they were carried out every 150.8 hours. The time between successive blasting works significantly increased in 2011 and 2012 compared to the 2009-2010 years.

The results of observations done in 2012 year are not well comparable with results obtained in years 2009-2011 since in 2012 the exploitation was constrained due to difficult mining and geological conditions (occurrence of vast stone zones); excavation was stopped at the beginning of October. The necessity to increase the number of faces in 2011 and 2012 might be connected with decreasing number of blasting works due to the aforementioned conditions and still high seismic hazard.

**Table 3.** Number of faces blasted in D-IE block in a given month in years 2009 and 2010.

| Month | I | II | III | IV | V | VI | VII | VIII | IX | X | XI | XII | Total Number of Faces | Faces in one blasting |
|---|---|---|---|---|---|---|---|---|---|---|---|---|---|---|
| **Year** | **2009** | | | | | | | | | | | | | |
| **Number of Faces** | 265 | 259 | 224 | 158 | 247 | 325 | 145 | 250 | 338 | 360 | 265 | 288 | 3,124 | 14.3 |
| **Year** | **2010** | | | | | | | | | | | | | |
| **Number of Faces** | 279 | 303 | 320 | 312 | 186 | 340 | 384 | 397 | 387 | 89 | 265 | 188 | 3,450 | 15.9 |
| **Year** | **2011** | | | | | | | | | | | | | |
| **Number of Faces** | 209 | 115 | 269 | 171 | 193 | 316 | 269 | 260 | 257 | 261 | 192 | 154 | 2,666 | 19.5 |
| **Year** | **2012** | | | | | | | | | | | | | |
| **Number of Faces** | 280 | 261 | 218 | 135 | 94 | 153 | 10 | 128 | 90 | 0 | 0 | 0 | 1,369 | 17.8 |

**Table 4.** Monthly average time between subsequent blast works in a given month in D-IE block in years 2009-2010.

| Month | I | II | III | IV | V | VI | VII | VIII | IX | X | XI | XII | per one year |
|---|---|---|---|---|---|---|---|---|---|---|---|---|---|
| **Year** | **2009** | | | | | | | | | | | | |
| **Time [h]** | 27 | 29 | 32 | 27 | 30 | 29 | 55 | 35 | 32 | 28 | 30 | 32 | 32.1 |
| **Year** | **2010** | | | | | | | | | | | | |
| **Time [h]** | 30 | 33 | 29 | 31 | 29 | 29 | 24 | 22 | 29 | 24 | 32 | 52 | 30.2 |
| **Year** | **2011** | | | | | | | | | | | | |
| **Time [h]** | 63 | 55 | 56 | 75 | 54 | 54 | 56 | 56 | 61 | 58 | 59 | 89 | 61.3 |
| **Year** | **2012** | | | | | | | | | | | | |
| **Time [h]** | 67 | 62 | 67 | 95 | 150 | 102 | 624 | 96 | 94 | - | - | - | 150.8 |

## 3.3 Effectiveness of group winning blasting in provocation of rock bursts and tremors

The assessment of the effectiveness of provoking seismic dynamic events by the group blasting works consists in determining the percentage of the number and energy of provoked events in the number and energy of all the recorded seismic phenomena. The group blasting works are the more effective, the higher the percentage of the number and the energy of the provoked seismic events is. The effectiveness of provocation related to energy is the most crucial factor because the release of cumulated energy in the rock mass reduces seismic hazard to a large extent. The effectiveness was assessed for the D-IE block and 2009-2012 years. The analysis covered rock bursts, tremors with energy of $E \geq 10^3$J and those with energy of $E \geq 10^5$J (i.e. high-energy tremors) (Table 5).

In years 2009-2012, in the entire mine, 15 rock bursts of $9.44 \cdot 10^8$J of total energy were provoked by blasting, which indicated 68% efficiency related to the number of rock bursts and 85% efficiency related to the energy of tremors causing the rock bursts. In the G-54 division blasting works provoked 8 ($3.16 \cdot 10^8$J), which posed 72.7% effectiveness for the number and 75.1% for energy. In the D-IE block 7 ($2.86 \cdot 10^8$J) rock bursts were provoked, which indicated 87.5% effectiveness for their number and 83.4% for energy.

In 2009, in the entire mine, the effectiveness of provoking rock bursts related to their number reached 44% and that for energy, 58.8%. In the G-54 division and the D-IE block, the effectiveness of rock bursts' provocation for their number was 66.7% and 100%

respectively while that for energy amounted to 75% and 100% respectively. In 2010, in the whole mine, the provocation effectiveness of rock bursts related to their number was 77,8% and that related to energy reached 91.5%. In the G-54 division, the effectiveness of rock bursts' provocation for their number was 80% and that for energy reached 76.2%. In the D-IE block, the effectiveness for rock bursts' number was 75% and for energy reached 73,2%. In 2011, in the entire mine, the effectiveness of provoking the rock bursts related to their number and energy reached 75% and 77% respectively. In the G-54 division, the effectiveness for rock bursts' number was 66.7% and that for energy reached 72.3%. In the D-IE block the effectiveness for both their number and energy amounted to 100%. In 2012 no rock bursts appeared. It can be stated that the provocation of rock bursts by group blasting works was highly effective.

In 2009, in D-IE block, the effectiveness of tremors' provocation in relation to their number was 17.9% and to their energy reached 84.4%. The effectiveness for high-energy tremors amounted to 17.2% for their number and 85.3% for their energy. In 2010, the provocation effectiveness for tremors' number was 21.1% and for their energy it amounted to 71.9%. The provocation effectiveness for high-energy tremors was 21.4% for number and 72.2% for energy. In 2011, the effectiveness of tremors' provocation in relation to their number was 20.4% and in relation to their energy it was 87.9%. The effectiveness for high-energy tremors was 27.8% for their number and 89.7% for their energy. In 2012, within first nine months, the effectiveness of tremors' provocation for their number was 38.2% and for their energy it was 16.9% while such effectiveness for high-energy tremors was 30.8% for their number and 7.8% for energy.

It can be noticed that in 2010 the effectiveness of provocation of tremors for their energy decreased and for their number increased compared to 2009 while in 2011 the effectiveness connected with number of tremors slightly dropped but energy provocation effectiveness increased largely compared to 2010. The provocation effectiveness associated with energy played the most decisive role in seismic hazard reduction. During 2009-2011 years in the D-IE block, such effectiveness was very high and the group blasting works fulfilled their aim completely. However, in 2012, the efficiency for energy decreased largely to 7.8% due to stopping the excavation; only liquidation operations were carried out.

**Table 5**. Effectiveness of blasting related to number and energy of tremors in D-IE block.

| Year | Tremors of Energy of $E \geq 10^3$ J | | Tremors of Energy of $E \geq 10^5$ J | |
|---|---|---|---|---|
| | Effectiveness related to Number of Tremors | Effectiveness related to Energy of Tremors | Effectiveness related to Number of Tremors | Effectiveness related to Energy of Tremors |
| [-] | [%] | [%] | [%] | [%] |
| 2009 | 17.9 | 84.4 | 17.2 | 85.3 |
| 2010 | 21.1 | 71.9 | 21.4 | 72.2 |
| 2011 | 20.4 | 87.9 | 27.8 | 89.7 |
| 2012 | 38.2 | 16.9 | 30.8 | 7.8 |

## 3.4 SRMS Index and seismic energy

The measurement of seismoacoustic activity induced with blasting works is used to assess the stress concentration in rock mass after blasting operations in KGHM mines.

The Microprocessor Meter of Crackles (MMC) records seismoacoustic activity occurring in the region of its installation site (up to approx. 50 m) before and after blasting operations and calculates the SRMS Index, which can vary from 1 to 10 class. The higher the value of the Index, the greater stress concentration in the rock mass. Any increase in the SRMS Index is treated as a growing disturbance of the rock balance due to an increase in stress concentration.

A significant rise of the Index, by 3 or more classes, may indicate an increased probability of a strong dynamic event occurrence. Moreover, a decrease in the number and energy of crackles may mean that the stress concentration zone is moving away, while the increase of these parameters may suggest approaching the region with enhanced stress concentration.

The SRMS Index was analysed in the D-IE block over the period of 2009-2011 years since the majority of seismoacoustic observations were then carried out and a large number of blasting works were regularly done. 996 measurements of the SRMS Index were performed (in 2009-380, in 2010-354, in 2011-262). In 2009, three significant growths of the SRMS Index were recorded; they took place in March, April and August. In March, the seismic activity was low, only three seismic tremors occurred, while in April there was a rock burst and in August the energy of tremors was significant. Thus, the increase in the Index properly indicated the destabilization of the rock mass twice. In July, there was a rock burst but the Index was stable and low. In 2010, there were six large growths in the Index, but only in December this growth was connected with a rock burst. In April, October and December, rock bursts occurred but the SRMS Index remained low and stable. In 2011, there were six large growths in the SRMS Index, but only in December the increase was connected with the rock burst. In years 2009-2011, fifteen significant growths of the Index were found and three of them were associated with rock bursts. The remaining five (63%) rock bursts were not accompanied by the substantial growth of the Index. It can be noticed that the stress concentration in the rock mass depicted by the SRMS Index was loosely connected with seismic and rock burst threat. The lack of signalling high-energy tremors with a clear increase in seismoacoustic activity recorded with the Microprocessor Meter of Crackles (MMC) may indicate another mechanism for provoking these tremors.

### 3.5 Effectiveness of provocation of seismic events in relation to parameters of group blasting works

The relationships between the effectiveness of the group blasting in the provocation of tremors and the number of blasting works, the time length between the subsequent blasting works and the number of blasted faces were determined. For this purpose, Pearson linear correlation coefficients and linear and polynomial regressions were calculated. The analysis was carried out for the D-IE block for high-energy tremors and for tremors of $E \geq 1 \cdot 10^3$J energy. The period 2009-2011 was selected for the analyses since the large number of regular blasting works and seismoacoustic observations were then performed.

No clear relationship was found between the effectiveness of tremor provocation and the number of blasting works and the number of blasted faces. The linear correlation coefficient between the number of blasting works and effectiveness was -0.13 (number of tremors) and -0.2 (energy of tremors). The correlation coefficient between the number of faces and the effectiveness was -0.1 (tremors' number) and -0.14 (energy).

The linear relationship between the effectiveness of tremor provocation and the length of time between successive blasting operations (frequency) was only found. The correlation coefficient of 0.48 (effectiveness related to number of tremors) and of 0.47 (effectiveness related to energy of tremors) were obtained for high-energy tremors. The correlation coefficient of 0.47 (effectiveness related to number of tremors) and of 0.50 (effectiveness related to energy of tremors) were obtained in case of tremors of $E \geq 1 \cdot 10^3$J energy. It can be said that the longer the time passed between successive blasting works, the greater their effectiveness in the provocation of tremors was. When the frequency of blasting was lower, the energy of provoked tremors was higher.

## 4 Conclusions

1. The seismic activity in the D-IE block over 2009-2012 years remained at a high level, which justified the application of the group blasting to reduce seismic and rock burst hazard. The total number and energy of rock bursts and tremors decreased in subsequent years, which may be connected with the high effectiveness of group blasting in provoking tremors and rock bursts.
2. The obtained results confirmed that the group blasting was the highly effective way of decreasing the seismic hazard in the D-IE block.
3. The seismoacoustic SRMS Index was not clearly related with the seismic activity and the associated rock burst threat in the D-IE block in 2009-2011; the lack of indicating high-energy tremors by means of an explicit growth of seismoacoustic activity might suggest other mechanism for generating these tremors. This would be an interesting issue for future research.
4. In 2009-2011 in the D-IE block both the number and energy effectiveness of provocation of seismic events did not depend on the number of group blasting works or blasted faces. However, they were inversely proportional to the frequency of blasting; the energy of induced tremors increased with decreasing intensity of blasting works. Future investigations are necessary to further develop and confirm these initial findings by analysis conducted in diverse geological and mining conditions and over extended time period.
5. The results indicate the necessity to assess the provocation effectiveness of blasting so that the seismic hazard could be controlled to a certain extent.

*The paper publishing was supported by the statutory research 0401/0125/17/21.*

## References

1. P. Kijewski (ed.), *Monografia, Zakłady Górnicze Polkowice-Sieroszowice (Monography Polkowice-Sieroszowice Mining Plant)*, KGHM Cuprum Sp. z o.o., Wrocław (2006) (in Polish)
2. W. Konopko, *Wieloźródłowość wstrząsów górotworu (Multi-origin of induced tremors)*, Przegląd Górniczy **65**, 72-75 (2009) (in Polish)
3. J. Butra, *Eksploatacja złoża rud miedzi w warunkach zagrożenia tąpaniami i zawałami (Exploitation of the copper ore deposit under conditions of rock bursts and roof falls)* KGHM Cuprum Sp. z o.o., Wrocław (2010) (in Polish)
4. A. Piestrzyński (ed.), *Monografia KGHM Polska Miedź S.A. (Monography of KGHM Polish Copper JSC)*, KGHM Cuprum Sp. z o.o., Wrocław (2007) (in Polish)
5. S. Bartlett, H. Burgess, B. Damjanović, R. Gowans, C. Lattanzi, *Raport techniczny dotyczący produkcji miedzi i srebra przez KGHM Polska Miedź S.A. w Legnicko-Głogowskim Okręgu Miedziowym w południowo-zachodniej Polsce, (Technical report on the production of copper and silver by KGHM Polish Copper JSC in the Legnica-Głogów Copper District in south-western Poland),* KGHM Polska Miedź S.A., Lubin (2013) (in Polish)
6. Z. Kłeczek, *Grupowe strzelanie przodków jako element profilaktyki tąpaniowej w kopalniach rud miedzi LGOM (Group blasting as a part of rock burst prevention in LGOM copper mines),* Górnictwo i Geoinżynieria **28**, 153-159 (2004) (in Polish)

# Comparison of the results of the seismic profiling and WAS-96/RMS seismoacoustic active method in an assessment of the impact of the overlying coal seam edge

*Sławomir* Olechowski[1], *Krzysztof* Krawiec[2,*], *Jakub* Kokowski[2], *Zbigniew* Szreder[3], *Paulina* Harba[2], and *Mateusz* Ćwiękała[1]

[1]PGG S.A. KWK ROW „Ruch Rydułtowy", Rydułtowy, Poland
[2]Mineral and Energy Economy Research Institute of the Polish Academy of Sciences, Cracow, Poland
[3]JSW S.A. KWK "Budryk", Ornontowice, Poland

**Abstract.** In the study, a comparison between the results of seismic profiling and the active seismoacoustic method WAS-96/RMS has been presented. The methods were used to assess the impact of an overlying coal seam edge on the relative stress state of the test heading in the chosen hard coal mine in the Upper Silesian Coal Basin, Poland. The work presents the methodologies of measurement, processing and interpretation of both methods. In the research area there were two edges of the adjacent seams at vertical distances of 70 and 100 m. The obtained results allowed for the development of conclusions regarding the effectiveness of both methods. It was stated, that the seismic profiling method, as well as the WAS-96/RMS method allowed for the identification of anomalies in the area of impact of the overlying coal seams. Based on previous experience, a comparison of the advantages and limitations of the seismic profiling and the WAS-96/RMS method has also been presented. As a result, it was found that seismic profiling should be the basic method for assessing the impact of the edges and remnants of exploited seams, whereas the WAS-96/RMS method may be used as a complementary method if a confirmation of the rockbursts threat is required.

**Keywords:** hard coal mining, seismic threat, stress concentration, edges of coal seam, seismic profiling, seismoacoustic method

## 1 Introduction

The seismic threat in Polish hard coal mines is associated with many factors of natural and mining origin [1–6]. One of the basic ones is stress concentration in coal seams which occurs in particular as a result of the impact of the edges and remnants of adjacent coal seams [7–

* Corresponding author: k.krawiec@meeri.pl

9]. The magnitude of such impacts is commonly recognized by seismic profiling [1, 10–18]. Refracted P-wave velocity profiling used to be performed according to the Dubiński method [10] updated in the work of Dubiński and Konopko [1]. Szreder et al. [15] made significant progress in the field of seismic data processing and its interpretation.

The assessment of the seismic threat in coal mine excavations may also be carried out using the active seismoacoustic method WAS-96/RMS developed by Kornowski et al. [19]. This method allows for the assessment of the seismic threat on the 10 m long section of the coal seam.

The aim of the study is to compare the results of both methods - seismic profiling and WAS-96/RMS in identical conditions of impact of the overlying edges of coal seams. Both methods are based on different parameters. In seismic profiling, the refracted P-wave is determined, while in the WAS-96/RMS method the $Z_{RMS}$ parameter is determined on the basis of energy intensity and the activity of the induced seismoacoustic emission.

The work presents the methodologies of measurement, processing and interpretation of both methods. The research was carried out in one of the hard coal mines in the Upper Silesian Coal Basin (USCB). In the area of research, there were two edges of adjacent seams at vertical distances of 70 and 100 m. The obtained results allowed for the development of conclusions regarding the effectiveness of both methods.

# 2 Methodology

## 2.1 Seismic profiling method

*Measurement methodology*

Depending on the conditions of propagation of the refracted P-wave in a coal seam [18]:

- The length of spread for 24 geophones may reach up to 115 meters adjusted to the energy of the wave source. The geophone interval should be assumed to be from 2 to 5 meters depending on the possibility of identifying refracted P-waves.
- The 40 Hz geophones are usually fixed on short anchors a few centimeters long in coal seam that is not intensively fractured or separated fragments.
- The seismic wave is excited with the use of a stroke with a sledgehammer of 5 kg.
- For improvement of the signal to noise ratio, at least six fold stacking should be applied.
- The seismic records are legible even in longer sections in the case of a low level of seismic noise and high amplification in the test equipment of the order of over 100 dB.
- The recording time and signal sampling always have to be tested in a specific investigation site. On the basis of previous experiments, one should select a sample interval of 0.125 ms and recording length of 0.5 sec.

*Methods of processing and interpretation*

In the interpretation stage, a two-layer model of a medium consisting of a fracture zone (plastic zone) and a solid zone (elastic zone) is assumed [15]. The reciprocal travel times method is used to calculate the velocity model and width of the fracture zone. The velocity model is corrected by the reverse analysis method. By changing the position of the seismic boundaries of the model, the calculated hodographs are adjusted to the observations. The accuracy of the calculations is verified by minimizing the average square error. In the final stage, velocity changes were compared with the reference velocity $V_0$ determined from the Dubiński empirical equation [10]:

$$V_0 = 1200 + 4.83H^{0.76} \tag{1}$$

where:
H – depth of survey area [m].

The reference velocity $V_0$ and the measured velocity $V_P$ were used to calculate the percentage anomaly A according to the formula:

$$A = \frac{V_P - V_0}{V_0} \cdot 100\ [\%] \tag{2}$$

The calculated anomaly is applied to evaluate the relative stress state based on the seismic scale, determined for USCB conditions (Tab. 1) or measured velocity in the region outside of anomaly. The error of the measured velocity did not exceed 50m/s [18].

**Table 1.** Seismic scale of relative stress changes in Upper Silesian Coal Basin conditions (on the basis of [1]).

| Degree of relative stress increase (decrease) | Positive (negative) seismic anomaly [%] | Scale of relative stress increase (decrease) | Probable increase (decrease) in relative stress [%] |
|---|---|---|---|
| 0 | < 5 (>-7.5) | Lack / very low | < 20 (<25) |
| 1 | 5 – 15 (-7.5 ÷ -15) | Low | 20 – 60 (25 ÷ 55) |
| 2 | 15 – 25 (-15 ÷ -25) | Medium | 60 – 140 (55 ÷ 80) |
| 3 | > 25 ( < -25) | High | > 140 (>80) |

## 2.2 Method of WAS-96/RMS

*Measurement methodology*

The measurement of seismoacoustic emission (SE) is carried out using two geophones (A and B) spaced symmetrically at a distance of 5 m from the blasthole in the coal seam (Fig. 1). Sensors should be placed as near as possible to the middle of the coal seam. The depth of the blasthole is 3 m, while the geophones holes are 1.5 m deep. A 1 kg charge of explosives is blasted.

The registration of SE is carried out in two-minute intervals and it lasts for 40 minutes. During the first 10 minutes, the SE noise is observed, and for the next 30 minutes the induced SE is observed as a result of blasting.

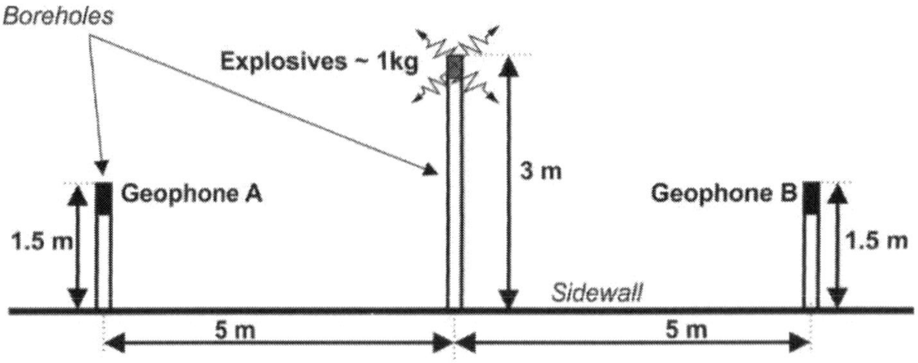


**Fig. 1.** The scheme of WAS-96/RMS measurement.

The measurement is carried out with the use of WLIS apparatus - multichannel seismoacoustic pulse counter or an equivalent apparatus from the point of view of the WAS-96/RMS method. WLIS enables the automatic counting of seismoacoustic pulses and the calculation of their energy. The apparatus records pulses with a sampling frequency of 5 kHz and dynamics of 60 dB for each channel.

Making measurements using the WAS-96/RMS method is possible only in conditions of seismic silence, without disturbances generated by mining works. Research in the freshly exposed sidewall may be undertaken 4 hours after the completion of mining.

*Methods of processing and interpretation*

The average pulse energy released in the two-minute intervals measured in Joules is given by the formula [19]:

$$\overline{\mathrm{E}} = 10\frac{(E_{WLIS}^{A} + E_{WLIS}^{B})}{40N} \tag{3}$$

where:
$E_{WLIS}^{A}$ and $E_{WLIS}^{B}$ – pulses energy from channels A and B in Joules,
N – total number of pulses on both channels.

Seismoacoustic activity and energy density in the time domain are estimated from formulas [19]:

$$n(t) = n_{-} + B \exp(-\beta t) + \eta_1 \tag{4}$$

$$E(t) = n(t)\overline{\mathrm{E}} + \eta_2 \tag{5}$$

where:
$n_{-}$ – median of seismoacoustic activity before blasting,
$B = n_1 - n_{-}$ where $n_1$ is seismoacoustic activity in the first two minutes after blasting,
$\beta$ – attenuation factor after blasting,
$\eta_1, \eta_2$ – constants of random disturbances.

The values of the parameters: $n_{-}$, $B$, $\beta$ i $\overline{\mathrm{E}}$ fully characterize the measurement results. On the basis of measured parameters, the plots of activity and energy density changes for sensors A and B are made separately (Fig. 2). The graphs should be calibrated, so that the maxima of energy density and activity overlap with each other. The graphs are used to determine the time of increase in seismoacoustic activity after blasting ($T_A$ and $T_B$).

The rockburst threat assessment is carried out on the basis of the $Z_{RMS}$ parameter, which is calculated using the formula [19]:

$$Z_{RMS} = C_1 + C_2 n_{-} + C_3 B + C_4 \beta^{-1} + C_5 \overline{\mathrm{E}} \tag{6}$$

where:
$C_1, C_2, C_3, C_4, C_5$ – weighting constants specifying the correlation between the measured parameters and the rockburst threat. They are recalculated after every survey in the same site.

The calculation of the $Z_{RMS}$ parameter is performed for each channel separately, and then a greater value of the $Z_{RMS}$ parameter is selected for the analysis. The criteria for assessing the threat of rockbursts using the WAS-96/RMS method are given in Table 2.

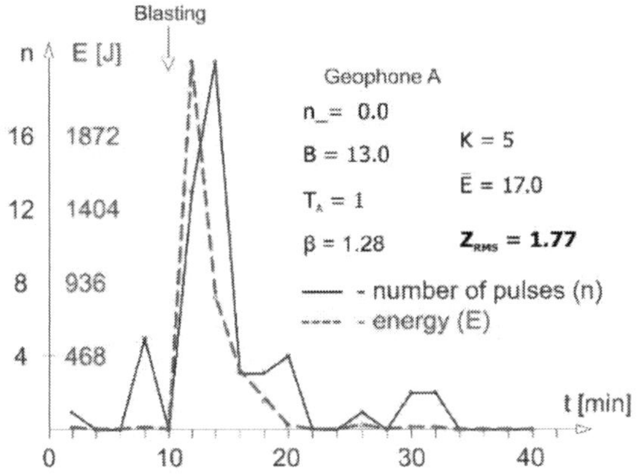


**Fig. 2.** Graph with parameters calculated in point P5 using WAS-96/RMS.

**Table 2.** Criteria for assessing the rockburst threat with the WAS-96/RMS method.

| $Z_{RMS}$ | Rockburst threat degree |
|---|---|
| $Z_{RMS} \leq 0{,}5$ | No occurrence – a |
| $0.5 < Z_{RMS} \leq 1.5$ | Weak – b |
| $1.5 < Z_{RMS} \leq 2.5$ | Medium – c |
| $2.5 < Z_{RMS}$ | High – d |

## 3 Research area and geological - mining conditions

The measurements were conducted in the region of a planned A longwall in the X0 coal seam (Fig. 3). The seismic profiling and seismoacoustic measurements were conducted in the A1 heading at a depth of about 1200 m. The measurements were located in the impact zone of the overlying edges of the X1 and X2 coal seams, lying about 70 m and 100 m above the X0 coal seam.

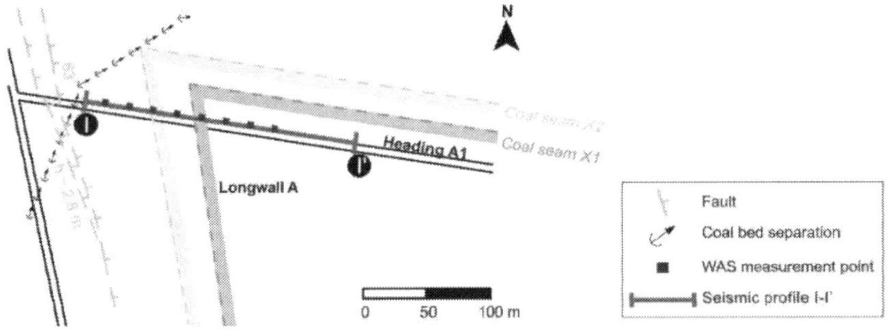


**Fig. 3.** Location of the seismic profile and WAS-96/RMS measurement points on the mining map with overlying coal seams edges and tectonic disturbances.

In the research area, the X0 coal seam has a mean thickness of about 3.6 m. The slope of the coal seam is between 0 and 5° in the N direction. The X0 coal seam is surrounded by mudstone layers of 3.9 m in thickness as the immediate roof. Above the mudstones there lies a sandstone layer of about 20 m in thickness. On the 20th meter of the heading A1 from the west side, there is a 10 meters wide fault zone with a maximum throw of 2.8 m and approximate angle of dip 63°/NEE. Separation of X0 coal seam occurs in the vicinity of the fault zone.

Seismic profiling was performed along the I-I 'profile (Fig 3). Geophones of 40 Hz were spaced out every 5 m over a distance of 200 m along the northern sidewall of the A1 heading. They were fixed approximately in the middle of the coal seam. The seismic wave was excited by a 4 kg sledgehammer at points located between the geophones. A sampling time of 0.125 ms and a registration time of 0.5 s were used. Six-fold vertical stacking was applied at each blasting point.

WAS-96/RMS measurements were made using the WLIS apparatus. Eight blasting boreholes spaced 20 m from each other were drilled. An explosive charge of 1 kg was blasted. Two 10 Hz geophones were symmetrically installed at each measuring point from both sides of the blasting borehole. The observation time of SE at a given point was 40 min.

## 4 Results and analysis

Figure 4 presents the results of seismic profiling and the active seismoacoustic WAS-96/RMS method together with the mining situation.

*Seismic profiling*

The velocity of the refracted P-wave on the profile varies from 2346 to 2457 m/s.. Two anomalous increases in the velocity of the P-wave are observed. The first anomaly $A_1$=4.9% is on the initial section of the profile up to approx. 25 m. It is probably associated with close proximity to the fault zone and the coal bed separation zone. The second anomaly $A_2$=8.4% occurs on the section from approximately 60 to 130 m of the measurement profile. The anomaly value was calculated for the reference velocity of $V_0$=2260 m/s. The $A_2$ anomaly is weak, which corresponds to an increase in stress in the range of 20 to 60%. Presumably, the anomaly $A_2$ is the effect of the impact of the edge of the X1 seam, and to a lesser extent the impact of the X2 edge.

The seismic profile is in large part underneath the gobs of the X1 and X2 seams. The effect of the stress reduction of these gobs is weak due to the impact of the edges of the X1 and X2 seams parallel to the profile (Fig. 3).

*WAS-96 / RMS measurements*

The value of the $Z_{RMS}$ coefficient characterizing the rockburst threat level varies from 0.84 to 2.23 (Fig. 4b). The greatest value of the coefficient was obtained for the P1 measuring point, which may be related to the proximity of the fault zone and the X0 coal seam separation zone. At the next three measuring points, the values of the coefficient range from 0.84 to 1.21 indicating a weak rock burst hazard threat. At point P5, located directly under the edge of the X1 seam, the value of the $Z_{RMS}$ coefficient increases and amounts to 1.79, this corresponds to the medium threat degree. In points P6 - P8, the values of the coefficient are smaller and do not exceed the value of 1.37.

Figure 4c shows graphs of the average energy of pulses and energy density in two-minute intervals. The character of both graphs is similar to the ZRMS coefficient graph. The maximum mean value of the energy pulse amounts to En=17 J and was measured at P5.

Diagram 4d shows two graphs of the average number of pulses over two-minute intervals. The first graph concerns the registration after blasting within 30 minutes. The second graph also includes pulses from 10 minutes registration before blasting. The greatest values of the average number of pulses were measured at points P1 and P2 and were respectively 6 and 10 pulses per 2 min. Increased seismoacoustic activity is probably related to the proximity of the fault zone and the X0 coal seam separation zone. The number of pulses measured at the remaining points is smaller and amounts to approximately 1 - 3 pulses per 2 min. There was no increase in activity at the edge of the X1 seam.

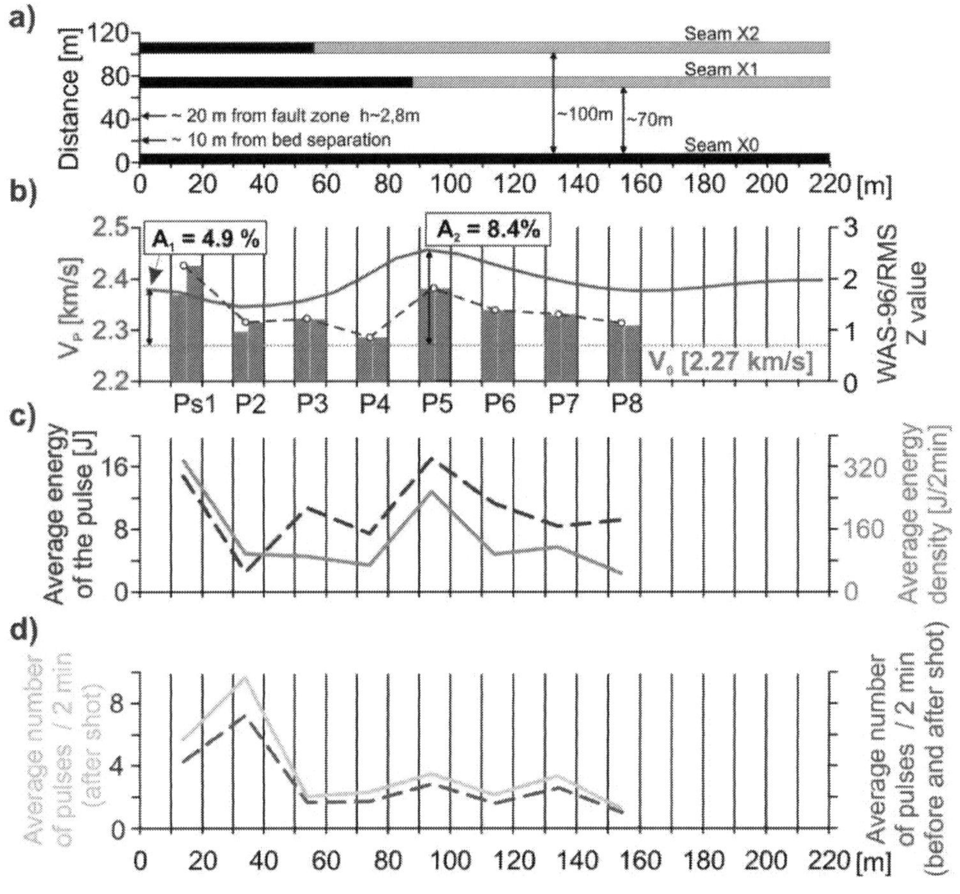


**Fig. 4.** Comparison of the results of seismic profiling and active seismoacoustic method WAS-96/RMS; a) coal seams edges location; b) refracted P-wave velocity in coal seam (solid red line – measured value, dashed blue line - reference value) and $Z_{RMS}$ parameter from WAS-96/RMS measurements (grey bars - value for each geophone); c) average energy of the pulse and energy density; d) average number of pulses obtained from WAS-96/RMS measurements.

When comparing the results of the measurements of seismic profiling and the WAS-96/RMS method, a similarity may be observed in the presence of anomalous zones in the area of impact of the overlying coal seams. The anomalous increases in measured parameters: P-wave velocity, ZRMS parameters and average energy are clearly correlated. In contrast, the course of the seismoacoustic activity is different. The smaller number of pulses may explain why the stronger pulses appeared in the zone of greater impact of the overlying edges.

On the basis of up to date studies, a comparison of the advantages and limitations of the seismic profiling method and the WAS-96/RMS method is presented (Tab. 3 and 4).

**Table 3.** Advantages and limitations of seismic profiling.

| Advantages | Limitations |
|---|---|
| • Changes in the P-wave velocity in the coal seam correspond to changes in relative stress.<br>• Measurement is carried out in-site on the threatened section of excavation.<br>• Survey is non-destructive.<br>• The acquired information may be considered as continuous.<br>• Measurement is relatively short in time (2 h for 100 m profile according to the described methodology).<br>• Measurement cost is relatively low. | • High impact of mining disturbances on wave field registration.<br>• Complicated wave image requiring experience in data processing and interpretation.<br>• Great influence of the disturbed zone in the sidewall of excavation at the site of sensor installation on the recording quality.<br>• Reduction of increments of the refracted P-wave velocity with increasing stress (problem at greater depths).<br>• Necessity to maintain the seismic silence during the measurement in the research area (about 20 minutes per 100 m of profile).<br>• Disturbances in the registration of seismic waves associated with complex tectonics and lithology (faults, seam thinning, washouts, rock partings, dirt bands).<br>• The use of non-intrinsically safe measuring equipment is possible in an atmosphere of up to 0.5% of methane in accordance with applicable regulations. |

**Table 4.** Advantages and limitations of WAS-96/RMS method.

| Advantages | Limitations |
|---|---|
| • Changes in the density of energy of seismoacoustic emission correspond to stress changes in the coal seam.<br>• Measurement is carried out in-site on the threatened section of excavation.<br>• Processing and interpretation of seismoacoustic data is automated.<br>• Adaptive algorithm is used to improve the calculation of the $Z_{RMS}$ parameter. | • Zonal valid results of up to 10m wide.<br>• Destructive test requiring blasting of 1 kg of explosives.<br>• Relatively long time of measurement associated with the preparation of boreholes, loading explosives and blasting.<br>• Necessity to stop mining work in the research area.<br>• The measurement costs are relatively high.<br>• Performing the measurement is limited by the possibility of the use of explosives in conditions of natural and mining hazards in accordance with applicable regulations. |

# 5 Conclusions

In the study, a comparison of the results of seismic profiling and the WAS-96/RMS method carried out in identical measurement conditions of the impact of the overlying coal seam edges has been presented. Based on the analysis, the following conclusions were made:

1. Both the seismic profiling and the WAS-96/RMS method allowed for the identification of anomalies in the area of impact of the overlying coal seam edge (seismic anomaly $A_2$=8.4% - low degree and $Z_{RMS}$=1.79 - medium degree).
2. At the beginning of the measurement profile, both methods indicate an increase in the threat related to the vicinity of the fault zone and the coal seam separation zone. The anomalous increase of velocity $A_1$ is enough clear, but this is only a part of this anomaly. In the case of the WAS-96/RMS method, the degree of threat is more evident. This increase is marked clearly on the energy density graph (Fig. 4c). In this case, it is difficult

to indicate which of the methods has shown the seismic threat more correctly. The problem will be explained during the operation of the planned longwall A.

3. The seismic profiling method seems to provide more clear results. Measured velocity values provide approximately a continuous graph of their changes, which corresponds to relative changes in stress. In the case of the WAS-96/RMS method, we gain a discrete graph associated with the zonal measurement from the 10 m long section. The method proposed by the Kornowski et al. [19] also allows a 60 m long section, 30 m each side of the blasting hole, in the case of homogeneous conditions in the coal seam .
4. Costs and time of the measurements are more favorable for seismic profiling. The measurements of 200 m of seismic profiling takes about four hours with the engagement of approximately 4-5 people of the measurement group together with the mining superviser. In the case of the WAS-96/RMS method, seven measurements of approx. 30 m each over a distance of approx. 200 m requires at least two work shifts, with a group of at least seven people engaged: drill and blast miners, measurement operators and the mining superviser.
5. The results of the research indicate that seismic profiling should be the basic method for assessing the impact of the edges and remnants of exploited coal seams. The WAS-96/RMS method may be used as a complementary method if a confirmation of the rockbursts threat is required.

## References

1. J. Dubiński, W. Konopko, *Tąpania – ocena, prognoza, zwalczanie* (*Rock bursts – assessment, prognosis, defeating)* Wyd. GIG, Katowice (2000) (in Polish)
2. J. Dubiński, Z. Pilecki, W. Zuberek, *Badania geofizyczne w kopalniach – przeszłość, teraźniejszość, i zamierzenia na przyszłość (Geophysical research in mines - past, present and future plans)* Wyd. IGSMiE PAN, Kraków (2001) (in Polish)
3. Z. Pilecki, *Dynamic analysis of mining tremor impact on excavation*, Proc. Int. FLAC Symp. on Num. Model. in Geom., Detournay & Hart, 397–400 (1999)
4. Z. Pilecki, *Metoda oceny zachowania się masywu skalnego wokół wyrobiska podziemnego (A method of assessing the behavior of the rock mass around the underground excavation)* Studia, Rozprawy, Monografie **59,** IGSMiE PAN, Kraków (1999) (in Polish)
5. Z. Burtan, A. Zorychta, J. Cieślik, D.D. Chlebowski, Arch. Min. Sci. **59**, 3, 691–704 (2014) doi: 10.2478/amsc-2014-0048
6. Z. Burtan, Z. Rak, J. Stasica, Gospod. Surowcami Min. **24**, 1, 185–200 (2008) (in Polish)
7. M. Kwaśniewski, J. Wang, *Analiza numeryczna deformacji górotworu wywołanych eksploatacją górniczą (Numerical analysis of mining-induced deformations of the rock mass)*, Proc. of the School of Underground Mining, 153–187 (1996) (in Polish)
8. Z. Pilecki, *Modelowanie numeryczne pola naprężenia w górotworze naruszonym wielopokładową eksploatacją węgla kamiennego w warunkach silnego zagrożenia sejsmicznego (Numerical modeling of stress field influenced by multi-seam hard coal mining in strong seismic threat conditions)* The Bulletin of The Mineral and Energy Economy Research Institute of the Polish Academy of Sciences **80**, 93–102 (2011) (in Polish)
9. K. Krawiec, R. Czarny, E3S Web Conf. **24**, 01007 (2017) doi: 10.1051/e3sconf/20172403001
10. J. Dubiński, *Sejsmiczna metoda wyprzedzającej oceny zagrożenia wstrząsami górniczymi w kopalniach węgla kamiennego (Seismic method of shock hazard assessment in hard coal mines)* Central Mining Institute, Katowice (1989) (in Polish)

11. A. Tor, P. Chmiel, Z. Pilecki, A. Jakubów, R. Skatuła, Z. Szreder, J. Kłosiński, *Kontrola efektywności strzelań torpedujących za pomocą profilowań sejsmicznych (Control of the effectiveness of torpedo blasting using seismic profiling)*, Bezpieczeństwo Pracy i Ochrona Środowiska w Górnictwie, Miesięcznik WUG **142**, 6, 62–64 (2006).
12. Z. Pilecki, Z. Szreder, *Możliwości wykorzystania współczynnika tłumienia do oceny stanu naprężenia i deformacji za pomocą profilowań sejsmicznych w warunkach dużych głębokości (Possibilities of using the damping factor to assess the state of stress and deformation by means of seismic profiling in conditions of great depth)*, Prace Naukowe GIG **4**, 197–203 (2007) (in Polish).
13. Z. Pilecki, J. Kłosiński, *Profilowanie tłumienia fali sejsmicznej w pokładzie węgla (Profiling of seismic wave attenuation in hard coal seam),* Proc. 11th Conf. On Warsztaty Górnicze, 381–393 (2007) (in Polish)
14. Z. Szreder, Z. Pilecki, J. Kłosiński, *Comparison of Profiling Results of Attenuation and Velocity of Refracted P-wave in Coal-seam*, Proc. 14th EAGE European Meeting of Environmental and Engineering Geophysics (2008) doi: 10.3997/2214-4609.20146265
15. Z. Szreder, Z. Pilecki, J. Kłosiński, *Effectiveness of recognition of exploitation edge influence with the help of profiling of attenuation and velocity of seismic wave,* Gospod. Surowcami Min. **24**, 2/3, 215–226 (2008)
16. Ł. Wojtecki, *Wykorzystanie metod sejsmicznych w wersji kinematycznej do określania stanu naprężeń w górotworze podczas eksploatacji w Kopalni Węgla Kamiennego "Bielszowice"(The use of seismic method in the kinematic version to determine the state of stresses in rock mass during operation in the Coal Mine "Bielszowice")*, Prace Naukowe GIG, 578–597 (2011) (in Polish)
17. Ł. Wojtecki, G. Dzik, *Profilowanie sejsmiczne ociosów chodników węglowych w wersji kinematycznej i tłumieniowej (Seismic profiling along sidewalls in the kinematic and attenuation versions)* Mechanizacja i Automatyzacja Górnictwa **50**, 4, 29–37 (2012) (in Polish).
18. Z. Szreder, M. Barnaś, E3S Web of Conf. **24**, 01007 (2017) doi: 10.1051/e3sconf/20172401007
19. J. Kornowski, H. Sokołowski, M. Trombik, A. Waśko, *Zasady i zakres stosowania metody wzbudzonej aktywności sejsmoakustycznej WAS-96/RMS do oceny stanu zagrożenia tąpaniami (The principles and scope of application of the method of induced seismoacoustic activity WAS-96/RMS for the assessment of rock bursts threat)* Central Mining Institute, Instrukcje **4**, Katowice (1996) (in Polish)

# Recognition of gasogeodynamic zones in the rock mass using seismic tomography in Rudna copper ore mine

*Krzysztof* Porębski[1], *Eugeniusz* Koziarz[1], *Arkadiusz* Anderko[1], *Krzysztof* Krawiec[2,*], *Rafał* Czarny[2], *Jakub* Kokowski[2], and *Paulina* Harba[2]

[1]*KGHM Polska Miedź S.A., Lubin, Poland*
[2]*Mineral and Energy Economy Research Institute, Polish Academy of Sciences, Cracow, Poland*

**Abstract.** In this work, the results of four seismic tomography surveys are presented. The research was conducted to identify the zones exposed to the threat of gas and rock outburst. The changes to the dolomite layer stiffness in the mining excavation roofs were analyzed. The surveys were conducted in the Rudna copper ore mine in the field of XXVIII/1. The research area was about 0.21 km2. The seismic waves were generated by a small amount of explosive material (100 – 300 g) located and installed in short blast holes (1.5 – 2.0 m). The processing and the interpretation of the measurement data did not cause serious problems due to the more favourable elastic properties of the dolomite layer compared to the adjacent anhydrite and sandstone layers. As a result, the maps of parameters like the longitudinal wave velocity (P-wave), the shear wave velocity (S-wave), and the ratio of the P-wave velocity to S-wave velocity and the dynamic Young modulus were estimated. The results showed that the changes in the seismic parameters were relatively small over most of the research area. This may be evidence of the minor effects of gas and rock outbursts.

**Keywords:** copper ore mine, gasogeodynamic phenomena, seismic tomography, seismic parameters

## 1 Introduction

The gasogeodynamic threat in the Rudna copper ore mine is related to the accumulation of gas and water under high pressure in the highly fissured and porous areas of the dolomite layer [1–3]. A unique technology based on seismic tomography, drilling of control boreholes and borehole ground penetrating radar (BGPR) surveys was developed for the identification of these areas [4–6]. In the first stage of this technology a seismic tomography survey in the dolomite layer is conducted. The second stage consists of drilling control boreholes in the locations of anomaly indicated by the seismic tomography results. The drillings are conducted with special care, this is accomplished by registering all signs of fissuring and gas or water outflows. The rock quality design coefficient (RQD) is also calculated. In the last

* Corresponding author: k.krawiec@meeri.pl

stage the BGPR surveys are conducted in the drilled control boreholes in order to obtain more detailed results [7–9].

In this work the results of four series of surveys using the seismic tomography method are gathered. The surveys were conducted in the XXVIII/1 field in the geological and mining conditions of Rudna copper ore mine in the years 2013 – 2017.

## 2 Geological conditions

The copper ore deposits in the Rudna mine occur in three lithological types: the sandstone of the Weissliegendes formation, cupriferous clay slates and the dolomites of the Zechstein series. A characteristic feature of the deposit area is the non-uniform formation of the lithological structure and diverse mineralization [10]. The variability in the lithological structure of rock types is related to the diverse morphology of the topmost sandstone layer [10–12].

Over the entire area of the copper ore deposit, morphological elevations and depressions occur [13–14]. The elevations are the relics of sand dunes, which originated in desert conditions during the period of the Rotliegend. The marine transgression in the Zechstein period brought about the flattening of the dune elevations and the resedimentation of sand in the lowering of the interdune areas. During the current operation of the Rudna mine 5 elevations have been identified. They are 15–35 m high forms, with a maximum length of 25 km and a width of approximately 1 km. The northeast slopes of the elevations are very steep, and the southwest slopes are slightly inclined, largely due to the inclination of the main tectonic unit of the Fore-Sudetian Monocline.

However, the morphology of the slates and the lower part of the dolomite series is not strictly bound up with the structure of the elevations. The localization of the dolomite series between the more plastic series of sandstone and anhydrite has a significant impact on the tectonic processes occurring in the dolomites. The Zechstein dolomite series was subjected to tension and compression during the period of the formation of folded structures, overthrusts, and faults [15]. The fissures thus created were filled with material that had been transported in water.

Within the Zechstein dolomite series, in the areas of fissuring isolated from above by the layers of anhydrite, gas traps are observed. In all probability the gases could flow into the dolomite series through a system of fissures from the Carboniferous strata during the formation of the Fore-Sudetian Monocline [1–3, 11] (Fig. 1). The characteristics of the gas traps indicate that they are most likely to occur in the slopes of morphological elevations in zones of tensile stress. They may also be present in the weakly inclined top portion of the Zechstein formations, if in the dolomite series, some larger zones of fissures and faults were formed in the area of a drag fold structure or an overthrust [11]. It was assumed that the gas traps could also exist along the direction of major tectonic structures [2–3].

In the research area three lithological layers were analyzed: the sandstone layer in the lower part of the roadway, the layer of dolomite and the anhydrite layer in the roof. Figure 2 shows an example of a cross-section obtained on the basis of the core of a borehole drilled in the roof underneath the sidewalk. Based on the borehole data it was observed that the dolomite layer has a thickness of up to 12 m. The dolomite layer has the most favorable elastic properties compared to adjacent layers. For this reason, the identification of seismic waves in a wave image does not pose major difficulties. Based on the modeling results [6], it was found that the first arrivals of the P and S waves are related to the direct wave in the dolomite.

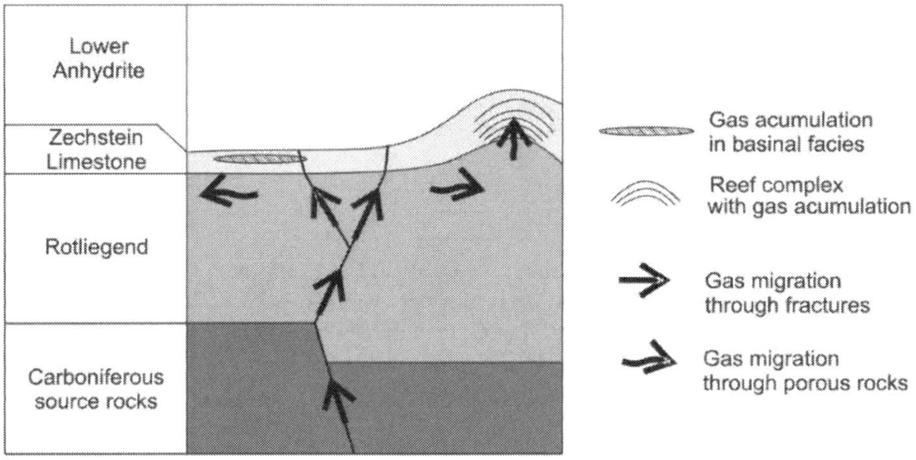


**Fig.1.** Schematic diagrams showing possible sources of the gas accumulation in basinal facies in the Zechstein Limestones in the Rudna Copper Mine [1].

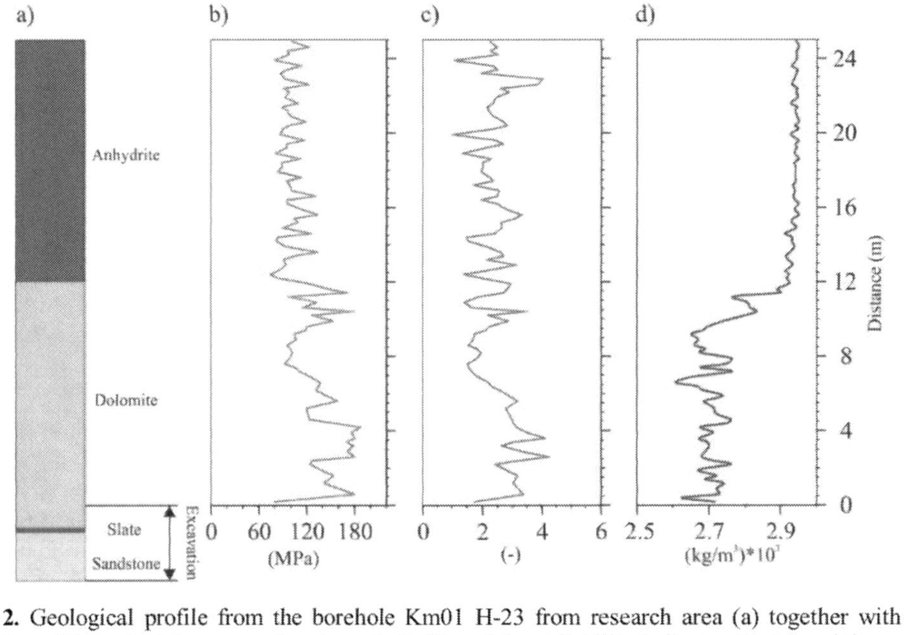


**Fig. 2.** Geological profile from the borehole Km01 H-23 from research area (a) together with the profiling of the uniaxial compressive strength $R_c$ (b) rock burst liability indicator $W_{et}$ (c) and the rock's volumetric density $\rho$ (d) (on the basis of KGHM Cuprum Research and Development Center data).

## 3 Methodology

*Measurement methodology*

The most important factor from the point of view of measurements in the seismic tomography method is the appropriate location of geophones and wave excitation points. The measurement scheme is being developed in order to obtain the best possible coverage with the seismic rays of the analyzed area. A significant influence on the geometry of the seismic tomography was the development of mining works and excavation availability.

In the mining conditions of the field XXVIII/1 in the Rudna mine in each of the four series of tests, the measurement profiles were located in two perpendicular drifts (Fig. 3). The drifts were about 500 m apart. The profile located on the NE side of the examined area consisted of 24 geophones. The profile installed on the SW side of the test area consisted of 72 geophones. The sensors were installed with an interval of 5 m in the roof section of the drifts in the dolomite layer.

A seismic wave was excited in blasting holes using explosives of 0.3 kg. The number of blast holes as well as the distance between holes for individual measurement series were variable. Typically, at least 19 blasting holes were used at intervals not exceeding 30 m. The length of the blast hole was 1.5 - 2 m.

The measurements were carried out using four sets of 24-channel Geode seismic apparatus. Geophones of 100 Hz natural frequency were used. For all registrations, the signal sampling rate was 0.125 ms, and the registration time was 2 s. The transmission of the signal with information concerning the timing of blasting was carried out using a two-core cable by breaking their loop on the explosive. The cable length was up to 1200 m.

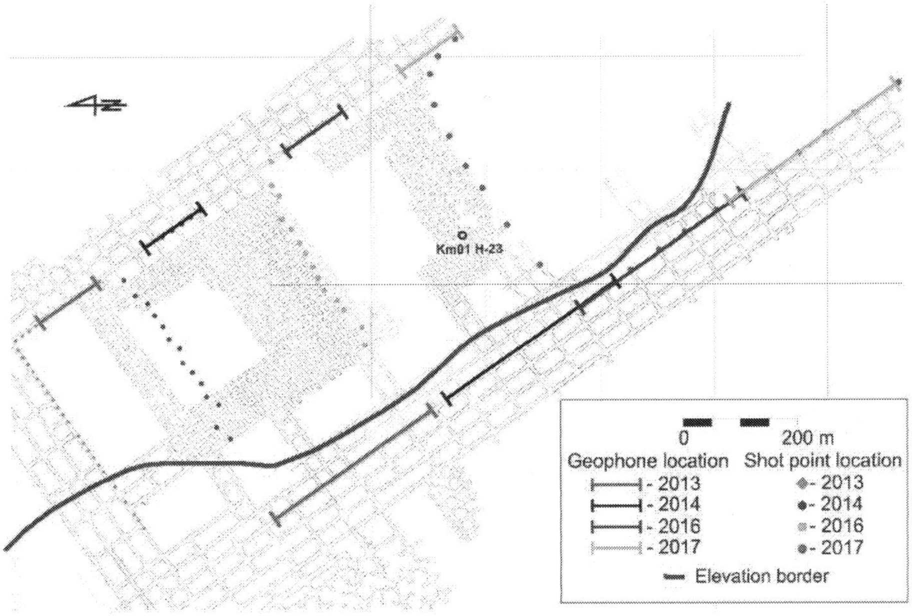


**Fig. 3.** Measurements scheme.

*Processing and interpretation methodology*

To determine the field of transverse and longitudinal wave velocity in the rock mass in the Rudna mine, the SIRT (Simultaneous Iterative Reconstruction Technique) algorithm was used [16]. Data processing consisted of [17]:

- frequency filtration and traces analysis for the identification of various types of seismic waves (Fig. 4),
- determining the times of the first brakes in the longitudinal and transverse wave,
- implementation of measurement geometry,
- proposal of the initial velocity model and discretization of the studied area to a finite number of cells.

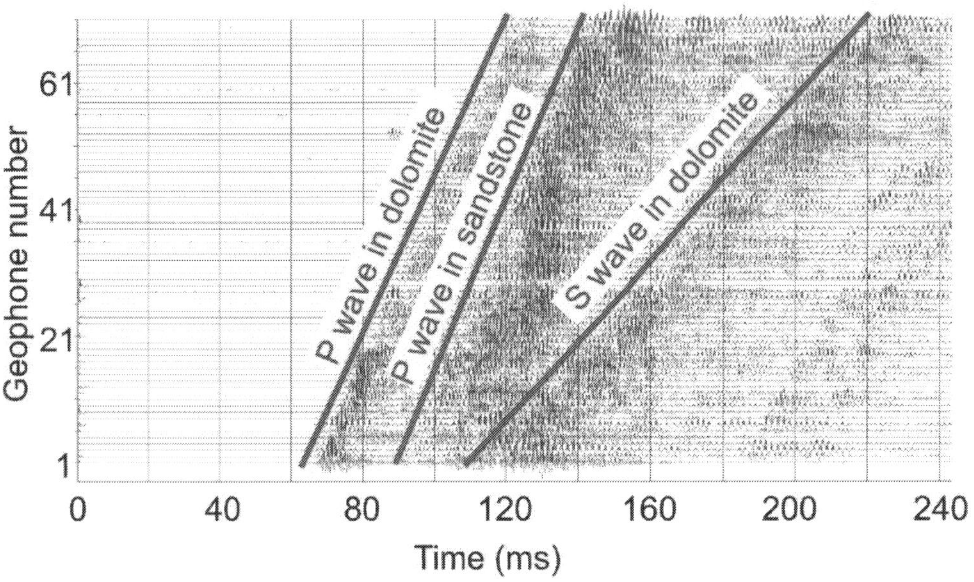


**Fig. 4.** Example of the seismic record from one of the blast point on the profile.

Next, an interpretation was carried out, which consisted of:

- inversion calculations for the curvilinear reconstruction of the seismic rays,
- calculations of matching errors between calculated and measured times for the resulting model,
- subsequent inversion calculations for improved data,
- developing a map of the variability of seismic parameters using interpolation procedures.

An example of the computational parameters of tomography for the longitudinal and transverse wave for the fourth measurement series is summarized in Table 1.

**Table 1.** Seismic tomography parameters used in the fourth measurement series in the Rudna mine.

| Parameter | P-wave | S-wave |
|---|---|---|
| Number of seismic rays | 914 | 921 |
| Mesh dimensions | 20x20m | 20x20m |
| The amount of iteration required to stabilize the final model | 20 | 21 |
| The sum of residual deviations | 7.8 | 7.23 |
| RMS error | 0.57 | 1.32 |
| The maximum mismatch between the calculated and measured times for the resulting model for a single seismic ray | 1.7 ms | 3.8 ms |
| The average value of the seismic coverage of a single cell | 72 | 46 |

## 4 Results and analysis

On the basis of seismic tomography surveys, maps of P-wave (Fig. 5) and S -wave (Fig. 6) velocity were developed for four measurement series in the dolomite layer in the field XXVIII/1 in the years from 2013 to 2017. The map of the dynamic Young's modulus was also developed based on changes in the P and S wave velocity values (Fig. 7). In order to compare the results of measurements, the maps are presented in the same color scale for each series of a given parameter. Larger anomalies are described by numbers.

The comparison of results from subsequent measurement series indicates that it is not possible to obtain the continuity of the isolines of seismic parameters. The secondary stress field associated with the progress of exploitation has a significant influence on the trajectory of the isolines.

On the maps of seismic parameters, two types of anomalies were distinguished: positive anomalies and negative anomalies. It is assumed that the negative anomaly arises in the zone of intense cracks and higher porosity, in which gas and water may also potentially accumulate. The positive anomaly is the effect of the tightening of existing cracks and pores.

On the map of the P-wave (Fig. 5), in the zone of the excavation front, clear positive anomalies marked with number 1 may be observed. The anomalies are associated with the load and deformation of direct roof layers in the exploitation zone. This influence is particularly visible in the north-eastern corners of the research areas due to the greater intensity of mining works. The bending of the roof layers results in the closing of pores and cracks and, as a consequence, an increase in the stiffness of the dolomite layer. These effects increase the velocity of the seismic waves. Negative anomalies are marked with number 2. They were identified on the P-wave velocity map in the results of the first and fourth series of measurements. The anomaly from the fourth series of measurements was located directly at the excavation front and could probably be associated with an excessive load of this fragment of the rock mass. The anomaly identified in the first series of tests, despite being located near the front, covers a much larger area compared to the fourth series. During the drilling of the control borehole, the presence of gas was found to be responsible for this anomaly, but its outflow was small. In addition, the obtained core from the borehole indicated the presence of an intense cracks zone, thus confirming the results of the seismic tomography [4]. In the other series of measurements, negative anomalies were found to be relatively small and the velocities slightly deviated from the values in the undisturbed rock mass. In the control boreholes carried out in this area, no gas outflow was detected, but only a water flow of 3 l/min.

On the collective map of the field of S-wave velocity changes (Fig. 6) anomalies are marked with numbers ranging from 1 to 4. Anomalies 1 and 2 have a similar range and value compared to anomalies marked on the map of P-wave velocity changes. In the third measurement series a greater negative anomaly marked with number 3 appeared. No gas outflows were recorded in the drilled control borehole as well as during the mining operations. Only on the side walls of the drift there was a slight moisture. A wider negative anomaly has also been identified in the fourth measurement series and marked with the number 4. S-wave velocities changes in this zone are small, however.

Figure 7 shows a cumulative map of the changes in the dynamic modulus of elasticity. The calculated anomalies coincide quite closely with anomalies identified on the map of S wave velocity changes. The elastic modulus values vary in range from approx. 68 - 90 GPa. A dolomite density of 2800 kg/m3 was assumed for the calculation of this parameter.

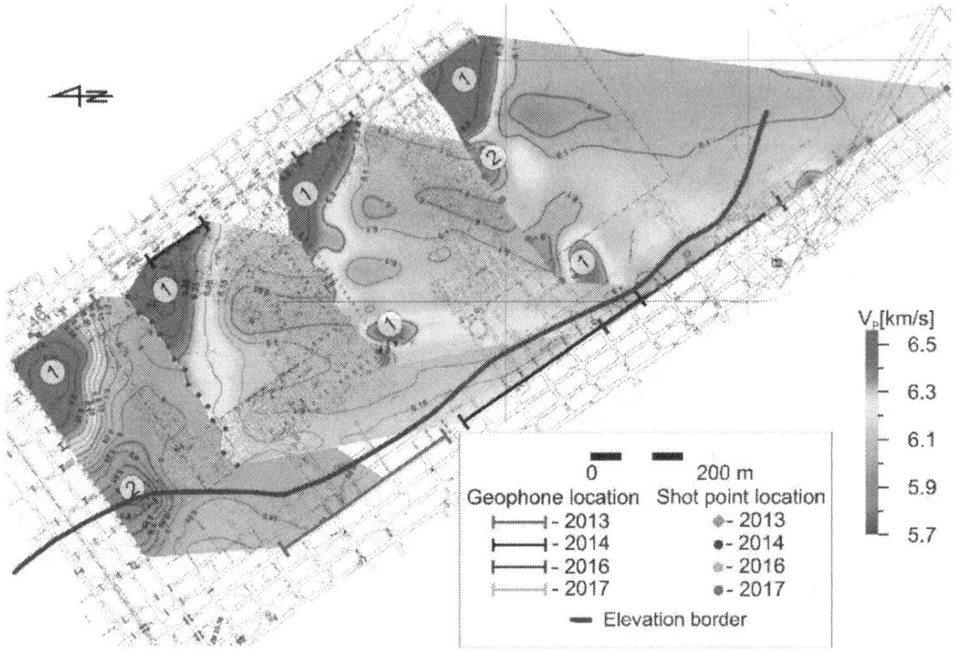


**Fig. 5.** Map of P-wave velocity changes.

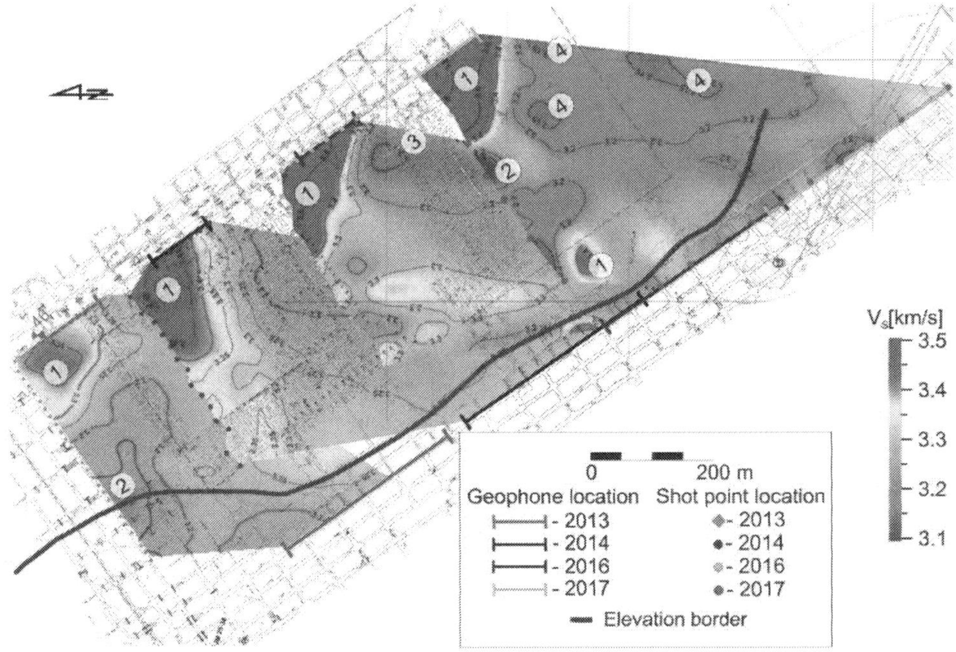


**Fig. 6.** Map of S-wave velocity changes.

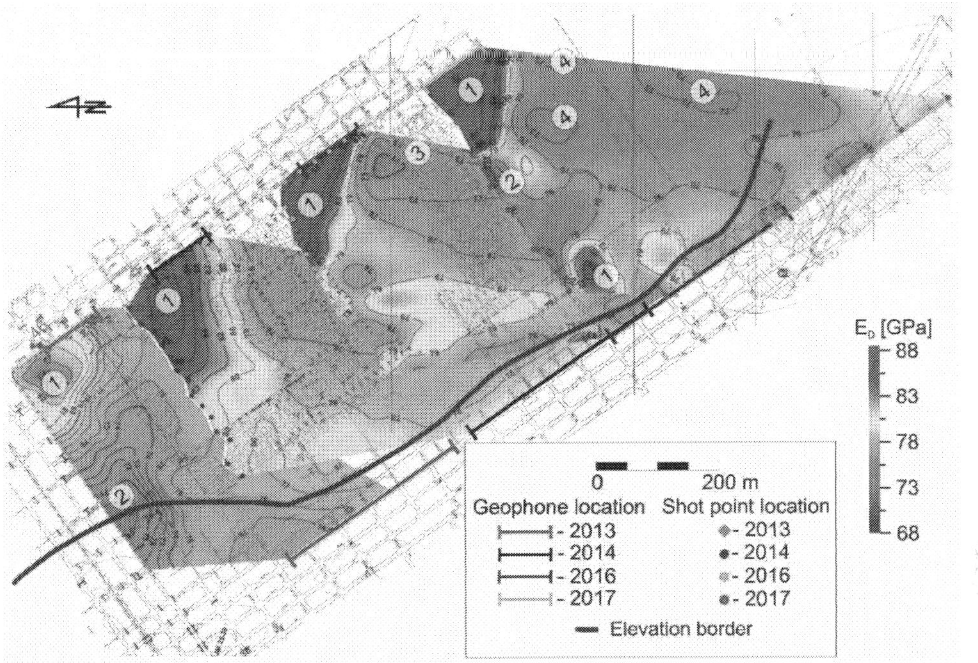


**Fig. 7.** Map of Young modulus changes.

## 5 Conclusions

In the study, we present the results of seismic tomography measurements carried out for the purpose of recognizing rock mass properties, in particular, the aspect of the occurrence of gasogeodynic phenomena in Rudna copper mine. The research was aimed at identifying changes in the stiffness of the dolomite layer in the roof of mining excavations due to the location of gas traps. It was carried out over a very large area in a field several hundred meters in length and width. The measurements were carried out using 96 geophones and four sets of seismic apparatus connected to each other. The source of the wave was a small explosive (about 300 g). The processing and interpretation of the measurement data did not cause major problems due to the more favorable elastic properties of the dolomite layer, compared to the adjacent layers of anhydrite and sandstone, high source energy and longer wave propagation time. As a result, maps of changes in seismic parameters such as P-wave velocity, S-wave velocity, Young's modulus of elasticity and others were calculated.

The use of the proposed methodology made it possible to formulate the following conclusions:

- As a result of the verification of seismic anomalies by means of borehole tests, the closest relationship existed for the S-wave velocity and the dynamic Young's modulus changes. It should be assumed that the other negative and positive anomalies, which were not controlled in the borehole tests, have analogical sources to the anomalies that were verified.
- Small changes in seismic parameters indicate the relatively favorable levels of gasogeodynamic risk in the rock mass, characterized by a weak variation in fracture intensity in the dolomite layer.
- In the vicinity of the front line of the exploited field, observed seismic anomalies are related to the influence of mining operations. The nature of the anomalies indicates that

they occur due to the tightening of existing pores and cracks. As a consequence, an increase in the stiffness of the dolomite layer is observed. The main reason for this is the additional load from the bending of the roof strata.

## References

1. A. Poszytek L. Dudek R. Rożek, J. Petrol. Geol. **41**, 1, 67–83 (2018) doi: 10.1111/jpg.12693
2. J. Dec, K. Pietsch, P. Marzec, Ann. Soc. Geol. Pol, **81**, 63–78 (2011)
3. J. Dec, K. Pietsch, Gospod. Surowcami Min. **28**, 3, 93–112 (2012) doi: 10.2478/v10269-012-0025-z
4. Z. Pilecki, M. Laskowski, A. Hryciuk, E. Pilecka, R. Czarny, J. Wróbel, E. Koziarz, K. Krawiec, CIM Journal **5**, 3, 194–202 (2014)
5. Z. Pilecki, M. Laskowski, K. Porębski, E. Pilecka, D. Jach, R. Czarny, J. Wróbel, P. Piasecki, K. Krawiec, E. Koziarz, P. Harba, *Rozpoznanie stref zagrożenia gazogeodynamicznego metodą tomografii sejsmicznej na przykładzie pola XXVIII/1 w ZG Rudna (Recognition of gaso-geodynamic threat zones using seismic tomography in XXVIII/1 field in Rudna copper mine)* Cuprum Ore Mining Scientific and Technical Magazine **4**, 77, 109–120 (2015) (in Polish)
6. Z. Pilecki, K. Porębski, E. Pilecka, D. Jach, R. Czarny, J. Wróbel, P. Harba, P. Piasecki, E. Koziarz, *Experiences in recognition of gaso-geodynamic zones in the rock mass of a copper ore mine using geophysical methods*, Proc. 24th World Mining Congress, 18–21 June, Rio de Janeiro, Brazil, 155–164 (2016)
7. T. Łątka, R. Czarny, K. Krawiec, M. Kudyk, Z. Pilecki, *Eksperymentalne badania położenia nieci agłości, pustek i stref rozluźnień w górotworze za pomocą georadaru otworowego, (Experimental researches of discontinuous, empties and fracture zones location in rock mass by borehole georadar)* Zeszyty Naukowe IGSMiE PAN – The Bulletin MEERI PAS **77**, 67–75 (2010) (in Polish)
8. T. Łątka, Z. Pilecki, *Zalety i ograniczenia otworowej metody georadarowej w badaniach geoinżynierskich (Advantages and limitations of borehole GPR method in the geoengineering investigations)* Zeszyty Naukowe IGSMiE PAN – The Bulletin MEERI PAS **89**, 19–31 (2015) (in Polish)
9. T. Łątka, Z. Pilecki, *Metoda oceny stopnia spękania ośrodka skalnego za pomocą georadaru otworowego (Method of assessing the degree of rock mass fragmentation using borehole georadar)* IGSMiE PAN, Kraków (2016) (In Polish)
10. A. Piestrzyński (ed.) *Monografia KGHM Polska Miedź S.A. (Monography KGHM Polska Miedź S.A.)* KGHM CUPRUM Sp. z o.o., Lubin (2007) (in Polish)
11. J. K. Błaszczyk, *Wpływ paleomorfologii stropu białego spągowca na zmienność facjalną serii złożowej w Zagłębiu Lubińskim (Palaeomorphology of Weissliegendes top as the control on facies variability in ore-bearing series of Lubin Copper-Field, southwestern Poland)* Geologia Sudetica **16**, 1, 195–217 (1981) (in Polish)
12. J. Ślizowski, Z. Pilecki, K. Urbańczyk, E. Pilecka, L. Lankof, R. Czarny, Adv. High Energy Phys. **2013**, 461764, (2013) doi: 10.1155/2013/461764
13. W. Kaczmarek, *Zróżnicowanie mineralizacji miedziowej a wykształcenie litologiczne białego spągowca w kopalniach LGOM (Differentiation of copper mineralization and lithology of Weissliegendes formations in copper ore mines)* Unpublished doctoral dissertation, University of Wrocław, Wrocław, Poland (2006)

14. J. Suchan, R. Rożek, A. Hryciuk, *Warunki sedymentacji i zróżnicowanie facjalne wapienia cechsztyńskiego a zagrożenie gazowe i gazogeodynamiczne w O/ZG „Rudna", KGHM „Polska Miedź" S.A. (Sedimentological conditions and facies diversity of the Zechstein Limesone to the hazard of gas and gasogeodynamic in O/ZG "Rudna", KGHM "Polska Miedz" S.A.)* Proc. 5th Polish Sedimentological Conf. POKOS V, 16–19 May, Żywiec, Poland, 1–8 (2013)
15. T. Jerzykiewicz, P. Kijewski, J. Mroczkowski, A. K. Teisseyre, *Geneza osadów białego spągowca monokliny przedsudeckiej (Origin of the Weissliegendes deposits in the fore-sudetic monocline)* Geologia Sudetica **11**, 1, 57–89 (1976) (in Polish)
16. P. Gilbert, J. Theor. Biol. **36**, 105–117 (1972)
17. B. Lehmann, *Seismic traveltime tomography for engineering and exploration applications*, EAGE, Houten, Netherlands (2007)

# 3D model of Sin Quyen iron oxide copper deposit in Lao Cai, North Vietnam

*Hao* Duong Van[1,*], *Chau* Nguyen Dinh[2], and *Władysław* Zygo[2]

[1]Hanoi University of Mining and Geology, Hanoi, Vietnam
[2]AGH University of Science and Technology, Cracow, Poland

**Abstract.** The paper presents a 3D model of ore bodies of the Sin Quyen iron oxide copper-gold deposit in Lao Cai province, North Vietnam. To build 3D model we used the computer software MineScape and archival geological, geophysical data recorded from 146 boreholes, 21 tunnels and 216 trenches, including the data obtained from chemical and neutron instrumental activation analysis of 50 ore samples collected in 2015 year. The modelling process was composed of three stages: first was checking and validating of the input data, second was data processing and the last building 3D model. The model shows that the ore bodies extend below level of -300 m, such information will be used for future exploration. Based on the obtained model, the reserves of selected metals were calculated and equal to 570 000 tonnes for copper and 190 tonnes for uranium. The obtained Cu reserve is comparable with that estimated and published in the archival report of the Vietnam Geological Department.

**Keywords:** 3D modelling, Sin Quyen deposit, geophysical data, MineScape software

## 1 Introduction

One of the applications of modelling software in geology is building a 3D model of geological structure for an investigated region [1-13]. Using the 3D models the potential mineral target can be predicted and evaluating mineral resources can be performed [13-20]. The aims of the study are to build 3D geometric model of morphology, ore-bodies and to calculate the ore reserves for the iron oxide copper-gold (IOCG) Sin Quyen deposit using the archival geophysical data and the recent chemical analysis. Interpretation of the obtained model enables to gain some additional data, which will be useful for future exploration.

## 2 The study area

The Sin Quyen deposit was discovered in the 70's of the 20 century by the radiometric surveys performed under the project of Federation of Vietnam Geologists. This deposit with area of 200 ha at 22°37'20''N latitude and 103°48'00''E longitude is located between Fanxipan Belt

* Corresponding author: haodnth@gmail.com

and Red River zones. The Fanxipan Belt is the boundary between South China block and Indochina block. This belt is dominated by Proterozoic rocks [21-22]. The Red River zone is the southeastern part of the Ailao Shan-Red River tectonic element [23, 24] (Fig. 1). According with Ta Viet Dung report [25] the Sin Quyen deposit is composed of 17 principal, steeply dipping and discontinuous ore bodies, occurring in form of lenses and veins enclosed, elongate with 100-200 m width, and approximately 2.5 km long. The copper ore bearing suite is the Proterozoic Sin Quyen metamorphosed sediment formation (SQF), which is built for amphibolites, migmatised granito - gneisses and schists trending in NW–SE direction between 280–320 degrees and dipping under 50-85 degrees [26-29]. The ore mineralization is strongly controlled by the Sin Quyen reverse fault, Pin Ngan Chai and Thung Sang faults, which are parallel to the Red River faults system in North Vietnam (Fig. 1).

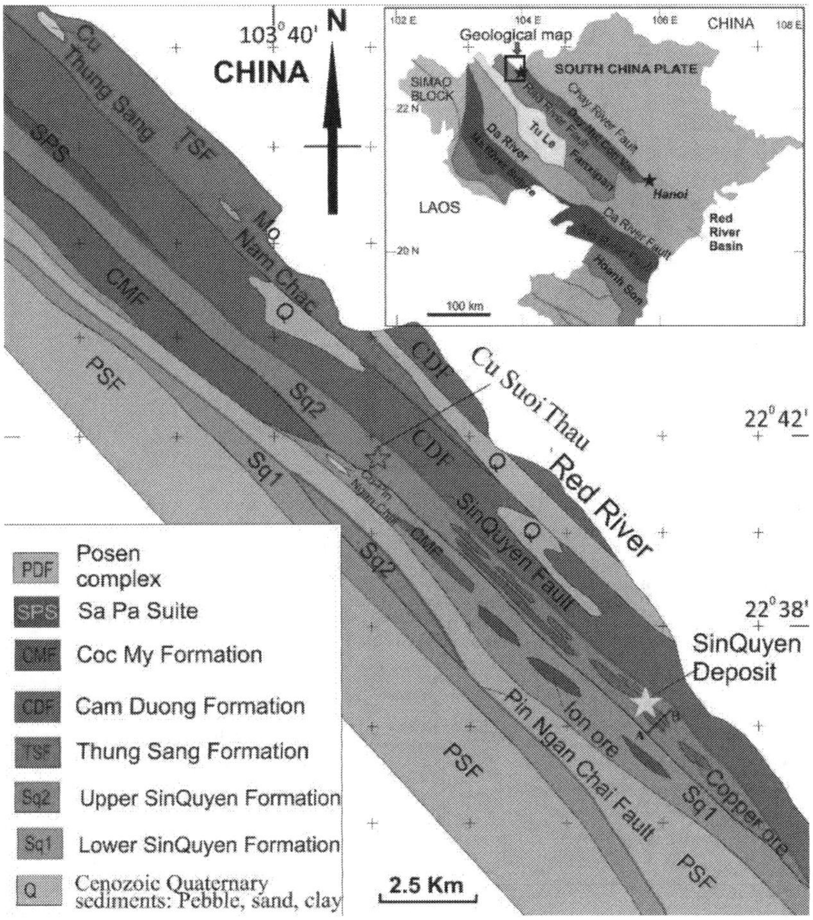


**Fig. 1.** The geological map of the region.

## 3 3D geometric modelling

### 3.1 Data for model

The archival data were obtained from Vietnam Geological Department under the permission "2971/ĐCKS-ĐC with 16/10/2015 of issue date". All the geophysical data and geographical coordinates (including depths) were collected from 146 boreholes, 21 underground galleries and 216 trenches distributed on the geological profiles designed over the studied deposit. The total measured metric lengths of boreholes, tunnels and trenches amount to 41 957, 7345 and 25 000 m respectively. The authors also used geological map and the chemical results analyzed at ACME Labs (certified laboratory in Vancouver, Canada); neutron activation analysis (NAA) as well as gamma spectrometry with HPGe detector at AGH UST for 50 solid samples. The correlation between uranium and copper contents of the 50 analyzed samples is presented on figure 2.

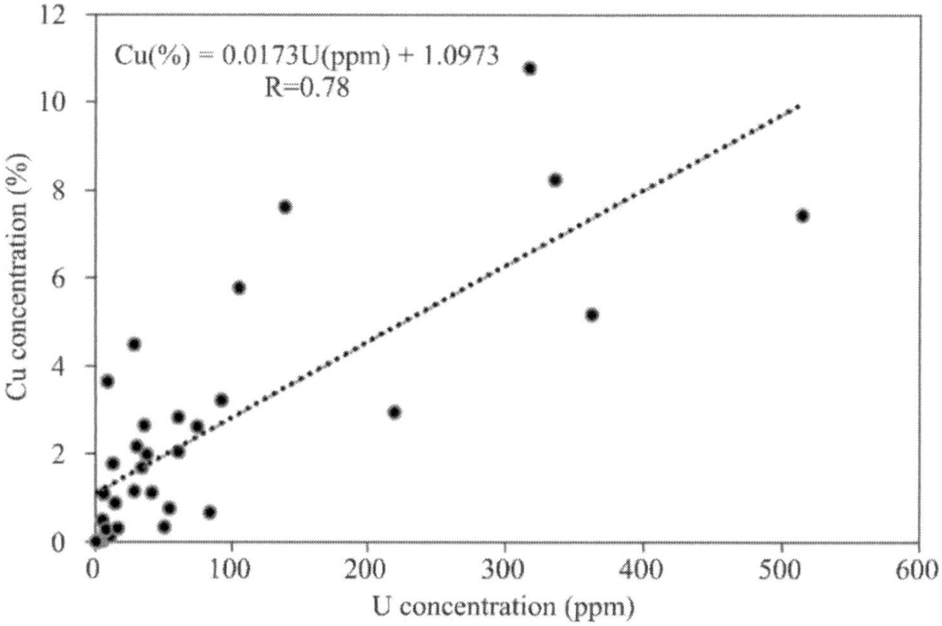


**Fig. 2.** The correlation between U – Cu concentration.

### 3.2 3D modeling methodology

The 3D model of the Sin Quyen deposit was built using software MineScape, product of the ABB™ Company. The scheme of the modeling process is presented on figure 3. The modeling procedure composes of three stages. The first stage was related to inputting and validating the data. The database was transformed into the digital data, then the coordinates of boreholes, cross-sections, the geophysical data record from the galleries and trenches, borehole logs together with the geochemical and lithological data where imported into the software.

The second step was data processing. Most of the archival materials required processing and validating, and the verified data were setup as the consistent data set. 28 vertical cross-sections were created based on the obtained data base. Every ore body presented on each cross-section was determined. The ore-bodies were characterized by high gamma intensity and low resistivity at the borehole logging curves (Fig. 4). Next the ore body elements on the cross-sections were identified using the characteristic geological parameters, structural direction and geophysical data. Every ore-body was drawn in an adequate polygon, which was encoded by a given ID and stored in the software. All the ID codes of the same ore body were arranged into a defined section.

The third stage comprised computation and creation of the 3D block model. The given ore body was built by the connection of the ore body's elements presented on the cross-surfaces, which are defined as the horizontal surface perpendicular to the projection of the ore bodies' trending direction on the XY surface (Fig. 5).

The new geochemical data together with the drill core analyzed results and lithology information were used as the input. The data were processed, correlated and interpreted.

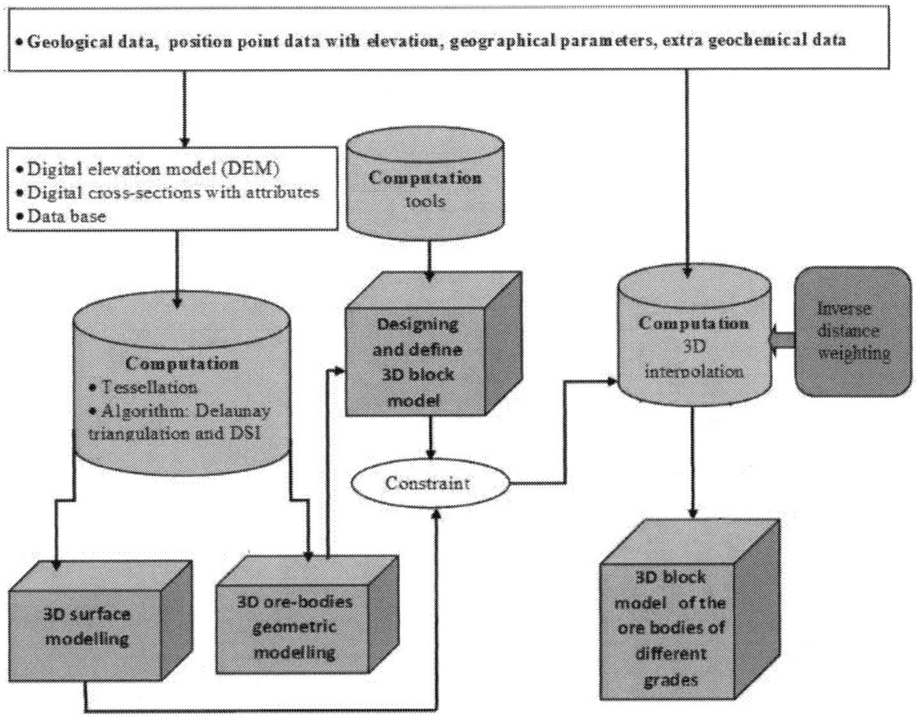


**Fig. 3.** The block scheme of 3D modeling process.

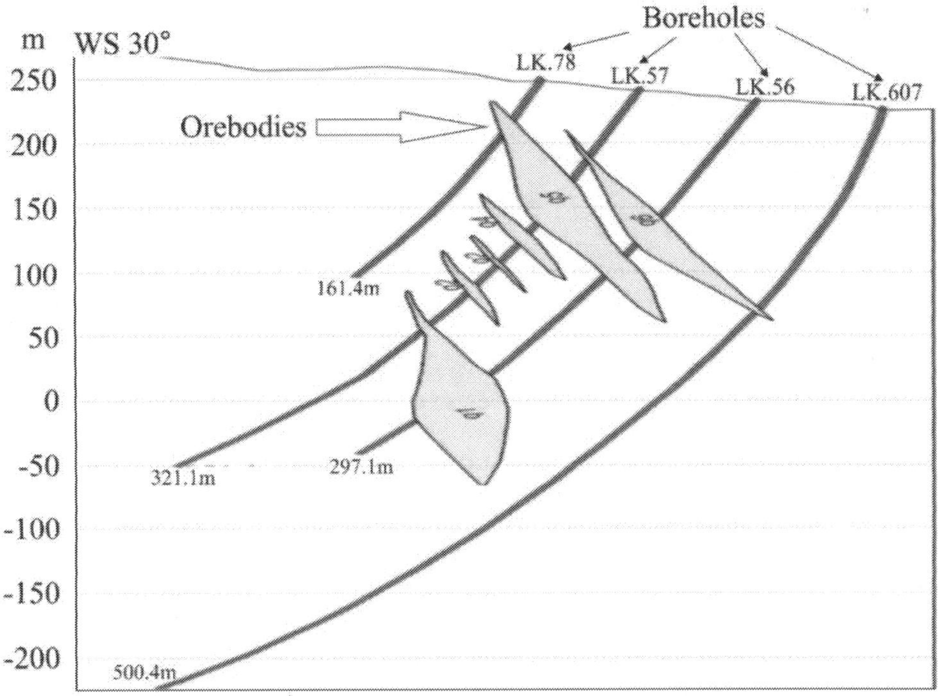


**Fig. 5.** Creating wireframes of the ore bodies based on borehole geophysics and structural data.

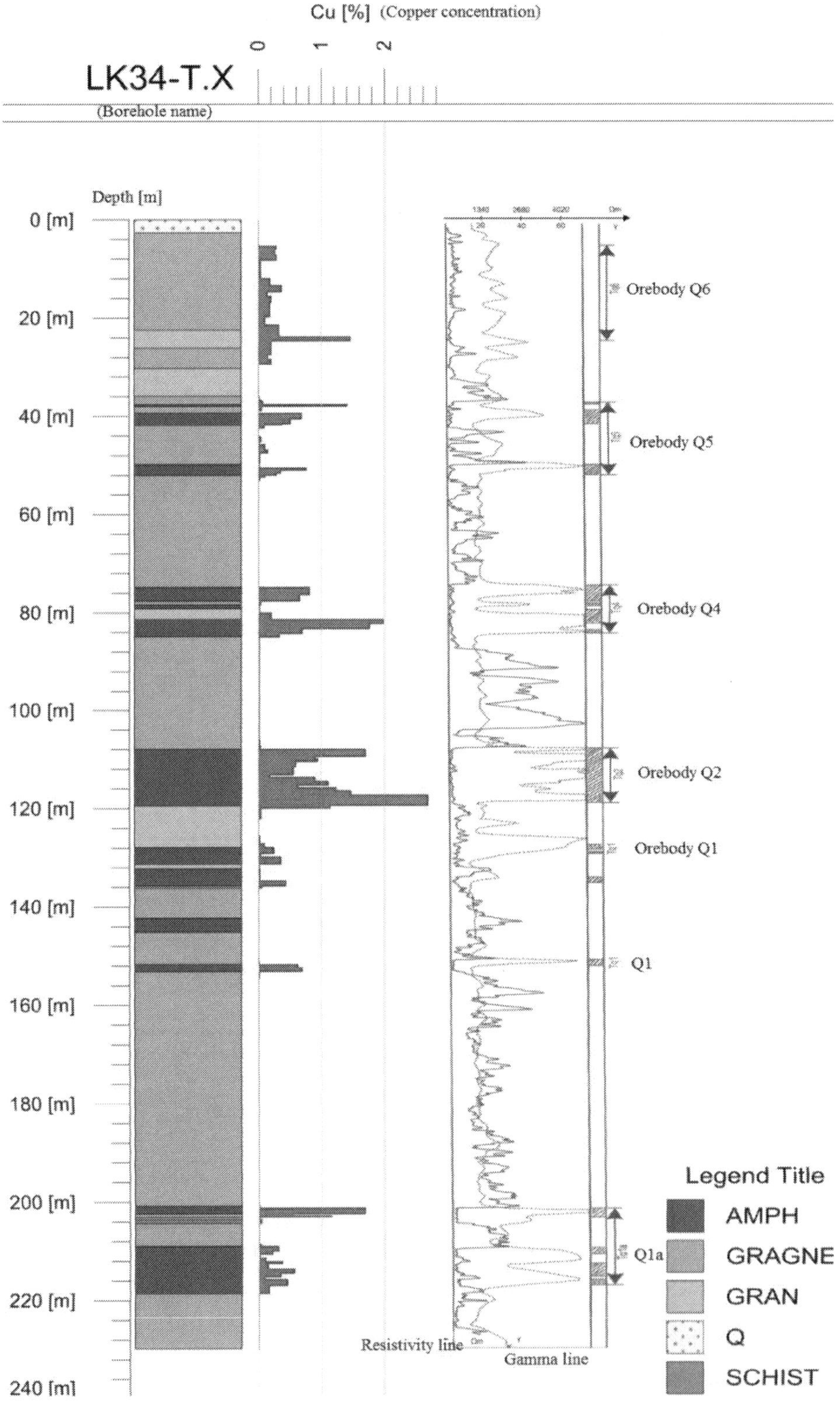


**Fig. 4.** The ore-bodies and Cu grades determined using natural gamma and resistivity curves.

The obtained results such as the fit curves, the geochemical behaviors of U and Cu elements and their distributions emerged from the interpretation. They served together with the coordinates of the well-known analyzed points as the data used to interpolate the points, where there was a lack of the analyzed data. The block model was created only inside wireframes prepared for all ore bodies. The interpolation was performed using inverse distance weighting, which background will be compendiously described in the next paragraph.

## 3.3 Interpolation

### *3.3.1 Inverse Distance Weighting (IDW)*

Based on distribution of samples and software predispositions for grade interpolation non-geostatistical (classical) method was chosen. One of the classical interpolation methods is inverse distance weighting, in which each sample is weighted according to some power of the inverse of its distance from the block to be estimated. In the IDW interpolation the search neighborhood controls both the selection of input points and the amount of weighting given to each point. In this method the search ellipsoid needs to be defined. The search ellipsoid and its parameters controls the process of interpolation. In the block model copper content was interpolated. Before this process samples were flagged to its own wireframes. This step prevents from mixing samples between different ore-bodies. The radius for the search ellipsoid was set on 200 m. After the interpolation, the obtained block model was validated. The set of temporary sections through the block model was created and validated against samples from drill holes.

### *3.3.2 Discrete Smooth Interpolation (DSI)*

The discrete smooth interpolation was used for estimation of the unknown nodes based on the known nodes on the network [30]. The known and unknown nodes are on a common network mesh of a given discrete geological object (e.g. a triangulated surface). The DSI is used to compute the location coordinates (elevation) of nodes and to measure the total "roughness" of various interpolating solutions using a specialized quadratic criterion. It allows also selection of the solution showing the least roughness [31]. At the each node the roughness is defined as the square of the difference between its elevation and the weighted average elevation of the neighbor nodes in the mesh. The DSI method is the useful approach, which is a weighted sum of a local geometrical roughness and the violation degree of the constraints used to this object.

### *3.3.3 Delaunay Triangulation*

The Delaunay triangulation is used to build a model of the terrain surface elevations, which can start with a number of sample points of known height. The Delaunay triangulation always attempts to maximize the amount of acute angle of the triangles in the triangulation. In particular, the Delaunay triangulation avoids sliver triangles. By this way the estimation can reach the highest precision.

## 3.4 The 3D surface model

The 3D surface modeling is built to back up the initial natural pattern of topography. The model shows elevated levels and all ore outcrops in the initial area. The surface modelling demonstrates various factors such as: topography, ore-body outcrops and boreholes localization (Fig. 6). In the Sin Quyen deposit the surface model indicates the outcrop of the copper ore bodies in the terrain along NW–SE strikes, which is rougher in comparison with the surrounding zones. In the depression center there is Ngoi Phat water stream formed by the geological faults trending on the northeast direction under 30-40 degrees of azimuth. The boreholes with core samples mark in yellow color are also presented on figure 6.

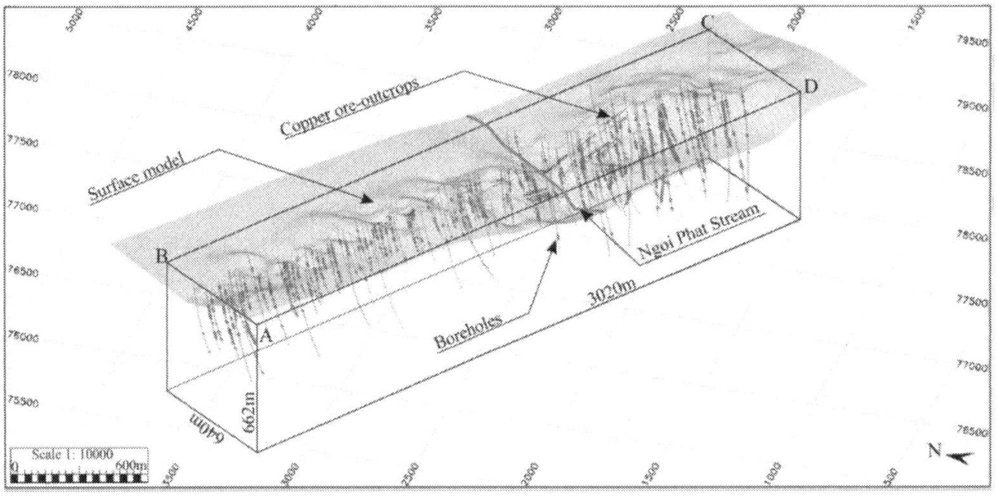


**Fig. 6.** The surface model of the studied area.

### 3.5 The 3D model of ore-bodies

The figure 7 presents the modeled projections of all ore bodies on the surface XY. Each ore body was built based on the connection of its fragments in the different boreholes. The fragments belong to an ore-body are characterized by the similar geophysical properties, related rock formation, and the coordinates of the ore-body. In order to present more clearly every ore body was marked with different color.

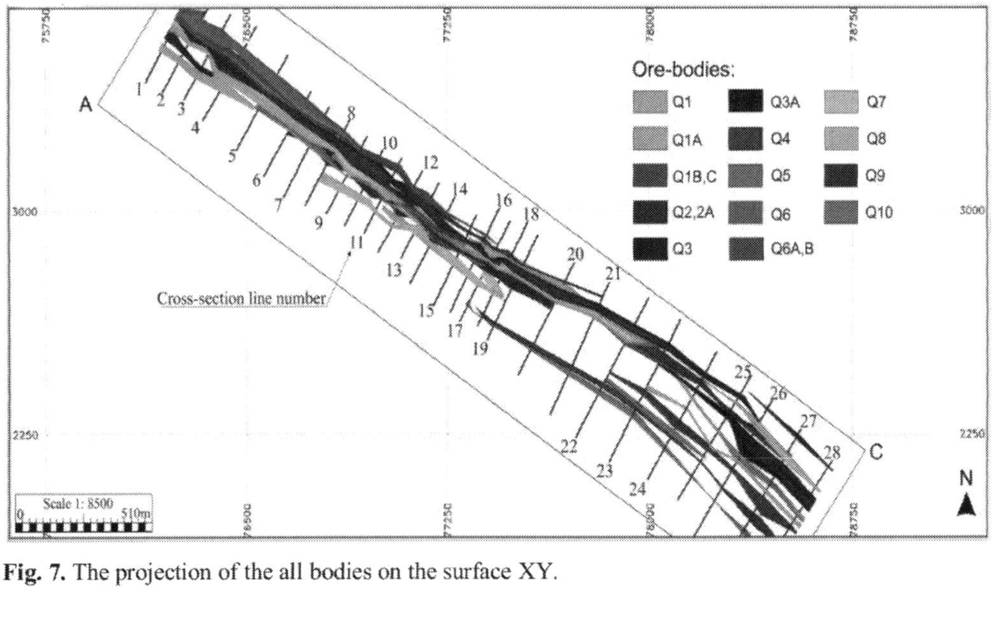


**Fig. 7.** The projection of the all bodies on the surface XY.

Based on the relation of U-Cu, the model of the distribution of different grades of Cu metal was built (Fig. 8).

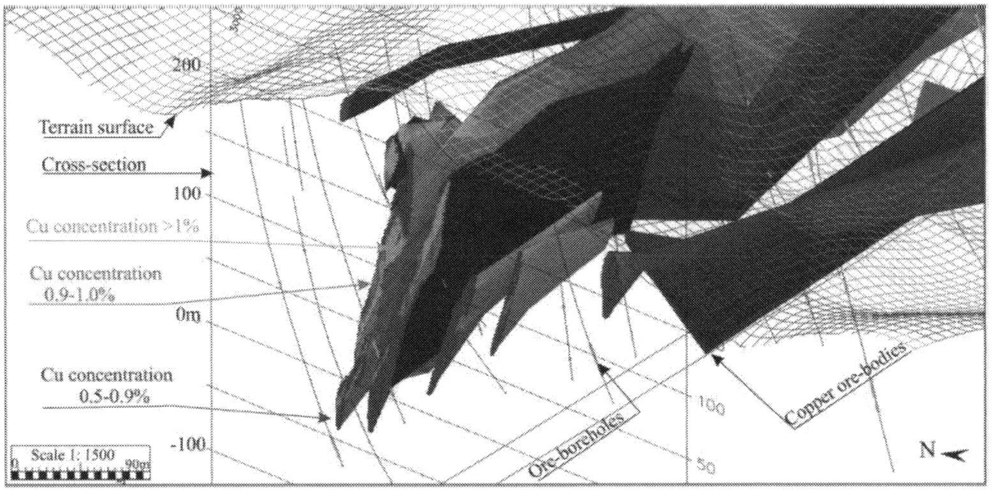


**Fig. 8.** View out of the slice of ore bodies with different grades.

### 3.6 Resources

Based on the block model, the Cu and U resources elements were estimated and amounted to 569 600 tons and 190 tons respectively. The calculated Cu resources are comparable with that given in the report prepared by Ta Viet Dung [29].

## 4 Conclusions

The chemical and geophysical data were useful to estimate the correlation and determine the geometrical parameters of the ore-bodies. Using the computer software MineScape 5.12 the 3D models of ore bodies were built. The 3D models enable to estimate the resources of Cu and other accompanying metals. The obtained 3D model enables us to follow not only the shape of ore-bodies but also the metal grade distribution in the study deposit. The accuracy of the 3D modeling is strongly depends on the conception model, used algorithms and the input data.

## Reference


1. J. P. Chiles, C. Aug, A. Guillen, T. Lees, *Proceedings of International Symposium on Orebody Modelling and Strategic Mine Planning, Perth, Australia, 22–24 November 2004.* 313–320 (2004)
2. E. Rodrigues Duran, R. Di Primio, Z. Anka, D. Stoddart, B. Horsfield, Mar. Petrol. Geol. **45**, 281-303 (2013) doi: 10.1016/j.orggeochem.2013.07.011
3. C. Fei, L. Jiangping, W. Jing, Z. Yuquan, Y. Mingyu, J. Appl. Geophys. **134**, 246–252 (2016)
4. Z. Guo, L. Sun, A. Jia, T. Lu, Petrol. Explor. Dev+ **42,** 83–91 (2015) doi: 10.1016/S1876-3804(15)60009-X
5. A.S. Høyer, F. Jørgensen, P.B.E. Sandersen, A. Viezzoli, I. Møller, J. Appl. Geophys. **122**, 94–102 (2015) doi: 1016/j.jappgeo.2015.09.004

6. J.L. Mari, B. Yven, Mar. Petrol. Geol.. **53**, 133-153 (2014) doi: 10.1016/j.marpetgeo.2013.10.014
7. M. Perrin, B. Zhu, J.F. Rainaud, S. Schneider, J. of Petrol. Sci. Eng. **47**, 89–104 (2005) doi: 10.1016/j.petrol.2004.11.010
8. S. H.H. Nielsen, F. Cunningham, R. Hay, G. Partington, M. Stokes, Ore Geol. Rev. **71**, 578–591 (2015) doi: 10.1016/j.oregeorev.2015.02.001
9. N.M. Sirakov, F.H. Muge, Comput. Geosci. **27**, 59–69 (2001) doi: 10.1016/S0098-3004(00)00055-8
10. J. Wu, Y. Sun, B. Wang, Y. Wang, L. Xu, C. Dai, Petrol. Explor. Dev+. **39,** 99–106 (2012) doi: 10.1016/S1876-3804(12)60020-2
11. Y.Y. Zhang, D.J.. Liu, G.H. Ai, M.J,Qin. J. Appl. Geophys.. **109**, 292–300 (2014) doi: 0.1016/j.jappgeo.2014.08.011
12. A.T. Zehnder, R.W. Allmendinger, J. .Struct. Geol.. **22**, 1009–1014 (2000) doi: 10.1016/S0191-8141(00)00037-7
13. A.M. Lemon, N.L. Jones, Comput. Geosci. **29**, 547–555 (2003) doi: 10.1016/S0098-3004(03)00051-7
14. P. Calcagno, G. Courrioux, A. Guillen, J.P. Chiles, Phys. Earth Planet. In. **171**, 147–157 (2008) doi: 10.1016/j.pepi.2008.06.013
15. G. Wang, S. Zhang, C. Yan, Y. Song, Y. Sun, D. Lo, F. Xu, Comput. Geosci. **37**, 1976–1988 (2011) doi: 10.1016/j.cageo.2011.05.007
16. P. Gumiel, M. Arias, A. Martín-Izard, Geol. J. **45**, 623–633 (2010) doi: 10.1002/gj.1192
17. A. Martin-Izard, D. Arias, M. Arias, P. Gumiel, D.J. Sanderson, C. Castañon, A. Lavandeira, J. Sanchez, Ore Geol. Rev. **71**, 457–476 (2015) doi: 10.1016/j.oregeorev.2015.06.006
18. M. Maxelon, P. Renard, G. Courrioux, M. Brandli, N. Mancktelow, Comput. Geosci. **35**, 644–658 (2009) doi: 10.1016/j.cageo.2008.06.005
19. O. Kaufmann, T. Martin, Comput. Geosci. **34**, 278–290 (2008) doi: 10.1016/j.cageo.2007.09.005
20. G. Wang, R. Li, E.J.M. Carranza, S. Zhang, C. Yan, Y. Zhu, J. Qu, D. Hong, Y. Song, J. Han, Z. Ma, H. Zhang, F. Yang, Ore Geol. Rev. **71**, 592–610 (2015) doi: 10.1016/j.oregeorev.2015.03.002
21. P.T. Hieu, F.K. Chen, N.T.B. Thuy, C.N. Quoc, S.Q. Li, J. Geodyn. **69**, 106–121 (2013) doi: 10.1016/j.oregeorev.2015.03.002
22. P.H. Leloup, R. Lacassin, P. Tapponnier, U. Scharer, D. Zhong, X. Liu, L. Zhang, S. Ji, T.T. Phan, Tectonophysics. **251**, 3–84 (1995) doi: 10.1016/0040-1951(95)00070-4
23. H. H. Huang, Z.J. Xu, Y.M. Wu, X. Song, B.S. Huang, L.M. Nguyen, Tectonophysics. **584**, 230–239 (2013) doi: 10.1016/j.tecto.2012.03.030
24. P. Tapponnier, R. Lacassin, P.H. Leloup, U. Schuarer, D. Zhong, H. Wu, X. Liu, S. Ji, L. Zhang, J. Zhong, Nature. **343**, 431-437 (1990) doi: 10.1038/343431a0
25. T. V. Dung, *Report of geological surveys and their results performed at the IOCG Sin Quyen deposit in Lao Cai, North Vietnam. Main Department of Geology of Vietnam*, 318 p. (1975) (in Vietnamese)
26. S. Ishihara, H. Hideo, H. Mihoko, P.N. Can, T.D. Pham, T.A. Tran, *Bulletin of the Geological Survey of Japan*. **62**, 197 – 209 (2011)
27. R.N. McLean, *PGC Publishing, Adelaide*. **2**, 293–301 (2001)

28. J. Pieczonka, A. Piestrzyński, L.K. Phon ., N.D.Chau, P. Jodłowski, *Viet-Pol 2015 second international conference on scientific research cooperation between Vietnam and Poland in Earth Sciences*. 331-353 (2015)
29. I.V. Gaskov, T.A. T.T. Hoa. Tran, T.D. Pham, P.A. Nevolko, N.C, Pham, Russ. Geol. Geophys+. **53**, 442–456 (2012)
30. J.L. Mallet, Comput. Aided Des. **24**, 178–191 (1992) Doi: 10.1016/0010-4485(92)90054-E
31. J.L. Mallet, Math. Geol. 29, 199–219 (1997) doi: 10.1007/BF02769628C. Le Carlier de Veslud, M. Cuney, G. Lorilleux, J.J. Royer, M. Jebrak, Comput. Geosci. 35, 92–107 (2009) doi: 10.1016/j.cageo.2007.09.008

# Application of PANTERA system for monitoring vibrations induced by blasting works in open pit mining plants and the role of supervisors

*Grażyna* Dzik*, *Adam* Mirek

State Mining Authority, Katowice, Poland

**Abstract.** Measurements of vibrations induced by blasting works in mining plants are one of the possibilities of controlling the operations of open-pit facilities by supervisory authorities. As part of these works, impact of vibrations surveys on buildings within the range of harmful impacts zone of seismic vibrations are carried out to guarantee the safety and protection of their residents. The article presents the possibilities of using digital apparatus for recording and processing vibrations induced by blasting works used during supervisory activities in open-pit mines from 2015 to 2017. Recording and size of events depends on mining conditions and geological structure of the exploited deposit and is closely related to detonation of explosive. The use of dynamic impacts SWD scale against the backdrop of the current norm no PN-B-02170:2016-12 are discussed and examples of assessments of harmfulness of vibrations transmitted by the ground to buildings are presented. It was found that bigger and bigger number of monitoring measurements is performed every year, causes reducing the number of complaints and incriminating letters on mine companies receive in effects of impact of vibrations.

**Keywords**: open pit mines, vibration monitoring, blasting works, ground vibrations, dynamic impact scales SWD

## 1 Introduction

Mining works utilizing explosives can produce harmful effects on buildings and other structures. Recording para-seismic vibrations induced by blasting works in open-pit mining plants using the PANTERA digital vibration recorder is therefore the basic method for mining supervision authorities to reliably measure the parameters of para-seismic vibrations in the aspect of their harmful effects on buildings and other structures situated in the closest vicinity of mining plants. With this, the correct operation of the mining plant, including the blasting works conducted there, are also verified. Hence the idea to create a device, the architecture of which would comprehensively support inspection and control personnel in its operations. One of the most important features of this device is its continuous monitoring and cyclic measurements functionality, which has proven useful in the course of inspection

* Corresponding author: g.dzik@wug.gov.pl

works. The device automatically converts digital signal and calculates earth accelerations based on the measurement data and using a third octave filter, and ultimately generates reports presenting the results in SWD (Dynamic Impact Scales), including a full set of information required to identify the event. These features of the device can be used to conduct ongoing monitoring of the safety of seismic works. The complete calculation process does not require complex systemic calculations, and the end result is the earth acceleration value, visualized in SWD scales. What is noteworthy is that the most decisive factors that contribute to the overall effects of blasting works are, firstly: the geological structure of the deposit and its environment, then: the technical condition of protected buildings and structures, the method of transferring vibrations from the groundwork to the building, and, ultimately, the weight of the EC in the borehole or a series of boreholes. These factors influence the result of measurement. However, it should be stressed that the building deterioration process is related to both increased ambient factors and the very structure of the building, and the duration of impact of vibrations on these buildings.

The purpose of the article was to present the possibilities of uses of paraseismic apparatus in the operations of mining supervision uses to identify threats in the movement of open-pit mining plants, and to guarantee the seismic safety of blasting works. In this way, the catalog of control activities was expanded to include measurements of para-seismic vibration parameters in the aspect of their detrimental effect on the environment of mining plants, and a basic analysis and documentation of vibrations recorded, in terms of their effect on protected buildings and structures.

## 2 Laws regulating the impact of blasting works in open-pit mining plants

Control operations often result from the need to minimize the impact of mining works on buildings and structures situated in the vicinity of the mining plant, and involve a determination of the safety conditions pertaining to the environment of blasting works, and documentation of this impact level. The Geological and Mining Law [1] regulates issues related with supervision and control under Article 154 and Article 153 section 1 in the operations of a mining plant extracting minerals from deposits. As regarding blasting works and measurements carried out amid mining operations and related to their safety, all proceedings must conform to the provisions of Appendix 4 to the ordinance of the [9] Minister of Energy of 9 November 2016. This appendix lays down the principles of designating hazard zones around the place where blasting works are performed, and the material expert is authorized to designate the range of the hazard zone: based on rock fragment dispersion, para-seismic vibrations of the rock mass (estimated calculations) and airborne shock wave.

## 3 PANTERA Digital Vibration Recorder used for observing the impact of para-seismic vibrations on the environment

The PANTERA Digital Vibration Recorder is a portable, autonomous vibration and explosion recorder and analyzer, used for observing the impact of para-seismic vibrations on the environment. The instrument is also used to analyze low-frequency ground and building vibrations in areas threatened by para-seismic vibrations induced by human activity - i.e. blasting works. Any vibrations detected are recorded by accelerometers, and then saved in the recorder's memory, from where they are transferred by the user to computer memory, together with the processing software installed. This software is used to archive and process event records according to the Polish standard applicable to assessments of the harmful effects of vibrations on buildings according to the effective Dynamic Impact Scale (SWD). The CRD PANTERA system comprises three-component accelerometers (x, y - horizontal components, z - vertical component) which measure the acceleration of seismic vibrations, a GPS module for synchronizing the time and locating the device, a measurement system and a power supply system. Digital vibration recording is carried out by writing the values of "a" amplitudes in time "t" in the form of binary files saved in the recorder's memory. One of the characteristic features of the device is its automatic spatial orientation and time synchronization with any random time standard. Accelerometers measure the acceleration of vibrations in a range from 1 to 800 Hz, and the operator can configure the device and its parameters to suit the measurement range, the duration of recording and the recording release level. After completing the transfer of detected and recorded vibrations from the device, the operator can perform an amplitude and frequency analysis of the recording, filter signals, and then generate an event measurement report, archive data and, ultimately, assess the impact of vibrations on buildings and structures as per PN-B-02170:1985 [6]. It is noteworthy that measures have been taken to adapt the system to the latest and effective version of PN-B-02170:2016-12 [7]. However, these differences do no influence the result of measurement. The PANTERA recording and measurement device has a Calibration Certificate issued by a calibration laboratory (Laboratory of Technical Acoustic, Central Mining Institute) accredited by the Polish Centre for Accreditation, a signatory to EA MLA and ILAC MRA concerning the mutual recognition of calibration certificates.

**Fig. 1.** Digital Vibrations Recorder PANTERA [3].

Inspection and control works generally consist in verifying the safety of blasting works. Instruments recording vibrations caused by para-seismic shocks are used for this purpose. Correlated with data obtained for the blasting works, the records created can be used to verify the measured parameters against true measurements performed in the field. The flexible waves created disperse when the explosion energy is released, posing a threat to mine pits, and, most of all, to the buildings and structures on the surface. Control operations assume continuous observation (monitoring) and cyclic measurements. Permanent, long-term supervision of blasting works carried out by the company provides a holistic image of the rock form processes in place and foundation reactions and their impact on blasting works in the buildings under observation. Vibration recorders are used to assess the threat to buildings and structures caused by para-seismic vibrations. By documenting the impact of vibrations on buildings and structures by generating event reports, this impact can be controlled in a systemic manner. Additionally, a database is created and provides unlimited access for data verification.

## 4 Assessment of impact of blasting works on buildings and structures

Until November 2016, mining plants adhered to PN-B-02170:1985 [6] "Assessment of harmful effects of ground-transferred vibrations on buildings", a standard developed by a

Krakow University of Technology team headed by Prof. Roman Ciesielski [2]. The standard admitted the application of Dynamic Impact Scales (SWD) in the approximate assessment of vibration impact on buildings. In more than 30 years, sufficient knowledge and experience has been gathered through tests to modify it [10]. The outcome of these works is the amended standard PN-B-02170:2016-12 [7] bearing the same name, which has superseded the 1985 version. What is noteworthy is that one of the most significant amendments is the introduction of SWD for acceleration values, and the withdrawal of displacement scales which have proven redundant due to the type of sensors used. Therefore, acceleration and speed were adopted as the basic values to be recorded. Assessment is currently performed on the basis of the maximum value of the parameter, on third-octave bands. The descriptions of criteria applicable to assessments of vibrations transmitted onto buildings were specified to include SWD - i.e. zone and SWD limit descriptions, as well as descriptions of requirements for dynamic measurements - device installation requirements. In general terms, after recording the vibrations (Fig. 6), performing calculations (Fig. 7) and marking the results on nomograms (Fig. 8), the harmful effects of the vibrations recorded can be identified. In conclusion, the standard is applied to recognize the harmful effects of vibrations on existing buildings, as well as in designing buildings, if the building is expected to be prone to vibrations transmitted by the groundwork.

The criteria and scales used to assess the human-induced effects. SWD consider not only the intensity of vibrations or their frequency, but also the type, structure and dimension of buildings.

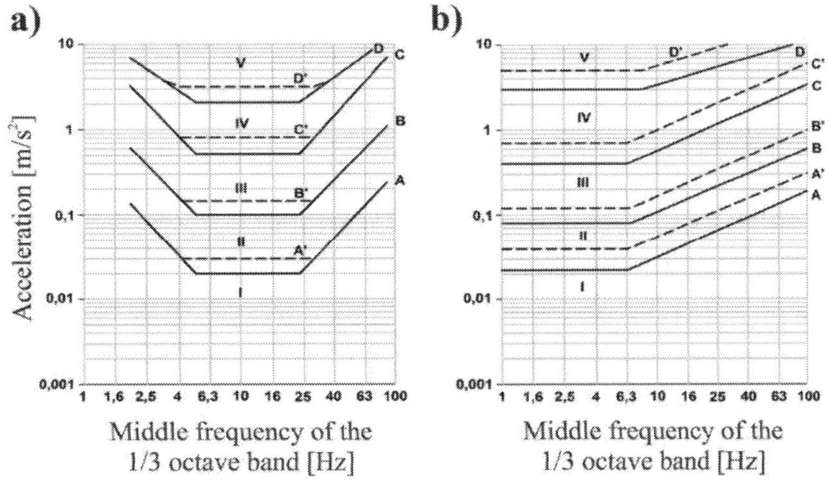


**Fig 2.** Logarithmic dynamic effect scales in the function of acceleration [ms$^{-2}$]: a) SWD-I and b) SWD-II, including descriptions of building vibration impact zones where x axis - middle frequency of the 1/3-octave band, y axis - maximum vibration acceleration amplitudes in the band (PN-B-02170:2016-12). Zone I - negligible vibrations, Zone II - harmless vibrations, Zone III - harmful vibrations, Zone IV - high-frequency vibrations, harmful to the building. Boundary lines of the SWD scale for the criteria of fulfilling the condition: A - perceptibility, B - stiffness, C - strength, D - stability. [7].

According to Fig. 2, SWD scales have five zones separated by borderlines which describe the level of harmfulness of vibrations on the buildings. SWD-I includes tight-structured buildings, of small external dimensions in the horizontal projection (<15m), one- or two-story, the height of which does not exceed any of the horizontal-projection dimensions. In turn, SWD-II includes buildings up to 5 above-ground stories, the height of which does not exceed the double lowest width of the building, as well as low buildings (up to 2 above-ground stories) which do not fulfill the requirements specified for SWD-I [7].

To obtain correct results for the vibration values measured, one needs to know the geological structure of the deposit and the environment, the technical condition of protected buildings, the conditions for the dispersion of waves between the groundwork and the building, and, ultimately, the weight of the explosive in the borehole or its series [10].

## 5 Test results and its analysis- impact of vibrations induced by blasting works and their effect on the environment.

Fig. 3 presents data for individual types of measurements performed in the course of controls implemented in open-pit mining plants relative to half-year time frames (with the exception of 2015, for which data was consolidated). The chart presents the number of various measurements performed in a time frame, i.e. a measurement performed for a residential building and the company building on the ground, a measurement in the harmful vibration zone, and monitoring. Until the end of 2016, the majority of measurements were performed for residential buildings and company buildings. However, following consultations concerning the device and the measurement methodology, measurements have been performed predominantly for residential buildings for a year now. The correctness of this methodology is proven by the mere small number of incriminating letters received by mining authorities or complaints concerning sensations or building damages occurring during or after blasting works carried out by companies. In this respect, we have noticed that, after 2016, following a series of intensive seismic safety inspections in companies conducting blasting works, a trend consisting in reducing the number of measurements has emerged in 2017. These trends generally result from a smaller number of complaints received, broader knowledge of residents of buildings situated in the closest vicinity of mining plants, and broader knowledge of companies themselves, as well as more effective measures taken by inspectors controlling the activity of mining plants.

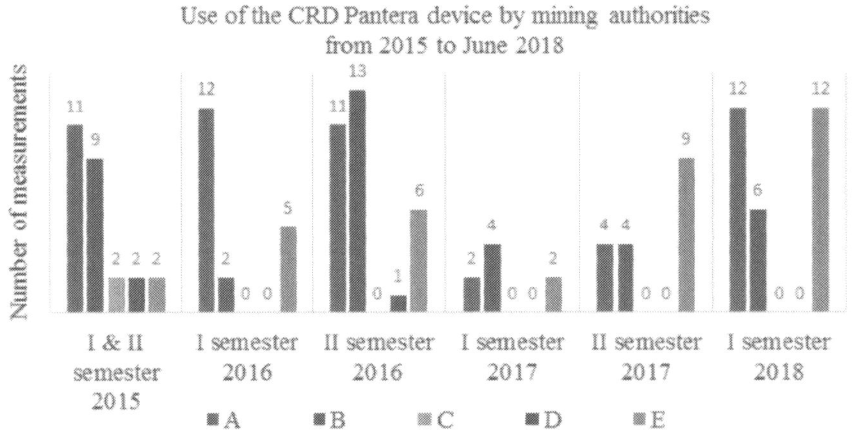


**Fig. 3.** Types and the number of measurements performed with the use of the PANTERA device by mining authorities from 2015 to June 2018 (A - Measurement on the residential building in connection with a complaint, B - Measurement on the entrepreneur's building in connection with the complaint, C - Measurement in the zone of harmful vibrations, D - Measurement on the ground, E - Monitoring).

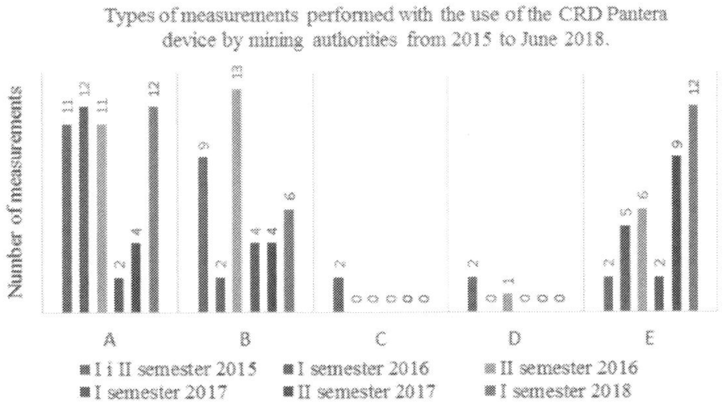


**Fig. 4.** Types of measurements performed with the use of the PANTERA device by mining authorities from 2015 to June 2018 (A - Measurement on the residential building in connection with a complaint, B - Measurement on the entrepreneur's building in connection with the complaint, C - Measurement in the zone of harmful vibrations, D - Measurement on the ground, E - Monitoring).

Fig. 4 presents the number of various measurements (performed for a residential building and the company building on the ground, a measurement in the harmful vibration zone, and monitoring) in half-year intervals, proving the legitimacy of measurements taken. Therefore, measurements performed in a residential building or with the company are the largest percentage of measurements carried out as part of supervision activities. According to Fig. 5, this is a total of 95% of measurements related to events recorded in open-pit mining plants. 40% of these measurements taken in 3.5 years were performed applying the monitoring method, using devices installed for a period of 2 weeks up to three months. Returning to Fig. 4. Note that the most measurements taken in the last year were continuous measurements. This practice was accepted not only by inspectors, but also by inquirers who demanded verification if the blasting works would be conducted safely, without effect on their buildings. Literature of the field includes multiple examples of monitoring activities [4, 5, 8]. It is therefore justified to strive to uphold this trend, as information from monitoring measurements is reliable, contrary to single, incidental event measurements which are often taken with the consent of the company. The next two columns contain measurements performed in the harmful vibrations' zone and on the ground, taken as part of tests only.

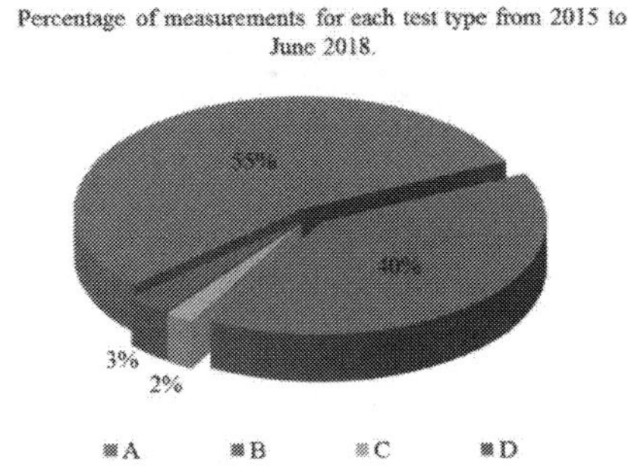


**Fig. 5.** Percentage of measurements performed for each test type from 2015 to June 2018 (A - Measurement on the residential building in connection with a complaint, B - Measurement on the entrepreneur's building in connection with the complaint, C - Measurement in the zone of harmful vibrations, D - Measurement on the ground).

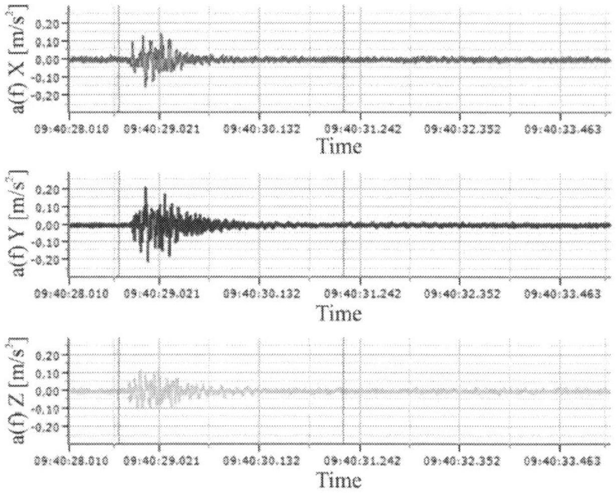


**Fig. 6a.** Example of seismogram for para-seismic vibrations recorded for event 1 as a result of blasting works involving a total charge of 450kg, distributed in a system of boreholes. The maximum acceleration at the sampling speed of 1600 Hz is 0.152 $m/s^2$ for axis X, 0.215 $m/s^2$ for axis Y, and 0.13 $m/s^2$ for axis Z.

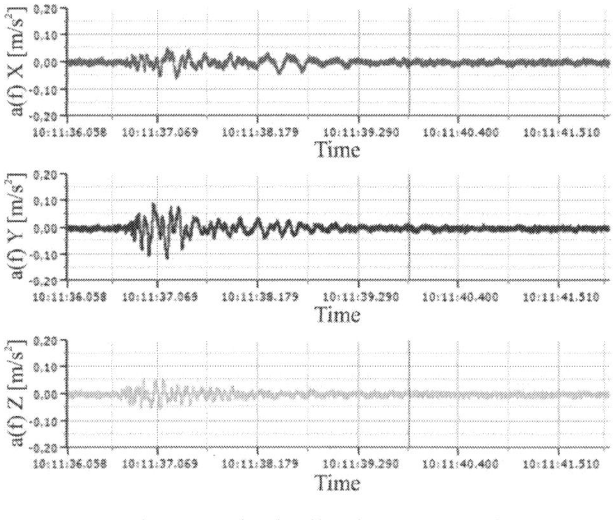


**Fig. 6b.** Example of seismogram for para-seismic vibrations recorded for event 2 as a result of blasting works involving a total charge of 250kg, distributed in a series of boreholes. The maximum acceleration at the sampling speed of 1600 Hz is 0.06 $m/s^2$ for axis X, 0.117 $m/s^2$ for axis Y, and 0.06 $m/s^2$ for axis Z.

Fig. 6a and Fig. 6b present an exemplary seismogram for vibrations recorded in the measurement point, i.e. in a residential building. A basic analysis of the seismogram is reduced to determining the maximum values of the measured parameter, i.e. acceleration in this case, compared to the values of vibration frequencies. An analysis applying a third-octave filter can be used to precisely assess the effect of vibrations induced by blasting works on buildings situated in the vicinity of the mine (Fig. 7a, Fig. 7b). The assessment is carried out by use of a dynamic impact scale SWD (Fig. 8a, Fig. 8b), where the maximum values of the acceleration amplitude are shown. As a result, an information document - a report is generated. For further analysis, the data is subject only to expert judgment. However, as current norm PN-B-02170:2016-12 [7] says, short-term vibrations with parameters fitting in the upper zone, if the given conditions from Table 4, column 3 of the above norm, effects can be attributed to the lower zone, provided that the appropriate conditions are met in the norm.

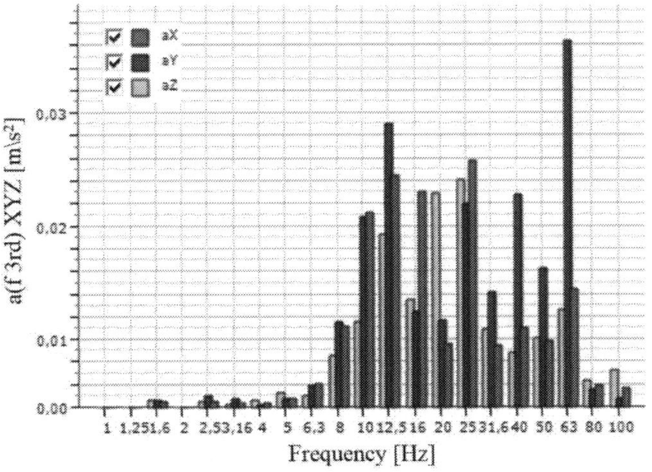


**Fig. 7a.** Diagram following a third-octave analysis for event 1, recorded for a residential building (component X - frequency [Hz], component Y – acceleration [m/s$^2$]).

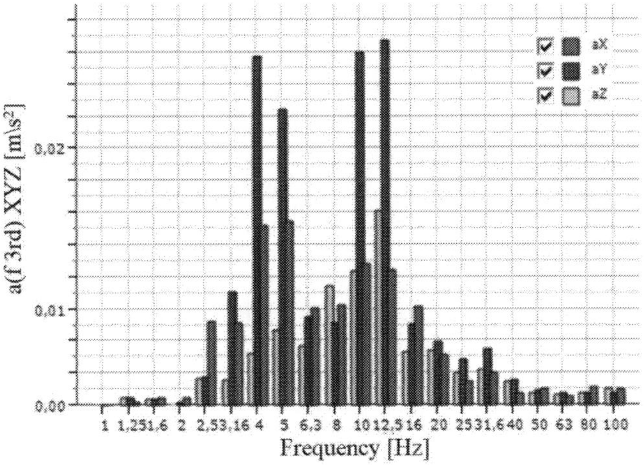


**Fig. 7b.** Diagram following a third-octave analysis for event 2, recorded for a residential building (component X - frequency [Hz], component Y – acceleration [m/s$^2$]).

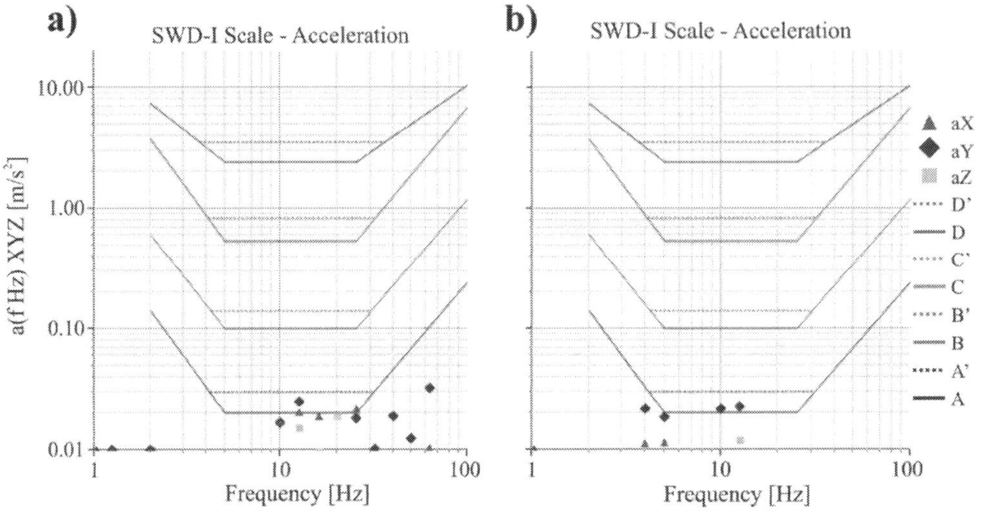


**Fig. 8.** Diagram of assessment of the impact of vibrations on a building, applying a SWD-I scale for a) event 1 and b) event 2, indicating maximum acceleration values for individual thirds (component X - frequency [Hz], component Y - acceleration[m/s$^2$]).

## 6 Conclusions

Based on the analysis of data and experience gathered from mining authorities, which used the PANTERA device for the last 3.5 years, the following conclusions were presented:

- The digital PANTERA system can be used to conduct ongoing control of the impact of blasting works on buildings situated in the vicinity of open-pit mining plants through monitoring or cyclic measurements. With reference to the standard, it allows to study the harmonic character and long-lasting vibrations, i.e. considering material fatigue.
- As a result of the subsidized enterprise, in 2015-2017, significant improvement in the quality of supervision performed by mining authorities and control of open-pit mining plants was witnessed. This situation contributes to achieving the anticipated effects consisting in better identification of threats to the environment and general safety.
- A data analysis indicated that, between 2015 and June 2018, 55% of control measurements were carried out in residential buildings, as recommended and according to the idea of para-seismic measurements following official complaints. The remaining 40% were control measurements carried out company buildings, and 5% were trial measurements (related to the initial phase of use of the device).
- The clear decrease in the number of complaints and incriminating letters received in the subject of general impact of vibrations on buildings correlates with the largest number of monitoring measurements completed.

Projects titled "Furnishing mining supervision bodies with equipment used to monitor mine operations", subsidized under the National Fund for Environmental Protection and Water Management allowed mining supervision bodies to purchase the CRD PANTERA system used to assess para-seismic vibrations. The PANTERA system was built in accordance with PN-B-02170:1985 [6]. It is currently adapted to the amended version of the standard PN-B-02170:2016-12 [7] "Assessment of harmful effects of ground-transferred vibrations on buildings". Many academic centers or licensed material experts from all over the country have switched to the new standard. Therefore, in our pursuit of global standards [10], we try to develop in terms of measurements and new possibilities of controlling the impact of blasting works on buildings. It must be noted, however, that, apart from creating databases, measurements performed with the use of para-seismic vibration detection devices used by mining supervision bodies are for the bodies' internal purposes only. This data undergoes expert analyses (performed by material experts) to present evidence in court cases. In conclusion, the most important role of recording the harmful effects of vibrations is the possibility of ordering the company to verify its blasting parameters.

## References

1. Act of 9 June 2011 – Geological and Mining Law (Journal of Laws.2017.2126)
2. R. Ciesielski, *Problemy diagnostyczne skutków wstrząsów górniczych na budowle naziemne. (Diagnostic problems of the effects of mining shocks on ground construction)* Górnictwo Odkrywkowe **5-6**, 13–19 (2004) (in Polish)
3. *Operation and Maintenance Documentation for the Digital Data Recorder*, instruction manual, EMAG-SERWIS
4. E. Maciąg, *Ocena wpływu wstrząsów górniczych na budynki (Assessment of the impact of mining shocks on buildings)* [in:] Mat. sympozjum „Warsztaty", IGSMiE PAN, Krakow, 297–317 (2000) (in Polish)
5. Z. Onderka, R. Biessikirski, J. Sieradzki, J. Winzer, *KSMD - System do monitorowania drgań powodowanych robotami strzelniczymi w otoczeniu kopalń odkrywkowych (KSMD – System for monitoring vibrations caused by blasting works in the surroundings of open pit mines)*, Warsztaty z cyklu „Zagrożenia naturalne w górnictwie" Mat. Sympozjum,199–211 (2002)
6. PN-B-02170:1985 Assessment of harmful effects of ground-transferred vibrations on buildings.
7. PN-B-02170:2016-12 *Assessment of harmful effects of ground-transferred vibrations on buildings*
8. J. Pyra, A. Sołtys, J. Winzer, *Monitoring drgań jako skuteczne narzędzie do kontrolowania oddziaływania na obiekty (Vibration monitoring as an effective tool for controlling the impact on objects)*, Mat. Konferencyjne. Technika Strzelnicza w Górnictwie i Budownictwie, Ustroń, 25-27.04.2013, 229–243 (2013)
9. *Ordinance of the Minister of Energy of 9 November 2016 on detailed requirements for the storage and use of blasting means and blasting equipment in mining plant operations* (Journal of Laws 2017.321)
10. J. Winzer, A. Sołtys, J. Pyra, *Oddziaływanie na otoczenie robót z użyciem materiałów wybuchowych (Influence on the surroundings of works with the use of explosives)*, Wyd. AGH, Krakow (2016)

# Problematic aspects of interpretation of seismic signals induced by detonation of explosives

*Józef* Pyra[*], and *Michał* Twardosz

AGH University of Science and Technology, Faculty of Mining and Geoengineering, Cracow, Poland

**Abstract.** The article presents problems concerning an interpretation of seismic signals registered in the ground and foundations of buildings which are induced by the detonation of explosives. In the course of blasting in opencast mines, harmful effects may occur. The primary impacts which usually occur are as listed, induced vibrations, airblast, acoustic wave and the flyrock. There are technical and technological methods to reduce them, but minimising one can enhance the impact of the other. Furthermore, impacts such as vibration, acoustic wave, and airblast may overlap - the additive effect, which might cause problems with the unambiguous identification and interpretation of the recordings. The paper presents the results of tests carried out on a laboratory scale of the detonation of varying weight explosives samples. Also, examples of seismic signals recorded during field measurements in the vicinity of open-pit mines are presented. Also, examples of seismic signals recorded during field measurements in the vicinity of open-pit mines are presented. These examples confirm doubt as to the applicability of PN-B-02170: 2016-12 to assess the effects of vibrations on the building structures in the event of a significant impact of the airblast.

**Keywords:** shock wave, airblast, ground vibration

## 1 Introduction

Detonation of explosives causes in addition to positive effects (e.g. fragmented rock, broken up foundations, creating a borrow pit for leveling etc.) negative impact in the form of ground and building vibrations, airblast waves, acoustic waves, post-blast fumes, dustiness, and many others depending on the specifics of conducted works using explosives. The blast induced vibrations, airblast wave and acoustic wave can be superimposed because as a physics phenomenon they are categorised as waves that propagate in different media (ground, water, air) with varying velocity. The coexistence of these interactions may cause an intensity change of one impact and reduce the effectiveness of prevention methods of other.

The problematic aspects of the impact of vibrations induced by the detonation of explosive charges on the surroundings is widely discussed around the world. The situation is slightly different when it comes to the issue of the airblast because it is a series of tests that

[*] Corresponding author: pyra@agh.edu.pl

are carried out for military purposes, in which the study of the closest blast zone impact is critical. The research mentioned above is carried out primarily for the protection of vehicles and military facilities. In the case of the use of explosives for civilian purposes (demolition works, blasting works in opencast mines, etc.), they should be designed in such way that they do not cause any damage in the environment. The problem may arise in the situation when the airblast wave affects the construction, and both the Polish standard [1] and the guidelines used in other countries [2, 3, 4] clearly state that the vibrations taken into account for the impact assessment are transmitted through the ground, and they are not caused by an airblast propagation.

Chengqing and Hong [5] state that in the case of detonation of explosives on the surface at small distances the decisive impact is registered from the airblast, at greater distances the effect of ground vibrations starts to outweigh, and at long distance, both phenomena can be now analysed separately.

Gupta, Roy and Singh [6] explains, however, that the vertical axis of vibrations of the rock medium causes oscillation equivalent to the velocity of sound, which causes vibrations in the air at the same frequency as the vertical vibrations of the rock medium, which may result in a disturbance of seismic signals recording. On the other hand, Alcudia [7]and Babcock [8] analysed the possibilities of noise reduction transmitted by air in seismic signals. For this purpose, microphones were used to record sound, so that at a later stage these recordings would serve as a kind of filter, thanks to which it is possible to separate the seismic signal. Moreover, in Faramarzi et al. [9], Kabwe and Wang [10], the issue of the impact of vibrations and airblast, caused by the detonation of explosives, is discussed in the aspect of protecting the environment and people.

The primary objective of the preliminary results of this study was to illustrate problematic aspects of assessment of the signals recorded by the geophones, and which are induced by the detonation on the surface of explosive charges. Incorrect analysis of signals may lead to misinterpretation, especially in the aspect of the impact assessment of vibrations. Another stated problem is the conversion of the airblast in the acoustic wave which should be analysed in different ways using appropriate measurement procedures and analysis.

## 2 Methodology

In order to observe the phenomenon, tests were carried out in field laboratory conditions, during which measurements of ground and building foundations vibrations were performed (sensors were mounted at ground level both inside and outside on the same supporting wall) and changes in air pressure (each test stand was equipped with a microphone). In the course of the research a total of 8 blast series were initiated in which the manner of locating a small mass of explosive charge was changed (about 3 g) starting from detonation on the surface, moving on to the suspension of the charge at 0.6 m height, and ending with a dug in charge at a 0.2 m depth and covering it with a steel sheet (thickness of 0.06 m). The article presents an analysis of the extreme cases, a suspended charge detonation, and an underground detonation.

Due to the fact that the tests were performed in the Republic of Polish, experimental studies were carried out in accordance with the following guidelines:

1. In the field of vibration measurement - Polish standard PN-B-02170:2016-12 Assessment of harmfulness of vibrations transmitted by the ground to buildings [1]. The standard specifies the location and method of mounting sensors for vibration measurements and the minimal technical parameters of vibration sensors. In the assessment using dynamic impacts (SWD I and SWD II scales), seismograms (full waveforms) of horizontal components of vibration should be used, ie with x and y (transversal and longitudinal) directions recorded at the measurement point towards the vibration source, at the rigid

structure node - at the intersection of load-bearing walls - located on the foundation of the building or in a rigid node on the wall of the underground storey at the ground level of the surrounding area. The seismograms should be analysed in one-third bands (tertiary analysis), obtaining in each of the bands the maximum (peak) values of acceleration or velocity of vibration. It means that the assessment of the recordings, according to the SWD scales requirements is to use entire waveforms of recordings of horizontal components. The analysis of the full entries for the x and y-axes is carried out by filtering the signal with a tertiary filter. The results obtained this way, presented as a histogram of peak particle velocity (PPV) values in a given frequency band, are plotted on the SWD scale assigning the effects corresponding to a given zone. It is an analysis method, based on which, in the case of short-term vibrations (pulses) is made the final assessment. As short-term vibrations, one should understand vibrations, the duration of which in a day does not exceed 3 minutes.

2. In terms of measuring the airblast - there is no guideline in regards to the methodology. The measurements were carried out following the recommendations contained in the [11, 12]. The information provided by the Regulation of the Minister of Economy, Labor and Social Policy of 9 July 2003. On occupational health and safety in the production, intra-factory transport and trading of explosives, including pyrotechnic articles [13], in particular in Annex 2 - Requirements regarding the location of facilities containing explosives classified as class 1, subclasses 1.1 and 1.5 or class 4.1 can be used to estimate the safety zones when explosives are detonated on the surface.

The UVS 1608 and VIBRALOC equipment from the Swedish company Nitro Consult AB equipped with broadband ribbon microphones were used to study the vibration intensity and pressure of the airblast. The vibration measurement apparatus meets the requirements of PN-B-02170: 2016-12.

Measuring stands were located as follows (Fig. 1):

Position 1 (st. 1, M1) - the closest to the detonation site - three-component sensor + microphone;

Position 2 (st. 2, M2) - away by 15 m from st. 1 - three-component sensor + microphone;

Position 3 (st. 3, M3) - on the supporting wall of the building (mounted outside) - three-component sensor + microphone;

Position 4 (st. 4, M4) - on the supporting wall of the building (mounted inside) - three-component sensor + microphone.

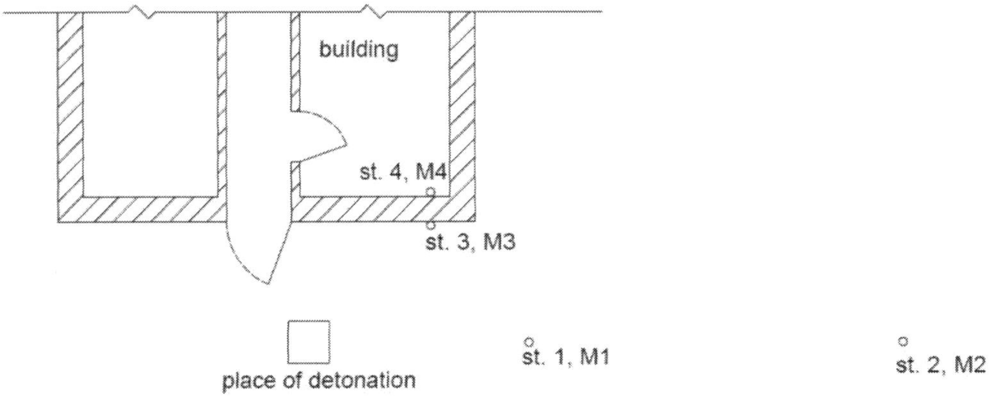


**Fig. 1.** The sketch of location of measurement stations and explosives detonation.

# 3 Results and discussion

## 3.1 Analysis of the laboratory test results

Table 1 presents the results of the analysis of full waveforms in the form of peak particle velocity (PPV) of recorded vibrations ($u_z$, $u_x$ and $u_y$) and their corresponding frequency ($f_z$, $f_x$ and $f_y$), real vector ($\mathbf{u_{zxy}}$) and pressure change corresponding to its frequency.

**Table 1.** Results of vibration and air shock measurements on individual positions

| **Record number** | **Position** | **Distance, m** | **PPV, m/s** | | | **Frequency, Hz** | | | **Vector, m/s** |
|---|---|---|---|---|---|---|---|---|---|
| | | | $u_z$ | $u_x$ | $u_y$ | $f_z$ | $f_x$ | $f_y$ | $u_{zxy}$ |
| Seria 1 | | | | | | | | | |
| 048.852 | st. 1 | 6,7 | 0,00235 | 0,00255 | 0,00083 | 360,7 | 380,7 | 476,2 | 0,002831 |
| | M1 | 6,7 | 103,0 | Pa | | 109,0 | Hz | | |
| | st. 2 | 21,5 | 0,00053 | 0,00045 | 0,0001 | 269,8 | 126,6 | 125,0 | 0,000693 |
| | M2 | 21,5 | 87,2 | Pa | | 21,8 | Hz | | |
| 723.1958 | st. 3 | 4,8 | 0,0009 | 0,00128 | 0,00105 | 352,0 | 351,0 | 332,0 | 0,001455 |
| | M3 | 4,8 | 343,9 | Pa | | 4,0 | Hz | | |
| 722.0936 | st. 4 | 5,0 | 0,00104 | 0,00163 | 0,00182 | 216,0 | 191,0 | 362,0 | 0,00188 |
| | M4 | 5,0 | 5,9 | Pa | | 146 | Hz | | |
| Seria 2 | | | | | | | | | |
| 048.853 | st. 1 | 7,3 | 0,00055 | 0,00155 | 0,00045 | 66,7 | 89,2 | 82,0 | 0,001614 |
| | M1 | 7,3 | 23,2 | Pa | | 155,0 | Hz | | |
| | st. 2 | 22,2 | 0,00045 | 0,00008 | 0,00015 | 12,3 | 33,8 | 78,1 | 0,000451 |
| | M2 | 22,2 | 1,2 | Pa | | 75,8 | Hz | | |
| 723.1959 | st. 3 | 6,6 | 0,00015 | 0,00013 | 0,00011 | 58,6 | 71,2 | 24,9 | 0,000166 |
| | M3 | 6,6 | 11,8 | Pa | | 4,0 | Hz | | |
| 722.0937 | st. 4 | 6,8 | 0,00017 | 0,00016 | 0,00023 | 107,0 | 55,4 | 3,8 | 0,000256 |
| | M4 | 6,8 | 1,1 | Pa | | 0,1 | Hz | | |

In order to outline the problems as mentioned earlier, Figures 2-5 present examples of vibration recordings at the horizontal axis (x) and airblast. Also, the amplitudes of the Fourier transform were determined. A full analysis was analogously performed for all components.

Figure 2 presents the recording of ground vibrations at position 1 and 2 (difference of the distance between them was 15 m) the horizontal component directed perpendicular to the detonation site of the explosive charge which was fired at 0.6 m height, and airblast pressure record at the same positions (the microphone was placed directly above the ground position at the height of 0.6 m).

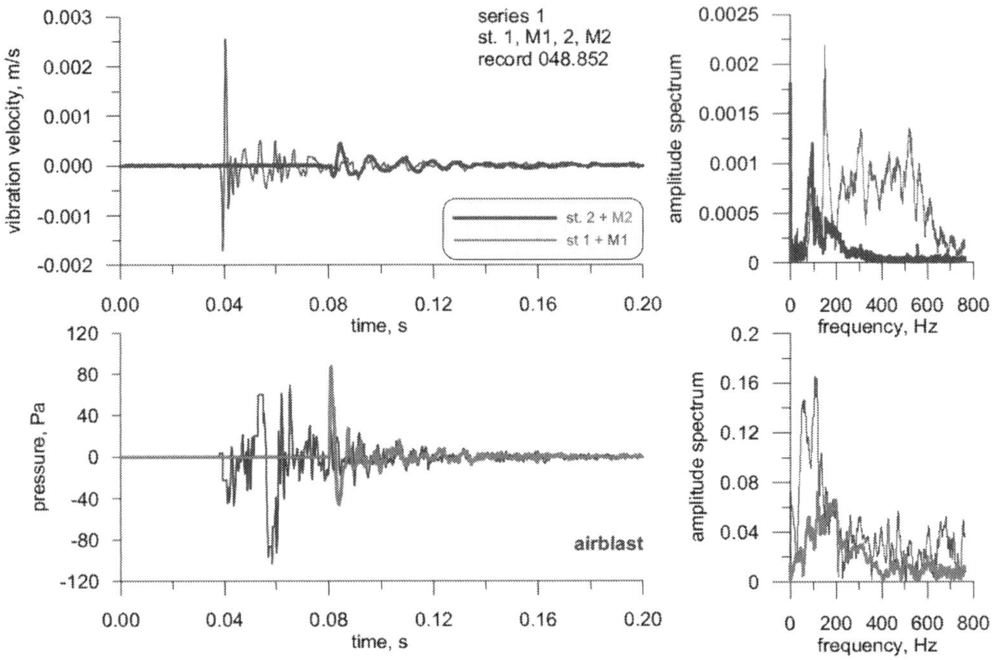


**Fig. 2.** Recording of ground vibrations and airblast pressure along with the amplitude spectrum (FFT) - series 1.

Registration at ground posts begun when the pressure changes reached the microphones, which probably indicates that the vibrations had been excited by air movement. When determining the difference between the entry time among microphones, the propagation velocity of the pressure change was calculated, which amounted to approx. 375 m/s (the air temperature during the measurements was about 15°C) which indicates that it was an airblast. Microphones, moreover, registered acoustic waves, which is confirmed by the frequency structure. During this test, vibration sensors and microphones were installed in the building object on the outer and inner side of the supporting (load bearing) wall at the ground level (the wall formed a mirror reflection for mounted sensors). The records of this study are shown in Figure 3.

Analyzing Fig. 3, it can be noticed that locking the microphone inside the facility (all doors and windows in the building were closed) resulted in a complete attenuation of the airblast. In the frequency structure of the recording from the microphone set outside, higher frequencies, distinctive for the acoustic wave, can be observed. What might be considered surprising, however, is the fact that the vibration sensors registered practically the same, except for the outside sensor which recorded a higher intensity in the first phase of vibration and the inside sensor in the second phase. The frequency structure is similar, the higher frequencies in the range from 200 Hz to 300 Hz dominate, which is not characteristic of the building object's natural vibrations. As a result of comparison Figure 2 and 3, it can be observed that the vibration sensors installed in the ground respond in an utterly different way compared to the sensors mounted on a building. This situation was repeated each time the explosive charge was placed in this specific manner. The other components of the sensor (vertical and transversal) behaved similarly, with the difference that the recording on the vertical axis is much longer than on the horizontal axes for stations 3 and 4.

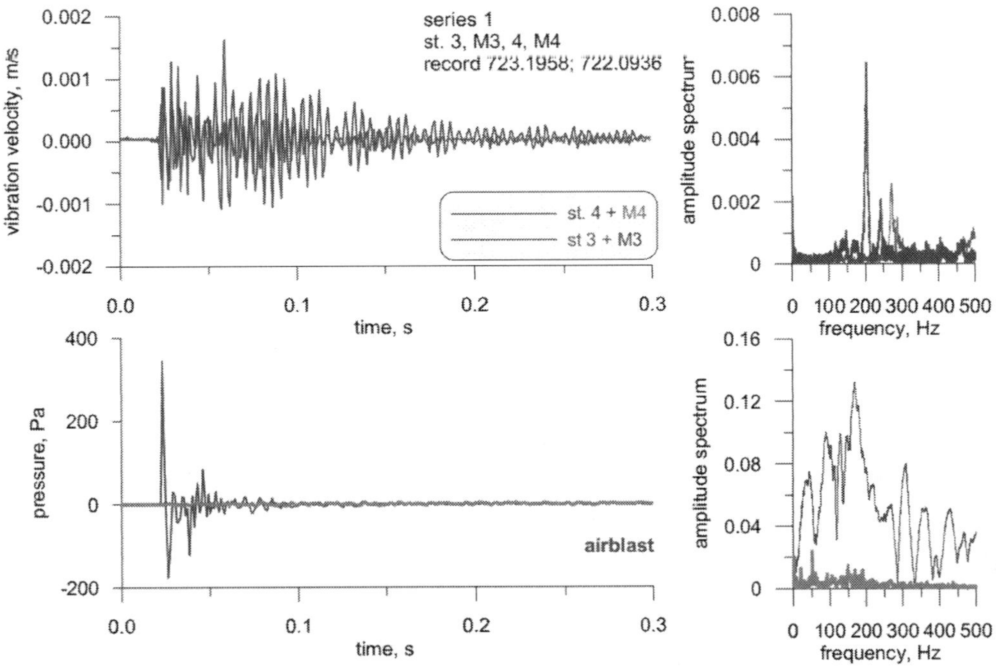


**Fig. 3.** Recording of structure vibrations and airblast pressure along with the amplitude spectrum (FFT) - series 1.

The next group of measurements was measurements carried out with buried and additionally covered with steel plate explosives charge. The results of those studies are shown in Fig. 4 for positions 1 and 2, as well as in Fig. 5 for test stands 3 and 4. When conducting the analysis of Fig. 4, it can be noticed that the records are different in comparison to the records presented in Figures 2 and 3. The shock wave has been completely suppressed. As a result of the detonation, the steel plate was slightly raised, which caused a recording from the fall in the second part of the entry (about 300 ms) at test station 1. At position two it is significantly reduced. By determining, in the same way as in the first series of tests, differences in the time of entry between microphones, the propagation velocity of the pressure change was calculated, which amounted to approx. 342 m/s (air temperature during measurements was about 15°C), which proves that it was an acoustic wave.

The excited vibrations in the ground were suppressed at the transition to the building, where vibrations were recorded, which slightly exceeded the trigger threshold (0.25 mm/s). It is also confirmed by the frequency structure, where the entire spectrum of frequencies ranges from 1 Hz to 500 Hz (noise).

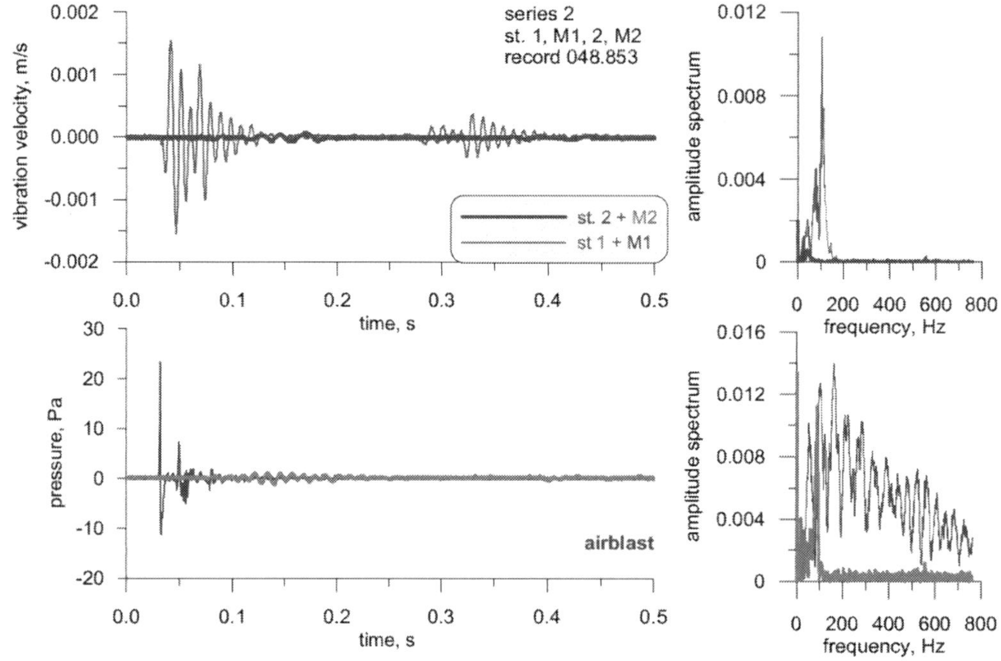


**Fig. 4.** Recording of ground vibrations and airblast pressure along with the amplitude spectrum (FFT) - series 2.

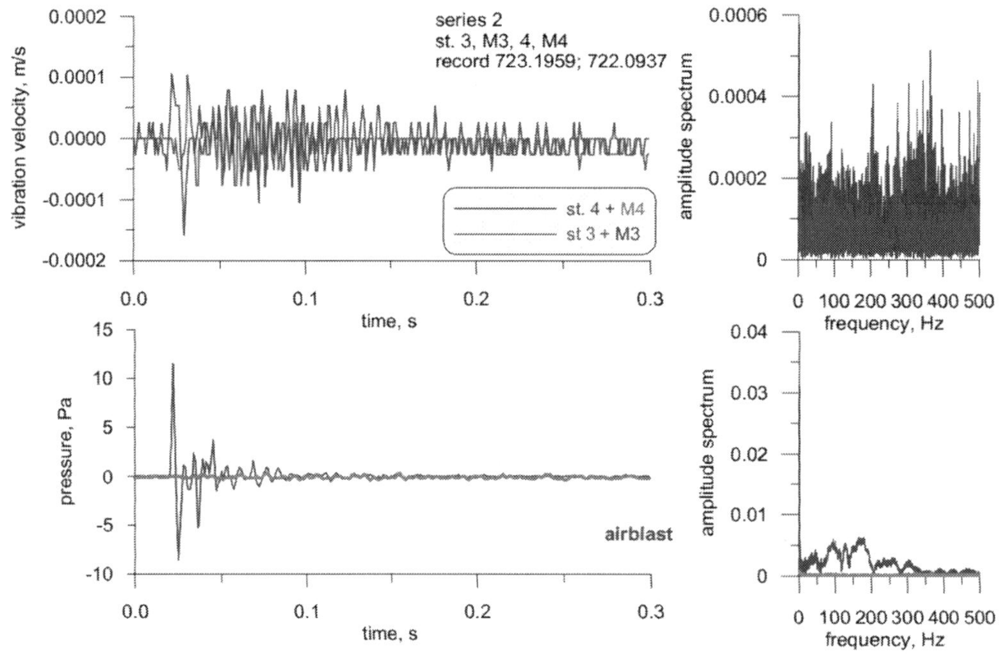


**Fig. 5.** Recording of structure vibrations and airblast pressure along with the amplitude spectrum (FFT) - series 2.

## 3.2 Analysis of selected results of field research

In the course of field measurements of ground's and building's vibrations in the vicinity of the chosen open-pit mine, where blasting works were carried out at the top floor of production levels, air shock waves were also recorded. Production blasting was carried out using the long hole method. Measurement positions were placed in the direction to the nearest buildings. Presented vibration and airblast registrations from the selected test stands, as shown in Fig. 6, were moved away from the blasting site, respectively: st. 21 and M21 - 630 m, st. 25 and 25 '- 1147 m.

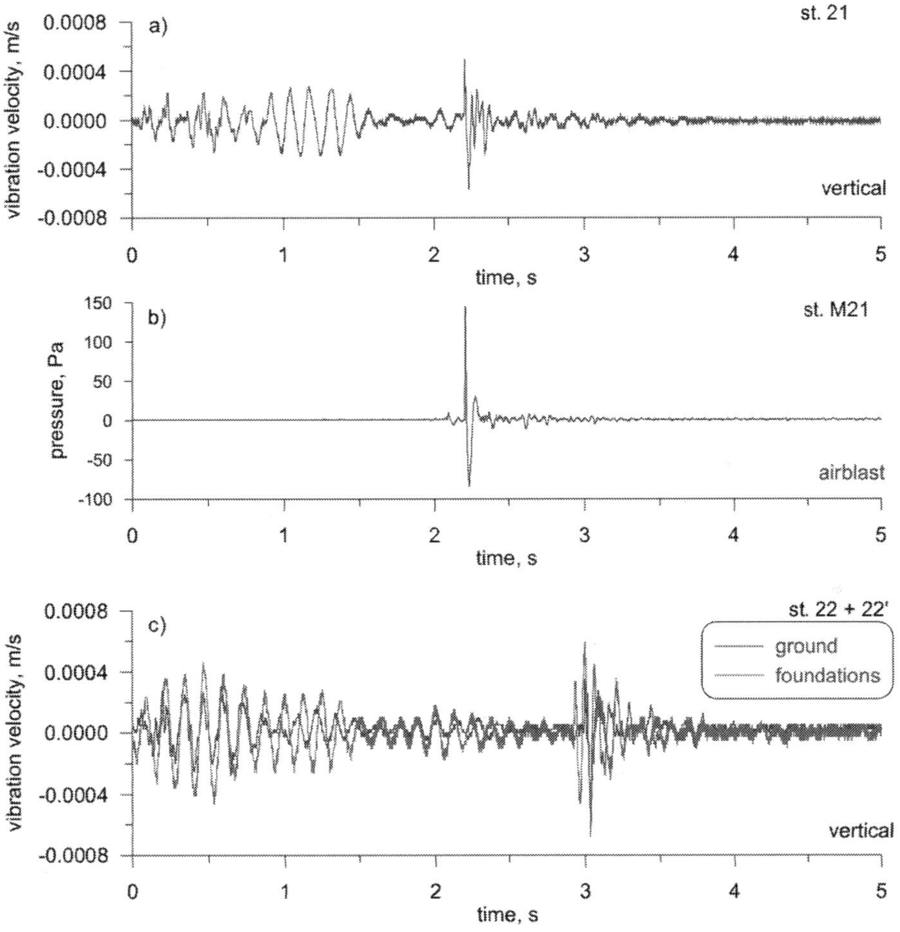


**Fig. 6.** Recordings of: a) ground vibrations; b) airblast; c) ground and building vibrations.

Vibration analysis of the records registered in the position 21 indicates the impact of airblast on vibration sensors, thanks to the installed microphone. Records from measurement stands 22 and 22' also clearly display previously mentioned impact, which caused higher intensity vibrations on the construction in comparison to vibrations transmitted by the ground. In order to correctly perform impact assessment for protected structures, moreover, do it in accordance with the requirements of the standard [1], it is necessary to eliminate from the signal the disruption (interference) caused by the airblast.

## 4 Conclusions

The article presents examples of signal measurements induced by the detonation of small charges of explosives placed in various ways, studies related to the distribution of selected physic fields. The perceived hazard is caused by the short-term and high-energy shock waves - acoustic, airblast (gust) and seismic. Based on the research and analyses carried out, it can be concluded that:

1. Doubts were confirmed as to the applicability of the PN-B-02170: 2016-12 standard, which concerns assessment of the vibration impact on buildings which are transmitted by the ground, and not by air (pressure change). The statement of this fact was possible thanks to the simultaneous measurement of vibrations and airblast.
2. In order to apply the standard [1] to assess the impact of vibrations on buildings, air-borne interference must be eliminated from the recordings.
3. There is still a doubt as to whether the recorded impact consisting of airblast and acoustic wave by the vibration sensors was an impact on the object and the substrate, or whether it was only affecting the sensor itself.

Research will be continued in order to develop a measurement and analytical methodology, so as not to misinterpret the recorded vibrations, and thus to incorrectly evaluate the impact on protected structures.

## References

1. PN-B-02170:2016-12– Assessment of the impact of vibrations transmitted by the ground on buildings (in Polish)
2. DIN 4015-1:2001-06 Vibrations in buildings - Part 1: Prediction of vibration parameters
3. DIN 4150-3:2016-12 Vibrations in buildings - Part 3: Effects on structures
4. Eurocode 8: Design of structures for earthquake resistance — Part 1: General rules, seismic actions and rules for buildings
5. W. Chengqing, H. Hong, International Journal of Impact Engineering **34**, 556–572 (2007) doi: 10.1016/j.ijimpeng.2005.11.003
6. R. N. Gupta, P.P. Roy, B. Singh, International Journals of Mining and Geological Engineering **6**, 15–26 (1988) doi: 10.1007/BF00881024
7. A. D. Alcudia, University of Calgary (2009) doi: 10.11575/PRISM/15296
8. N.M. Babcock, *Microphone suppression of air-noise on geophones*, (M. Sc. thesis, Faculty of the Department of Earth and Atmospheric Sciences University of Houston, 2012)
9. F. Faramarzi, M. A. Ebrahimi Farsangi, H. Mansouri, International Journal of Mining Science and Technology **24** (5) 663–669 (2014) doi: 10.1016/j.ijmst.2014.07.006
10. E. Kabwe, Y. Wang, Geomaterials **6**, 28–38 (2016) doi: 10.4236/gm.2016.61003
11. V. J. Stachura, D. E. Siskind, A.J. Engler *Airblast Instrumentation and Measurement Technique for Surface Mine Blast.* U. S. Bureau of Mines, Report of Investigations 8508 (2000)
12. J. Winzer, A. Sołtys, J. Pyra, *The impact of in use of blasting works on the environment*, AGH University of Science and Technology Press, Krakow (2016) (in Polish)
13. Regulation of the Minister of Labour and Social Policy of 9 July 2003 on health and safety during the production, in-house transport and circulation of explosives, including pyrotechnic articles., Journal of Laws of 2003, no. 163, item 1577 (in Polish)

# Nonlinear anisotropic diffusion techniques for seismic signal enhancing - Carpathian Foredeep study

*Mateusz* Zaręba*, and *Tomasz* Danek

AGH University of Science and Technology, Department of Geoinformatics and Applied Computer Science, Cracow, Poland

**Abstract.** The use of nonlinear anisotropic diffusion algorithm for advanced seismic signal processing in the complicated geological region of Carpathian Foredeep was examined. This technique allows for an improvement of seismic data quality and for more accurate interpretation by the recovery of a significant amount of structural information in the form of a correlating seismic reflections and by preserving true DHI indicators. It also allows searching for more subtle geological structures. Anisotropic diffusion is an iterative image processing algorithm that removes noise by modifying the data by solving partial differential equations. Moreover, it can reduce image noise without blurring the edges between regions of different chrominance or brightness. This filter preserves edges, lines, or other features relevant to the seismic structural and stratigraphic interpretation. The algorithm also enables noise reduction without removing significant information from a seismic section even for high dips values. For a better estimation of anisotropic diffusion structure tensor, the parameterization is done using the depth field and the calculations in the two-way travel time field. The presented research shows the results of using an anisotropic diffusion algorithm for post-stack and migration processing of seismic 3D data collected in Carpathian reservoir rocks of southern Poland.

**Keywords:** signal processing, seismic, anisotropic diffusion, Carpathian Foredeep, nonlinear algorithm, seismic interpretation

## 1 Introduction

Obtaining more accurate and detailed images of the subsurface based on seismic surveys is one of the core functions of seismic processing, especially when data is acquired in the region of complicated geology. Such conditions require more specific processing techniques to obtain an image that can be properly interpreted. The Carpathian Foredeep is undoubtedly an example of a region where complex geology is a challenge for seismic investigations. On seismic sections from this region, plenty of artefacts that are generated by different phenomena must be eliminated before a distinct reflection from particular geological

* Corresponding author: mattzareba@gmail.com

structures can be observed. The sedimentary profile of the Carpathian Foredeep is very specific and highly variable due to the mosaic type of flysch tectonic observed in seismic images [1]. In recent years, the understanding of the stratigraphy and lithology of The Carpathian Foredeep has increased, thanks to geophysical surveys and newly developed seismic and well-logging techniques [2]. Many studies have been done in this part of Poland because of its reservoir potential [3]. Seismic observations allowed the discovery of oil and gas reservoirs in the Carpathian Mountains and their Foredeep; however, despite many studies and progress in seismic techniques, proper processing and interpretation of seismic data is still a challenge. The importance of developing processing techniques, guided on improving the seismic image, has been widely discuses in many papers [4, 5, 6]. According to a recent geological modelling study of the inner structure of autochthonous Miocene deposits, a new seismic approach is essential for the recognition and proper localization of unconventional reservoirs in this region [7]. Another recent example of the critical importance of processing in The Carpathian Foredeep seismic interpretation was shown by Kubociński [8]. He presented new techniques of three-dimensional, angle domain pre-stack depth migration results for the discussed area. The importance of flysch rocks effects is studied in detail in [1]. These challenges of high values of dips and latterly changing velocities constitute an environment where poor signal quality, which degrades with depth, is not common and where the new techniques of seismic processing are needed. This paper is focused on one technique, namely non-linear anisotropic diffusion, which has been known and used for several years in digital image processing. Recently, the first application of this technique in the Polish Carpathian Foredeep seismic surveys was presented by Zaręba [9]. This paper presents the broadening results of the application of this technique in seismic data processing.

## 2 Anisotropic diffusion filtering

It has been observed that using filters based on anisotropic diffusion can simplify the digital image by filtering out signals that do not provide relevant information [10]. The most important advantage of the anisotropic diffusion is preserving the edges of the smooth structures [11]. A similar procedure can be applied in the case of a seismic signal defined as collection of samples gather in sets of particular traces. Anisotropic diffusion filtering can simplify a seismic section by reducing noise, which provides better quality horizons.

In case of seismic data, the role of pixels with high gradient magnitude can be replaced by amplitude changes in the time-space domain window. Usually a common depth point domain as a time-space representation is used. Enhancing seismic signatures can help in obtaining better sections for structural interpretation.

Anisotropic diffusion can improve fault interpretation in 3D seismic surveys and provides dip and fault inclinations and azimuth values (e.g., [9]). The second advantage of this algorithm is complex de-noising of seismic data while preventing the layer's borders and enhancing the signal as demonstrated by Baddari et al. [12]. Additionally, processing using this type of algorithm can improve the final results of depth migration. It is due to the fact that the velocity model for depth migration is created without perfect knowledge of the real position of structures. This procedure makes it possible to obtain the correct depth image of subsurface structures, which can be used for further stratigraphic interpretation (e.g., [13]). As a result, the obtained seismic section, after time migration with an enhanced signal to noise ratio and depth to time conversion, is more accurate.

A typical algorithm for anisotropy diffusion filtering is as follows (e.g., [14, 15]):

$$\frac{\partial U}{\partial t} = div(c\nabla U) \tag{1}$$

where c is a diffusion coefficient for filtering and U is the intensity function of the data.
Equation 1 shows typical convolution with Gaussian kernel. It helps preserve the edges of the image and deals with fuzziness. However, to prevent edges of a 3D structure, the diffusion coefficient from Equation 1 has to be replaced with tensors:

$$\frac{\partial U}{\partial t} = div(D\nabla U) \tag{2}$$

Diffusion coefficients are stored in Matrix D. It is simply the central difference disperse equation with a Neumann boundary condition. Weickert [16] proposed a use of structure tensor $J_\rho$ as powerful tool for examining coherence structures:

$$J_\rho(\nabla U_\sigma) = K_\rho^*(\nabla U_\sigma \otimes \nabla U_\sigma) \tag{3}$$

where $K_\rho$ is a Gaussian kernel, $U_\sigma$ is a Gaussian-smoothed version of the image, ρ is a scale factor, and σ is a noise scale.

Gradient vector $\mathbf{v_1}$ and the structure orientation $\mathbf{v_2}$, at scale ρ are represented by eigenvectors of $J_\rho$. To be able to use Equation 3 for 3D seismic needs, the structure tensor $J_\rho$ becomes:

$$J_\rho(\nabla U_\sigma) = K_\rho^* \begin{pmatrix} (\frac{\partial U_\sigma}{\partial x})^2 & \frac{\partial U_\sigma}{\partial x}\frac{\partial U_\sigma}{\partial y} & \frac{\partial U_\sigma}{\partial x}\frac{\partial U_\sigma}{\partial z} \\ \frac{\partial U_\sigma}{\partial x}\frac{\partial U_\sigma}{\partial y} & (\frac{\partial U_\sigma}{\partial y})^2 & \frac{\partial U_\sigma}{\partial y}\frac{\partial U_\sigma}{\partial z} \\ \frac{\partial U_\sigma}{\partial x}\frac{\partial U_\sigma}{\partial z} & \frac{\partial U_\sigma}{\partial y}\frac{\partial U_\sigma}{\partial z} & (\frac{\partial U_\sigma}{\partial z})^2 \end{pmatrix} \tag{4}$$

To be able to use Equation 4 in the algorithm, the iterative form is needed:

$$U^{n+1} = U^n + \Delta x\Delta y\Delta z\{\frac{\partial}{\partial x}(D_{11}^n U_x^n) + \frac{\partial}{\partial x}(D_{12}^n U_y^n) + \frac{\partial}{\partial x}(D_{13}^n U_z^n)\frac{\partial}{\partial y}(D_{21}^n U_x^n) + \frac{\partial}{\partial y}(D_{22}^n U_y^n) + \frac{\partial}{\partial y}(D_{23}^n U_z^n) + \frac{\partial}{\partial z}(D_{31}^n U_x^n) + \frac{\partial}{\partial z}(D_{32}^n U_y^n) + \frac{\partial}{\partial z}(D_{33}^n U_z^n)\} \tag{5}$$

Where $U^n$ and $U^{n+1}$ are the filter outputs at sample nΔs and (n+1)Δs, and $D^n$ is diffusion tensor.

Using the equations presented above, filtering is performed in the following steps:
1. Input seismic data are represented by $U^0$.
2. Structure gradient tensor $J_\rho(\nabla U^n)$ for $U^n$ and scale ρ is constructed using Equations 6, 7, and 8.

$$J_\rho(\nabla U^n) = K_\rho * J_0 = K_\rho * (U(\nabla U^n)^T) \tag{6}$$

$$K_\rho(\boldsymbol{r}) = (2\pi\rho^2)^{\frac{-3}{2}} \exp\left(\frac{|\boldsymbol{r}|^2}{(2\rho^2)}\right) \tag{7}$$

$$|\boldsymbol{r}|^2 = (x - x_0)^2 + (y - y_0)^2 + (z - z_0)^2 \tag{8}$$

3. Eigenvectors $\mathbf{V}_1$, $\mathbf{V}_2$, and $\mathbf{V}_3$ for $J_\rho(\nabla U^n)$ and eigenvalues $L_1$, $L_2$, and $L_3$ are calculated.
4. Using values from Step 3 and Equations 9 and 10, the diffusion tensor $D^n$ is evaluated.

$$\begin{cases} \vartheta_1 = \alpha + (1-\alpha)\exp(\frac{-c}{(L_1-L_2)^{2m}}) \\ \vartheta_2 = \alpha \\ \vartheta_3 = \alpha \end{cases} \tag{9}$$

$$D = (v_1 v_2 v_3)\begin{vmatrix} \vartheta_1 & 0 & 0 \\ 0 & \vartheta_2 & 0 \\ 0 & 0 & \vartheta_3 \end{vmatrix}(v_1 v_2 v_3)^T = \begin{vmatrix} D_{11} & D_{12} & D_{13} \\ D_{21} & D_{22} & D_{23} \\ D_{31} & D_{32} & D_{33} \end{vmatrix} \tag{10}$$

5. Using Equation 5, the value $U^{n+1}$ is calculated, until n=N-1.
6. The filtered signal is obtained as an output.

## 3 Local geology

The presented research was conducted on data from different parts of The Carpathian Foredeep (Fig. 1). 3D seismic data from different time periods were used. The main idea was to examine as many different stratigraphic levels as possible. In consequence - due to changing in stratigraphy - seismic data from different parts of the presented area have to be considered. Typical stratigraphy for this region consists of the oldest Precambrian complexes dated on the Neoproterozoic era. They are represented by strongly metamorphosed phyllites and meta-arqillites with different levels of tectonic involvement (dips over 60°) [17]. The higher Ordovician rocks are represented by grey-green, fine-grained, quartzite sandstones with glauconite, quartize conglomerates and crystal limestone [18, 19]. The Silurian complex is represented by loamed shales with limestone and marl interbreeds. Their thickness is about 20-300 meters [20]. The Devonian complex is made of carbonate rocks in the upper and middle part and terrigenous sediments in the lower part. Lower Devonian rocks are mostly argillaceous, grey sandstones, red or green clays, and mottled mudrocks. The thickness of this complex is variable. Upper Devon rocks of this complex manifest as limestone and dolomite breccias slightly successioning into carbonate rocks. For the last type of the mentioned complex, we can include dolomite, limestone, and micro-crystalic dolomitic limestone.

Triassic rocks are divided on three different complexes, i.e. the lower, consisting of sandstone, the middle, consisting of limestone (sediment of sea environment), and the Kajper, consisting of sandstone and mudrock [21]. The oldest Jurassic complexes of the Carpathian range are observed from the Middle Jurassic (Dogger). They have been deposited in the depression of Triassic sediments. Rocks of Middle Jurassic are represented by clastic deposits with thick intrabeddings of marl [22]. The Upper Jurassic is made of limestone complexes that are deposited on Middle Jurassic rocks or Paleozoic rocks. A Biochermas series is present by regular lines connected to the upheaval of the sea bottom. Their maximum thickness could be over several hundred meters [23]. The seismic signal inside these structures is highly attenuated.

The Lower Cretaceous period is represented by sediments similar to Upper Jurassic; however, they have strong facial differentiation. Their special range in the considered area is limited to the area between Radomyśl, Jastrząbka Stara and Zagorzyce [24]. On the morphologically differentiated surface of the substrate, the autochthonous Miocene deposits are present. Their thickness varies from 200m up to 1000m in the presented area. The lower Miocene series is filled with the erosive unevenness of the substrate. They often outcrop on slopes of depressions. Over the Sarmat complex, there are Quaternary deposits represented by clay, gravel, and ice and river-transported sands. Their thickness is less than 60 m on the considered area [25].

The presented geological structure of the area of investigation, with its complex tectonic and stratigraphy, various sediments and facies, is definitely a challenge for seismic processing and interpretation.

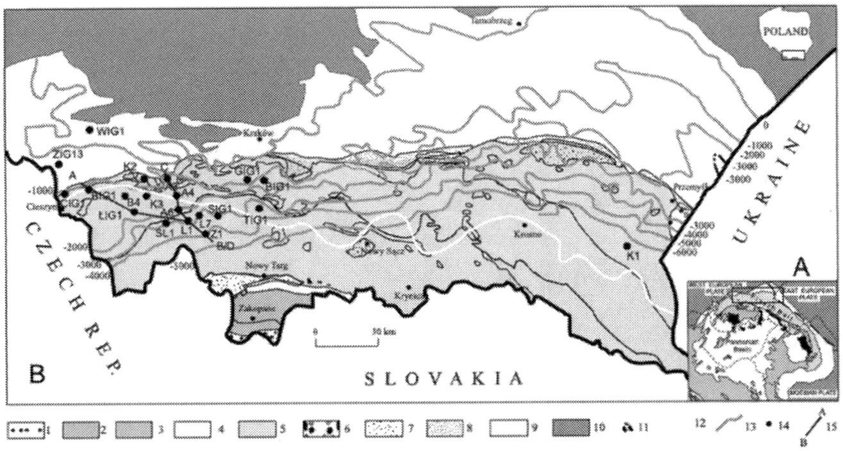


**Fig. 1.** Position of the Polish Carpathian Foredeep in the Carpathian -Pannonian Region. B. Sketch-map of the Polish Carpathians and their foredeep (after Oszczypko & Oszczypko-Clowes [26]); 1 – crystalline core of Tatra Mts., 2 – high and sub -Tatra units, 3 – Podhale Flysch, 4 – Pieniny Klippen Belt, 5 – Outer Carpathians, 6 – Stebnik Unit, 7 – Miocene deposits upon Carpathians, 8 – Zgłobice Unit, 9 -Miocene of the foredeep, 10 – Mesozoic and Palae-ozoic foreland deposits, 11 – andesites, 12 – northern extent of Lower Miocene, 13 – isobath of Miocene substratum, 14 – boreholes, 15 – geological cross-section.

## 4 Results of application in The Carpathian Foredeep

In this paper, data from different parts of The Carpathian Foredeep were used. The quality of recorded seismic data was relatively good with a high signal to noise ratio; however, some horizons had been poorly preserved in processed sections. The decreasing quality of the seismic signal in the Jurassic, Cretaceous, and autochthonous Miocene sediments, which are considered as reservoirs in The Carpathian Foredeep, have been noticed [27, 28]. As shown in Fig. 2, Mark 1, the signal in Miocene sediments has low amplitude with a lot of discontinuities. At a lower part of presented section, below strong anhydrite reflection (Mark 2), Craterous and Jurassic rocks are present (Mark 3). As is typical for anhydrite horizons, the decrease in amplitude below them is visible. To improve the quality of presented seismic section, anisotropic diffusion filtering was applied. Figure 2A shows the results of this operation. Improvement is clearly visible. Horizons inside the Miocene sediment complex were properly recovered and their continuity is highly improved. Moreover, the positive effect of the algorithm is also visible in the Mesozoic part, where interpretational value of data was highly improved.

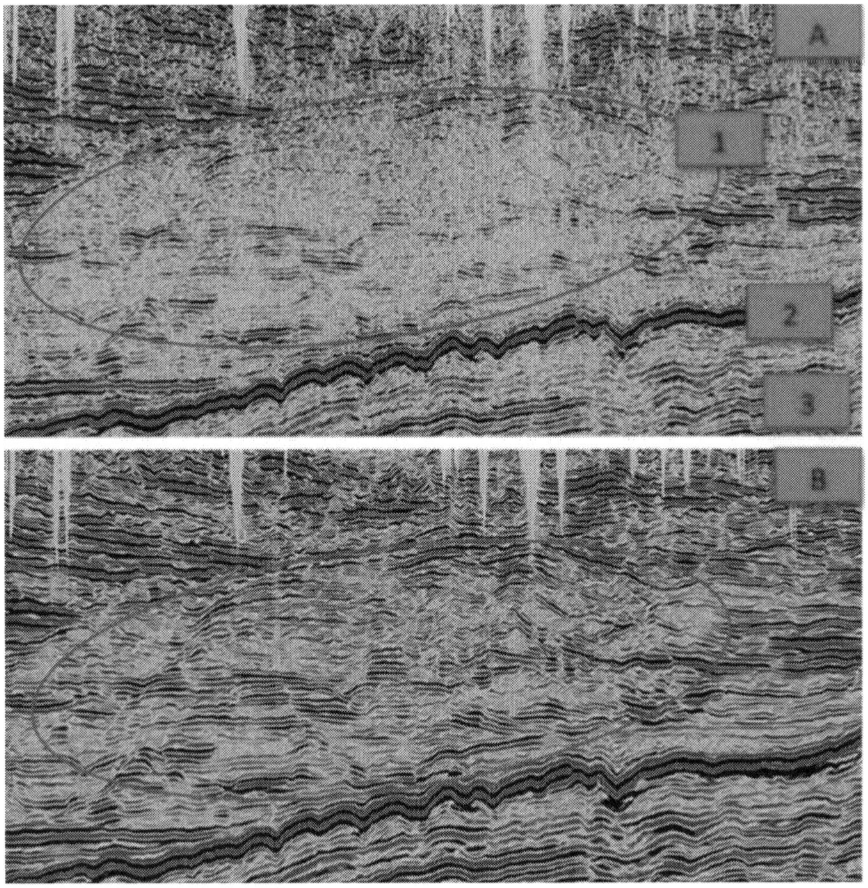


**Fig. 2.** A comparison of the seismic sections of 1 – Miocene, 2 – Anhydrite, 3 – Mesozoic sediments - before (A) and after (B) filtering application.

Research shows that mudstones and claystone facies of Miocene sediments are gas reservoirs [4]. Due to lateral changes in reservoir facies, it is important to obtain more detailed images of the subsurface. It is extremely important to figure out the genesis of seismic anomalies on sections. Dim outs, the chaotic, and anomalous signal in a particular region could be considered hydrocarbon indicators. However, it is possible that the mentioned phenomena are not connected with the presence of hydrocarbons. The history of using DHI (Direct Hydrocarbon Indicator) in The Carpathian Foredeep in Poland between years 1994 and 2003 shows that 32% of wells drilled in locations chosen according to DHI interpretation were dry [4]. Seismic processing techniques that preserve the "true" signal characteristic are necessary for the proper interpretation stratigraphy and DHI in the case of decreasing number of dry wells in The Carpathian Foredeep. It is easily noticeable that the seismic signal in Miocene sediments (Fig. 2 Mark 1) has been strengthened after the application of anisotropic diffusion filtering, but there is an area of carbonated structures (Fig. 3, orange circle) in which the signal has not been strengthened. It is a highly expected effect. Bioherms are, next to other reef structures, important in the case of carbonate reservoirs, even though they could be more oil-wet then water-wet in character when compared to similarly aged sandstones [29]. Bioherms can be identified on the seismic section as almost a reflection free zones [30] or zones with chaotic signals. They usually occur on uphill sections (as seen on Fig. 3). The correlation of Jurassic horizons after filtering (Fig. 3A) is clearly better, but the signal in the bioherm section is still chaotic and weak.

What is important relation between low and high amplitude parts of presented seismic sections is well preserved after anisotropic diffusion filtering. It is clearly visible that anisotropic diffusion filtering helps overcome problems that other structural filtering techniques do not. We obtained better structural information and preserved seismic anomalies that can indicate a hydrocarbon presence. Additionally, in Miocene sediments, the presented filtration helps to prevent the misinterpretation. An area in Fig. 2A, Section 1 without anisotropic diffusion filtering could be interpreted as a dim out. After applying anisotropy diffusion filtering, the signal is aligned over the entire area (Fig. 2B). The presented technique is independent of the observer, so it helps to obtain a more objective seismic image of the subsurface during the processing stage, which can be later properly interpreted.

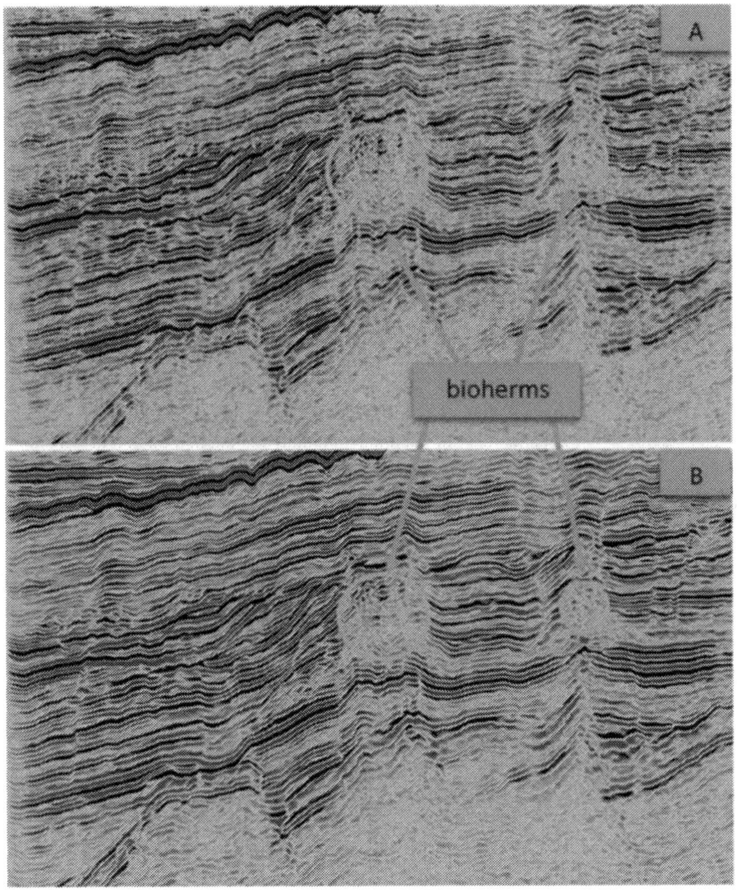


**Fig. 3.** A comparison of the seismic section of Mesozoic sediments before (A) and after (B) filtering application.

Another example of using the anisotropic diffusion algorithm comes from the region where Carpathian overlaps Miocene sediments. It is a very problematic region, because of the presence of dips. It causes problems with velocity analysis (because of significant lateral changes) and at later stages of time migration. In Fig. 4, the comparison of this region before and after filtering is presented. Dipping horizons have better correlation and are longer than their apparent image before filtration. Proper mapping of dips in the Carpathian range has always been a challenge for seismic processing. The ambiguous origin of problems should be resolved in the processing stage by using objective techniques (e.g., nonlinear anisotropic diffusion filtering). Only a proper image of subsurface can be correctly interpreted. We have to ensure that we strengthened true dips, and attenuated noise, migration artefacts, and other unwanted reflections during the processing stage (as shown in Fig. 4).

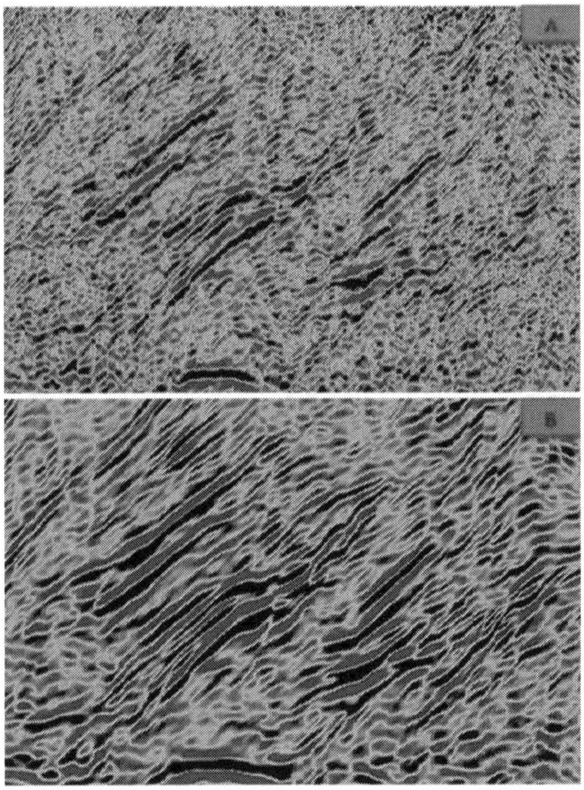

**Fig. 4.** A comparison of the seismic section of the Carpathian overlap before (A) and after (B) filtering application.

## 5 Conclusion

Presently, proper structural interpretation and the search for new hydrocarbons reservoirs require the use of precise techniques of geological structure imaging. The application of the digital image processing for enhancing 3D seismic surveys in areas where geological structures are very complicated and seismically "unfriendly" is very promising. The presented study shows the results of using a nonlinear anisotropic diffusion algorithm for post-stack processing of 3D seismic data collected in The Carpathian Foredeep region of southern Poland. Notably, the application of this algorithm provided a reliable seismic image of complicated geology, without any additional input from the interpreter.

It was shown that anisotropic diffusion can be used for advanced seismic signal processing. It allows for an improvement of seismic data quality and for more accurate interpretation by the recovery of a significant amount of structural information in the form of correlated seismic reflections. It is also a tool which allows searching for more subtle details of geological structures. According to the presented study, nonlinear anisotropic diffusion filtering is a powerful tool for the reliable reconstruction of the seismic image in the complicated geology region of The Carpathian Foredeep. It allowed proper structural interpretation and preserved seismic anomalies related to hydrocarbon reservoirs.

This research was supported by AGH - University of Science and Technology, Faculty of Geology, Geophysics and Environmental Protection as a part of the statutory project No. 11.11.140.613

## References

1. M. Myśliwiec et al., AAPG Memoir, **84**, 391 (2006), DOI:10.1306/985613M843073
2. W. Moryc, *Rozwój badan utworów miocenu w Karpatach Zachodnich na obszarze Bielsko - Kraków (Development of research on Western Carphatian Miocen deposits in the Bielsko-Kraków area)*, Geologia **31**, 7–8 (2005) (in Polish)
3. L. Jankowski, J. Probulski, *Rozwój tektoniczno-basenowy Karpat zewnętrznych na przykładzie budowy geologicznej złóż Grabownica, Strachocina i Łodyna oraz ich otoczenia (Tectonic and basinal evolution of the Outer Carpathians based on example of geological structure of the Grabownica, Strachocina and Łodyna hydrocarbon deposits)*, Geologia, **37**, 556–583 (2011) (in Polish)
4. M. Myśliwiec, *Poszukiwania złóż gazu ziemnego w osadach miocenu zapadliska przedkarpackiego na podstawie interpretacji anomalii sejsmicznych (Exploration for gas accumulations in the Miocene deposits of the Carpathian Foredeep using Direct Hydrocarbon Indicators*), Przegląd Geologiczny, **52**, 299–306 (2004) (in Polish)
5. S. Tlałka,73rd EAGE Conf. & Exhib. incorp. SPE EUROPEC (2011) DOI: 10.3997/22144609.20149535
6. M. Cyz, M. Malinowski, M. Mulińska, SEG Technical Program Expanded Abstracts, 692–696 (2017), DOI:10.1190/segam2017-17789092.1
7. B. Papiernik, M. Krzysztofik, S. Franczak, *Wstępne modele geologiczne miocenu autochtonicznego w wybranych strefach w pobliżu frontalnego nasunięcia karpackiego (Preliminary geological of autochthonous Miocene in the selected zones proximal to the Frontal Carphatian Overthrust)*, Cooperation of science and industry in hydrocarbon exploration and production, Conference magazine of the Oil and Gas Institute, Kraków, **209**, 101–105, (2016) (in Polish)
8. W. Kubociński, *Interakcyjna sektoryzacja i selektywne sumowanie kolekcji pełnego azymutu (Interactive sectoring and selective stacking of the 3D full azimuth directional gathers – the new tool to analyse structural image)*, Cooperation of science and industry in hydrocarbon exploration and production, Conference magazine of the Oil and Gas Institute, Kraków, **209**, 193–196 (2016) (in Polish)
9. M. Zaręba, *Zaawansowane metody zwiększania koherencji sygnału sejsmicznego (The use of advanced seismic techniques to enhance coherency of seismic signal)*, Cooperation of science and industry in hydrocarbon exploration and production, Conference magazine of the Oil and Gas Institute, Kraków, **209**, 201–204 (2016) (in Polish)
10. J. Weickert, *Anisotropic diffusion in image processing*, Department of Computer Science, University of Copenhagen, Copenhagen, Denmark (1998)
11. M. Black, G. Sapiro, LECT NOTES COMPUT SC, **1682**, 259–270 (2012) DOI: 10.1007/3-540-48236-9
12. K. Baddari, J. Ferahtia, T. Aifa, N. Djarfour, COMPUT GEOSCI, **37**, 4, 456–463 (2011) DOI:10.1016/j.cageo.2010.09.009
13. H. Kowalski, W. Kubociński, *Sejsmiczny obraz geologii: czasowy, głębokościowy, a może dyfrakcyjny? (Seismic image of geology: time, depth, or diffractive?)*, Cooperation of science and industry in hydrocarbon exploration and production, Conference magazine of the Oil and Gas Institute, Kraków, **198**, 247–251 (2014) (in Polish)

14. O. Lavialle, S. Pop, Ch. Germain, M. Donias, S. Guillon, N. Keskes, Y. Berthoumieu, J APPL GEOPHYS, **61**, 132–141 (2007) DOI: 10.1016/j.jappgeo.2006.06.002
15. E. Wang, G. Yan, L. Li, SEG Technical Program Expanded Abstracts, 1590–1594 (2013), DOI: 10.1190/segam2013-0564.1
16. J. Weickert, INT J COMPUT VISION, **31**, 111–127 (1999) DOI: 10.1023/A:1008009714131
17. W. Moryc, M. Jachowicz, *Utwory prekambryjskie w rejonie Bochnia-Tarnów-Dębica (Precambrian deposits in the Bochnia-Tarnów-Dębica area)*, Przegląd Geologiczny, **48**, 601–606 (2000) (in Polish)
18. W. Moryc, *Stratygrafia warstw w głębokim otworze Niwki 3 koło Dąbrowy Tarnowskiej (Stratigraphy of the geological formations in the deep borehole Niwki 3 near Dąbrowa Tarnowska)*, Geologia, **19**, 87–106 (1974) (in Polish)
19. H. Tomczyk, *Ordowik i sylur w podłożu Zapadliska Przedkarpackiego (Ordovician and Silurian in the basement of the Fore-Carpathian Depression)*, Rocznik Polskiego Towarzystwa Geologicznego, **33**, 289–320 (1963) (in Polish)
20. M. Jachowicz, *Mikroskamieniałości organiczne terenewu (kambr dolny) i późnego ediakaru (neoproterozoik) okolic Krakowa (The Terreneuvian and late Ediacaran organic microfossils from Kraków area)*, Biuletyn Państwowego Instytutu Geologicznego, **459**, 61–82 (2014) (in Polish)
21. H. Senkowiczowa, *Wapień Muszlowy na południowym zboczu Gór Świętokrzyskich między Czarną Nidą a Chmielnikiem (The Muschelkalk on the southern Slope of the Święty Krzyż Mts. between Czarna Nida and Chmielnik)*, Biuletyn Instytutu Geologicznego, **122**, 5–67 (1957) (in Polish)
22. W. Moryc, *Budowa geologiczna podłoża miocenu w rejonie Kraków-Pilzno. Cz. 2. Perm i mezozoik (Geological structure of Miocene substratum in Kraków-Pilzno region. Part 2. The Permian and Mesozoic period)*, Nafta-Gaz, **62**, 6, 263-282 (2006) (in Polish)
23. J. Gutowski, A. Urbaniec, Z. Złonkiewicz, L. Bobrek, B. Świetlik, P.Gliniak, *Stratygrafia górnej jury i dolnej kredy środkowej części przedpola polskich Karpat (Upper Jurassic and lower Cretaceous of the middle polish Carpathian foreland)*, Biuletyn Państwego Instytutu Geologicznego, **426**, 1–26 (2007) (in Polish)
24. A. Urbaniec, L. Borek, B. Świetlik, *Litostratygrafia i charakterystyka mikropaleontologiczna utworów kredy dolnej w środkowej części przedgórza Karpat (Lithostratigraphy and micropalaeontological characteristic of Lower Cretaceous strata in central part of the Carpathian Foreland)*, Przegląd Geologiczny. **58**, 12, 1161–1175 (2010) (in Polish)
25. J. Karnkowski, *Osady miocenu Zapadliska Przedkarpackiego (Miocene deposits of the Carpathian Foredeep)*, GEOL Q, **38**, 3, 377–394 (1994) (in Polish)
26. N. Oszczypko, M. Oszczypko-Clowes, ACTA GEOL POL **53**, 101–122 (2003)
27. M. Pawlewicz, *Total Petroleum Systems of the North Carpathian Province of Poland, Ukraine, Czech Republic, and Austria*, U.S. Geological Survey, Reston, Virginia (2006)
28. A. Shouket, J. Jarzyna, *Reservoir properties and facies characterization from core data and well logging: autochthonous Miocene sediments in the Carpathian Foredeep case study*, Geology, Geophysics & Environment, **40**, 4, 331–342 (2014)
29. P.C.H. Veeken, *Seismic Stratigraphy and Depositional Facies Models 1st Edition*, Academic Press, EAGE Publication, 313 (2013)
30. L. Buryakovsky, G. Chilingar, H. Rieke, S. Shin, Fundamentals of the Petrophysics of Oil and Gas Reservoirs, Wiley, USA, 47 (2012)

# Inversion of velocity parameters in multilayered elliptical anisotropy medium – synthetic data example

*Bartosz* Gierlach*, and *Tomasz* Danek

AGH University of Science and Technology, Department of Geoinformatics and Applied Computer Sciences, Cracow, Poland

**Abstract.** In the medium composed of parallel sedimentary layers, model of elliptic anisotropy is often better approximation of real conditions than simple layers with constant velocity. In such model a velocity is described with 3 parameters: $a$, $b$ (responsible for vertical velocity and its gradient) and $\chi$ – elliptic anisotropy coefficient, describing changes of velocity with direction of wave propagation. In this case, seismic rays are no longer straight lines but elliptic-shaped curves. In this contribution we present a tool for inversion of velocity parameters in multilayered medium considering model of elliptic anisotropy. Input data contains measured traveltimes of direct wave between all sources and receivers. The depth of boundaries and existence of elliptic anisotropy in specific layers are not subject of inversion. In order to obtain each synthetic traveltime, necessary for the target function, ray tracing must be carried out using Fermat's principle directly. The optimization in both parameters inversion and ray tracing is local and performed with Simplex algorithm from GSL library. The tests of the tool was conducted on the synthetic data with various types of start model containing one or more layers with or without elliptic anisotropy.

**Keywords:** seismic method, ray tracing, inversion, elastic anisotropy

## 1 Introduction

The basic model used in interpretation of seismic data is a set of parallel layers having constant velocity. In the real geological conditions there is a common occurrence of horizontal sedimentary layers in which velocity varies inside a layer. These changes can be approximated with a velocity gradient (linear change with depth) and an elliptical anisotropy (change with direction of wave propagation).

Description of the medium in such way demands additional parameters. It can be treated as an advantage in context of model accuracy, but on the other hand additional parameters may make the inversion process more difficult. The inversion of velocity parameters with

* Corresponding author: gierlach@agh.edu.pl

assumption of vertical gradient and elliptical anisotropy requires also optimization of ray path as their trajectory is arc-shaped.

The decision about optimal number of parameters should be made in the way that ensures balance between the fit of modelled traveltimes with measurements and the simplicity of the model. In this paper we used the Bayesian Information Criterion for choosing parameterization for specific dataset with respect to resolving power of the data.

## 2 Mathematic background

### 2.1 Model of elliptical anisotropy

In this paper, we consider model of Hookean solid described with 3 parameters per layer. A linear increase of velocity with depth is described with $a$ and $b$ parameters (velocity at zero depth and vertical velocity gradient, respectively), while elliptical anisotropy parameter $\chi$ connects vertical ($v_z$) and horizontal ($v_x$) velocities [1]:

$$\chi = \frac{v_x^2 + v_z^2}{2v_z^2} \tag{1}$$

The elliptical anisotropy model is simplification of model with Thomsen's constants (δ, ε) for the condition δ = ε as defined in [2] (see also [3]). It is characterized by elliptical wavefronts propagating from the source. The elliptical nature of the wavefront surface and the group velocity $v_{SH}$ as a function of group angle Φ is shown in the equation (4):

$$\frac{1}{v_{SH}^2(\Phi)} = \frac{\cos^2\Phi}{\beta^2} + \frac{\sin^2\Phi}{(1+2\gamma)\beta^2} \tag{4}$$

For the near-vertical incidence of seismic ray (short-spread approximation) one can use mentioned equations also to obtain velocity of P-waves propagating along the angle $\theta$ [4]:

$$v(\theta) = (a+bz)\sqrt{1+2\chi\sin^2\theta} \tag{6}$$

Thus, we obtain model with $ab\chi$ parametrization used for traveltime calculation. With the source at (0,0) and receiver at $(x,z)$ the traveltime of a signal propagating along the elliptic ray is given with formula[5]:

$$t = \frac{1}{b}\left\{\operatorname{atanh}\left[pbx - \sqrt{1-(1+2\chi)p^2a^2}\right] + \operatorname{atanh}\sqrt{1-(1+2\chi)p^2a^2}\right\} \tag{7}$$

where

$$p = \frac{2x}{\sqrt{\left[x^2+(1+2\chi)z^2\right]\left[(2a+bz)^2(1+2\chi)+b^2x^2\right]}} \tag{8}$$

The mentioned equations are proper for model composed of one layer. The analytical solution for traveltime in 2-layers model exists and is described in [5]. For the multilayered model, when one needs to find several points where ray crosses the boundaries, there is a need to perform optimization of ray path. Since the rays are not straight lines one cannot use solutions based on conventional Snell's law. In our method we use directly Fermat's principle (principle of least time) that the path taken between two points by a ray is the path that can be traversed in the least time.

### 2.2 Simplex optimization

A method proposed by Nelder and Mead [6] can be used for minimization of function of *n* variables. It does not use gradients thus can be used also for not differentiable functions.

Simplex in *n*-dimensional space of solutions is set of $n+1$ points creating polyhedron. During one iteration of the optimization function values at $n+1$ vertices of a simplex are calculated and vertex with the worst value is replaced by another point. The operations used to find new vertex of the polyhedron are reflection, expansion and contraction with respect to the symmetry centre of polyhedron, excluding discarded point. For more details, see [7].

### 2.3 Bayesian Information Criterion

In geophysical investigations statistical approach is often important for right interpretation of data. Traveltimes are always burdened with measurement errors and small-scale differences in elastic properties. Some of the velocity parameters can lose their importance for the result due to these errors which means that simpler model would be better to describe the medium.

One of the measures which allows one to choose optimal model is Bayesian Information Criterion (BIC) [8]. Under the assumption of normal distribution of the model errors or disturbances. The BIC value can be calculated as [9]:

$$BIC = n\ln(\sigma_e^2) + k\ln(n) \tag{9}$$

where $k$ is the number of parameters in the model ,$n$ is the number of data points and $\sigma_e^2$ is the sum of squared residuals. It should be minimized to ensure that assumed model gives the best balance between simplicity and fit to data. BIC value is larger either if the differences between measured and modelled values are larger or there is more parameters necessary to describe the model. Penalization for additional parameters is relatively large so BIC is more critical for the model over-parameterization if compared with another information criterions (i.e. Akaike) and is widely used in similar researches (see [10]).

## 3 Numerical example

Let us consider two scenarios – Model 1 is composed of three layers with the middle one anisotropic and Model 2 has five layers with forth layer anisotropic (see Table.1). The aim of the research was to test software for velocity parameters inversion and find the highest level of error for which adding elliptical anisotropy coefficient to the velocity model is justified.

### 3.1 Software description

In the research process we used our software written in C language and operating on Linux platform. The program has 3 working modes: forward modelling, single source and multiple source inversion. During the optimization process the number of layers and layers with allowed anisotropy and gradient are invariable.

**Table 1.** Velocity models used for the research.

| **Model 1** | | **Model2** | |
|---|---|---|---|
| Velocity | Bottom depth | Velocity | Bottom depth |
| v=2000m/s | 1500m | v=2000m/s | 1200m |
| a=2500m/s b=0.2s$^{-1}$ $\chi$=0.06 | 2600m | v=2500m/s | 1900m |
| v=4500m/s | | v=3200m/s | 2200m |
| | | a=3000m/s b=0. 2s$^{-1}$ $\chi$=0.04 | 2600m |
| | | v=4500m/s | |

Optimization of velocity model can be performed separately for each source or for several sources together. The target function is the sum of squared differences between modelled and measured traveltimes. For each velocity model and each pair of source and receiver the optimization of ray path is required. It is performed according to the Fermat's principle thus the target function is the traveltime for specific velocity model.

Both optimization of velocity model and ray path are carried out with Nelder-Mean simplex algorithm. The start steps of optimization are chosen arbitrarily and the stop criterion is based on distance between vertices.

### 3.2 Research outline

The first step of the procedure was generation of synthetic datasets basing on velocity models mentioned above and measurement sets composed of 40 geophones located inside the borehole at the depth in range between 2400m and 2900m below the surface. The waves were simulated by 10 sources at the surface located along the line with 200m spacing (between 200m and 2000km from the borehole). To simulate real conditions of measurements a random noise (normally distributed) was added to modelled traveltimes. The amplitudes of the noise was different, with standard deviations in range from 0.1% to 2% of mean traveltime.

For each dataset multiple source inversion was carried out with different number of model parameters. The simplest model has only constant velocities (3 or 5 parameters) while more complicated models has up to 9 or 11 parameters (isotropic layers treated as they would be anisotropic). Then BIC value was calculated for each variant of the start model to learn what is the optimal number of parameters for a given level of random noise.

### 3.3 Results

For each modelled set of traveltimes, inversion of velocity parameters was carried out successfully. Points where ray paths cross subsequent boundaries were calculated for each pair of source and receiver and can be displayed with our program (Fig. 1). Note that in the anisotropic layer path is displayed for simplicity as a straight line while actually it is arc-shaped according to the equation (4). Explicit formula for the ray path is presented in [10].

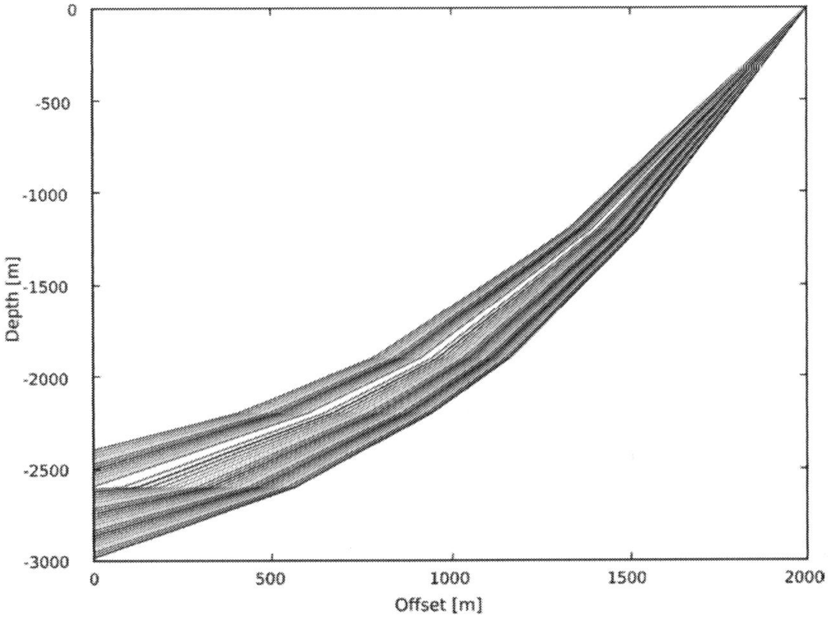


**Fig. 1.** Ray paths for source #10 and all receivers for 5-layers model.

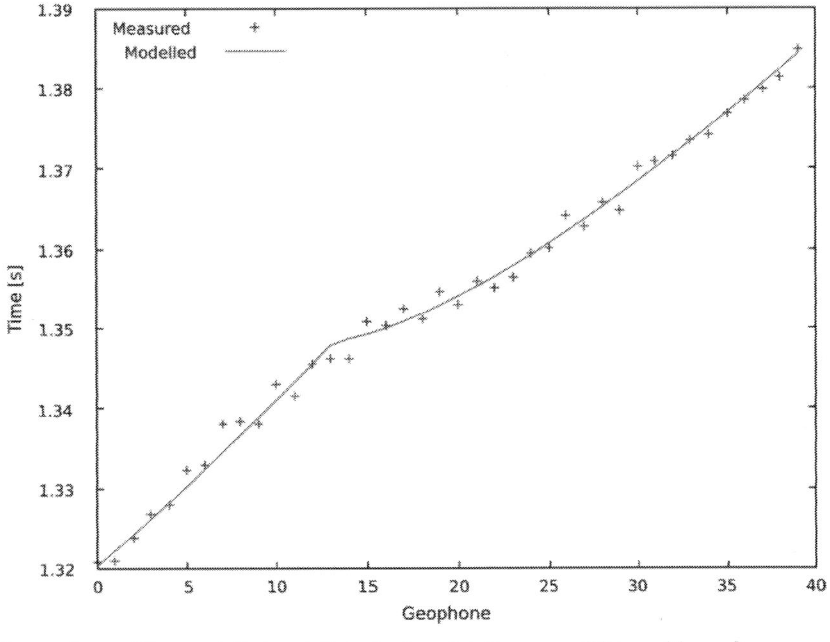


**Fig. 2.** Comparison of the modelled and disturbed model ("measured") traveltimes for source #10 with 0.1% noise (5 layers model).

The traveltimes obtained with inversion are in agreement with measured time for all variants (Fig. 2). For the number of parameters equal to that used for modelling (4 or 6 parameters) obtained velocity model is quite consistent with the simulated values. If there is less parameters, changes of other values compensate lack of anisotropy or gradient (though target function has greater value). If the number of parameters is greater than in simulation, the velocity model may contain some artificial anisotropy or gradient but it results from the added random deviations.

https://doi.org/10.1051/e3sconf/20186601017

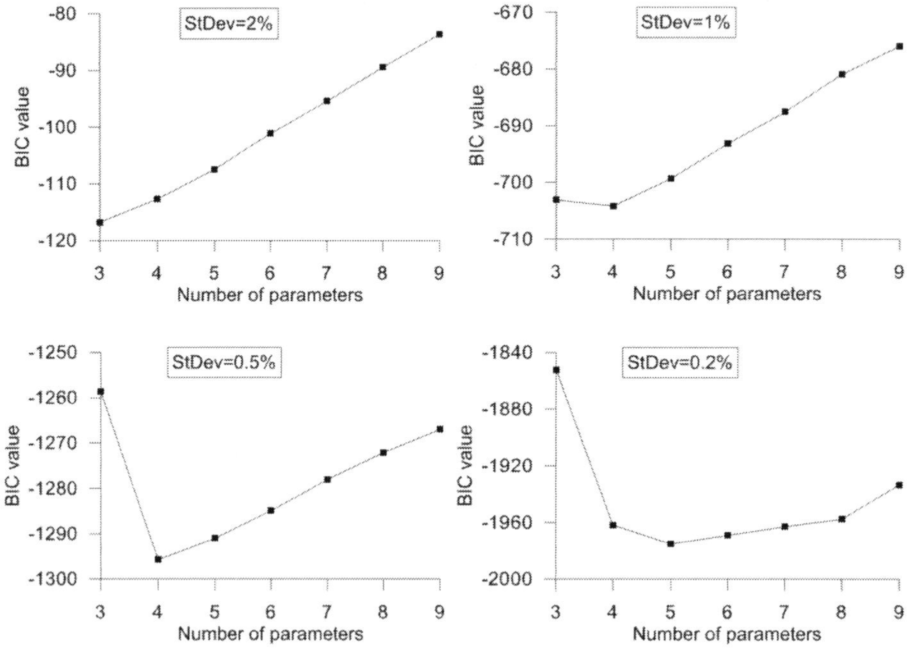


**Fig. 3.** BIC values for different levels of random noise for 3-layers model.

The next step was an analysis of obtained BIC values for each dataset to simulate decisions that should be made for real data without knowledge of the exact number of parameters.

For the 3-layered model (Fig. 3), analysis was conducted for 4 levels of random noise added. Noise with the standard deviation of 2% make model composed of 3 parameters optimal (all 3 layers isotropic and without gradient). If the noise drops to 1% or 0.5% velocity gradient in the middle layer should be considered as a forth parameter. Model including elliptical anisotropy coefficient is justified only if the noise has standard deviation equal 0.2% or less.

Analysis for the 5-layered model (Fig. 4) has shown that for the noise level of 0.2% or more, the additional parameters of forth layer are not justified and the model composed of 5 velocities is sufficient. Random noise not exceeding 0.1% gives basis to consider model with elliptical anisotropy and gradient in forth layer.

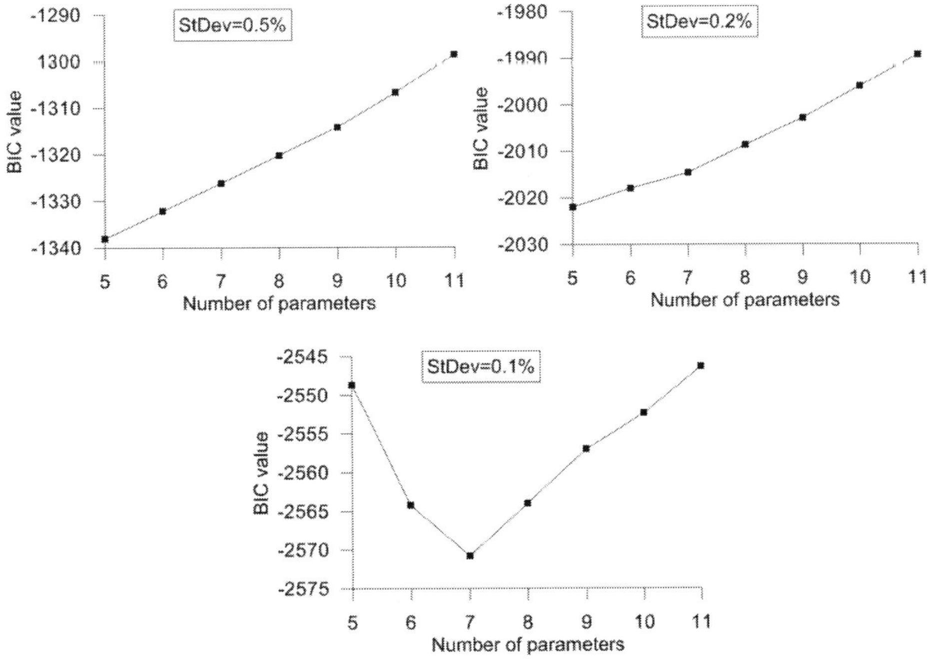


**Fig. 4.** BIC values for different levels of random noise for 5-layers model.

## 4 Conclusions

The proposed method for the inversion of the velocity model with assumption of elliptical anisotropy and vertical gradient allows one to obtain model with given number of velocity parameters. Additional parameters can improve similarity of modelled and measured traveltimes but with the low-quality data they may not carry useful information and make model too complex. The BIC analysis can help to choose optimal model parameterization for the specific dataset.

The analysis carried out on the synthetic data has shown that the method of velocity model optimization gives results consistent with the expected model. Knowing the exact traveltimes and location of sources and receivers (the scheme of vertical seismic profiling) one can use multiple sources inversion without problems with the solution stability.

Basing on the BIC analysis we can formulate two general conclusions. Firstly, the lower quality of data, the less parameters are adequate to describe the velocity model. With decreasing level of noise added to synthetic dataset minimum of BIC value is achieved by model with greater number of parameters (Figs. 3 and 4). Secondly, the greater total number of parameters in genuine model (i.e. number of layers), the lower level of noise allowing one to introduce additional parameter like gradient or anisotropy. The reason is that more parameters can compensate lack of this value and reduce the target function value.

This research was supported by AGH University of Science and Technology, Faculty of Geology, Geophysics and Environmental Protection as a part of the project No.15.11.140.194

## References

1. M. A. Slawinski, C. J. Wheaton, M. Powojowski, Geophysics **69**, 373-377 (2004) DOI: 10.1190/1.1707055

2. P. F. Daley, F.Hron, Geophysics **44**, 27-38 (1979) DOI: 10.1190/1.1440920
3. L. Thomsen, Geophysics **51**, 1954-1966 (1986) DOI: 10.1190/1.1442051
4. R. Brown, M. Lamoureux, M. Slawinski, R. Slawinski,. *Direct traveltime inversion of VSP data for elliptical anisotropy in layered media.* CREWES Research Report (2000)
5. Y. Rogister, M. Slawinski, Geophysics **70**, 37-41 (2005) DOI: 10.1190/1.2049347
6. J. A. Nelder, R. Mead, Comput. J. **7**, 308-313 (1965) DOI: 10.1093/comjnl/7.4.308
7. J. H. Mathews, K. D. Fink, *Numerical Methods Using Matlab* (Prentice Hall, New Jersey, 2004)
8. R. Kass, A. Raftery, J. Am. Stat. Assoc. **90**, 773-395 (1995) DOI: 10.1080/01621459.1995.10476572
9. M. B. Priestley, *Spectral Analysis and Time Series* (Academic Press. New York, 1981)
10. T. Danek, M. A. Slawinski, Geophysics **77**, 239-243 (2012) DOI: 10.1190/geo2012-0051.1

# Stability analysis of the Kosciuszko Mound using terrestrial laser scanner and numerical modelling

*Elżbieta* Pilecka*, and *Karolina* Tomaszkiewicz

Cracow University of Technology, Faculty of Civil Engineering, Cracow, Poland

**Abstract.** Landslides which form in anthropogenic soils are complicated from a geological engineering and geotechnical point of view. Each case requires a detailed investigation and the selection of effective reinforcements is a difficult project issue. The study presents the problem of the stability analysis of landslides occurring in the anthropogenic soils of the Kosciuszko Mound in Cracow. The previously performed protections are discussed to highlight their ineffectiveness and the current technical condition of the mound is also presented. By overlapping the results of displacement measurements made with a terrestrial laser scanner, a differential model of the terrain was created which made it possible to determine the size and direction of the deformation of the slopes of the mound and the tendencies for the development of landslide movements in this area. A cross-section, selected on the basis of the model, was numerically analysed using the finite element method (FEM) in the Midas GTS NX program. As a result of the analysis, the values of the displacements and strains occurring in the Mound were calculated. On the basis of the value of the safety factor obtained, it was also possible to assess the risk of landslide movements.

**Keywords:** Kosciuszko Mound, landslide, terrestrial laser scanning, numerical modelling

## 1 Introduction

Landslides formed from anthropogenic soils are complicated from a geological engineering and geotechnical point of view [1, 2]. The slope stability analysis and the selection of a method for protection against these landslides are a difficult design issue. An important aspect is also the monitoring of landslides, which helps to enhance the observation of the deformation of the area with time, determine the range of the occurrence and the risk of landslide development [3]. Landslide monitoring may be realized through different kinds of geological engineering and geophysical methods, including telemetric methods [3 – 8]. Such deformations may also be modelled by numerical simulations [9, 10].

The purpose of the article is to present the results of studies carried out on the periodically active landslides at the Kosciuszko Mound, which is an anthropogenic soil mound. Based on

* Corresponding author: epilecka@pk.edu.pl

the terrestrial laser scanner monitoring of the displacements of the Mound block, it was possible to locate the most deformed section. After that, it was subjected to a stability analysis in the Midas GTS NX program, the results of which allowed for the determination of the stability coefficient value and the risk of landslide movements.

The Kosciuszko Mound was built in the years 1820-1823 to commemorate the national hero Tadeusz Kosciuszko. It was built from soil covering the slope (loess formed in the form of dust and dust with loam). The mound is located on a thin layer of quaternary native soils. These soils are located on limestone, whose top is leaning towards the east, which also affects the direction of deformation. In the years 1850-1854, the base of the Mound was surrounded by a brick retaining wall and the chapel of St. Bronislawa was built in the western part of the wall [11].

Landslide processes on Kosciuszko Mound have taken place continuously from the moment it was built (the first mention dates from 1830-1840). Among the causes of the landslide processes in this area are the following: geological structure (the mound is made of an anthropogenic soil, mainly dusts and silty clays, having a tendency to settle under load and which are very sensitive to water - after dampening they show a decrease in strength parameters), the steep sloping angle of the slope of the Mound (around 46-51°, at the base of the embankment, locally 40°), penetration of waters from precipitation and thaws into the embankment leading to the moistening and plasticizing of the soil that makes up the Mound, incorrect execution of reinforcements to the Mound. The soil conditions occurring in the Kosciuszko Mound and its surroundings, due to the presence of active geodynamic processes have been described as complicated [12].

The year 1997 was the turning point for landslides on the Kosciuszko Mound because of long-lasting heavy rains and floods. As a result, soil landslides were created, and cracks made it possible for water to penetrate into the Mound. The paths were completely destroyed, and the platform bent in a north-west direction, threatening to shift the commemorative boulder on top of the Mound into the chapel. The mound began to fall apart from the top, breaking in half - the state of the Mound was defined as the state of a construction disaster.

In the years 2000-2002, the largest renovation of the Kosciuszko Mound was made - its cone was reconstructed by completely dismantling the delaminated 9-meter part of the embankment. On the whole surface of the slope below, injection bolts were put into place and the surface was secured with a geomat. An HDPE membrane was placed on the exposed portion. The embankment was rebuilt from soil which had been mixed with soil from the dismantling of its central part and a gravel mix with geogrid reinforcement in the outer part. In 2002-2003, depth and surface monitoring networks were set up (inclinometers, piezometers).

Many protection and repair works (drainage system, protection in the form of nets, foil, a geomembrane, micropiles, soil anchors) undertaken over the years after each occurrence of a landslide have proven to be insufficient and ineffective. This may have been due to inaccurate geological diagnosis - all of the tests performed included only the surface layer of the embankment, no deeper layers of soil that make up the Mound were sampled, neither were the soil layers of native soils and rocks. Soil samples were taken via penetration, which caused a disturbance of the soil structure and humidity. In 2012, a detailed determination of the soil of the Kosciuszko Mound was performed. On the basis of these results, presented in [13], this article analyses the stability of the Mound's embankment and offers a proposal for the protection against landslides.

## 2 Geotechnical condition of the Kosciuszko Mound

Up to the last occurrence of a Kosciuszko Mound landslide which occurred in May 2010 due to heavy rains and floods, 50% of the Mound's area has been destroyed. Temporary safeguards in the form of geogrids, ropes and anchors were put in place, which slowed the landslide processes, but did not ensure the stability of the cone. The current technical condition of the Mound is described as bad. If no further measures are undertaken to secure the escarpment, there may be a construction disaster.

By area, the largest deformations have appeared on the southern and eastern sides. There are many traces of landslides, an unevenness of the surface of the embankment in the form of swellings of the soil, collapse of the soil in the vicinity of concrete bands surrounding the paths. Safety nets are exposed in several places. There are zones of dried grass on the surface of the slope, which shows the presence of areas with reduced humidity in these places.

The subsidence of the land has led in many places to moving concrete bands and damage to reinforced concrete drainage canals. Many transversal and longitudinal cracks are visible on the stone walking paths around the Mound. The damage described above constitutes a significant threat to the stability of the structure. The biggest problem is the possibility of penetration of water into the Mound. It causes a continuous deterioration in the geotechnical parameters of the soil, which may lead to further settlements and deformations of the mound.

In addition, the disruption of the drainage channels has been observed on the southern slope of the Mound - the damaged installation not only does not fulfil its function, it may also cause the Mound to be irrigated by water that has not been drained.

There are visible cracks in the balustrades surrounding the observation platform at the top of the Mound [phot.2]. There are cracks and other damage on the wall surrounding the Mound in the eastern part [phot.1]. On the northern side of the wall, there is a visible white coating on the bricks, in some places the bricks are slightly darker, which may be a sign of the increased humidity of the soil zones.

**Phot. 1.** Fragment of the landslide on the eastern slope of the Kosciuszko Mound [phot. K. Tomaszkiewicz].

**Phot. 2.** Balustrade damage of the viewing platform of the Kosciuszko Mound [phot. K. Tomaszkiewicz].

## 3 Deformation monitoring of the Kosciuszko Mound using terrestrial laser scanning

Measurements of the area of the Kosciuszko Mound by a terrestrial laser scanner were performed twice, the measurements were performed over a year apart - in October 2015 and December 2016. By overlapping two numerical models of the surface area, it was possible to create a differential model, which showed that almost the whole corpus of the Kosciuszko Mound was deformed over a year between the measurements.

The slope is mainly deformed on the eastern side, which is slipping (displacement differences range from -400 mm to +400 mm, the bottom part of the slope has deformed the most). On the south-eastern side, the largest displacements (-200 mm) occur near the lowest walking path and over the retaining walls. Slope increments on the southern side are on average 50 mm, and in the lower part they vary from -200 mm locally to even +130 mm. Large displacements may be observed near the drainage system - displacements even reach up to +200 mm. The Northeast, North and West sides are dominated by displacement increments of +100 mm. The part of the Mound below the top of the observation platform is the least deformed, local surface displacements occur only in the northern part (within +80 mm). Slope displacements increase as you approach the base of the embankment.

**Fig. 1.** Laser scanning differential model of the Kosciuszko Mound - vertical displacements of the south-west slope [14].

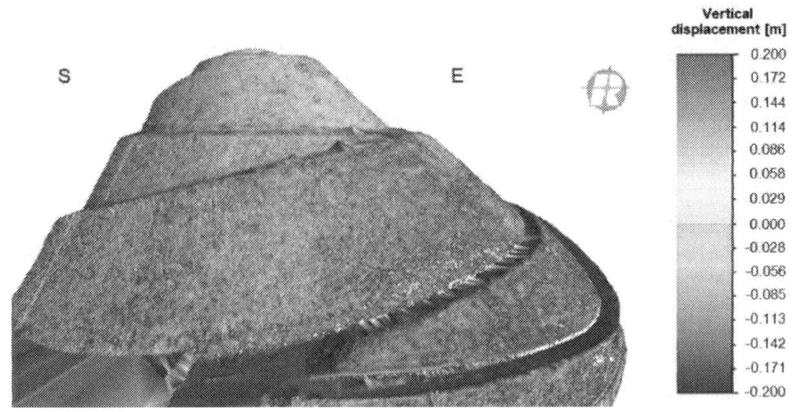


**Fig. 2.** Laser scanning differential model of the Kosciuszko Mound - vertical displacements of the south-eastern slope [14].

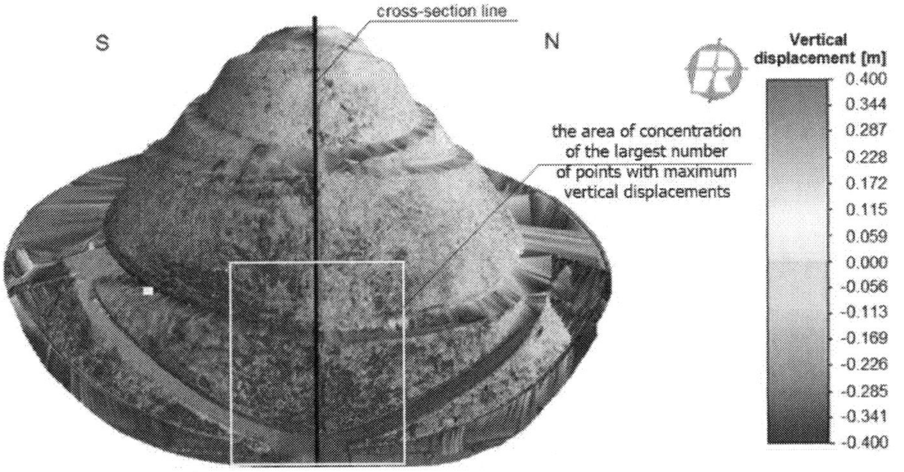


**Fig. 3.** Laser scanning differential model of the Kosciuszko Mound - vertical displacements of the eastern slope with a selection of the cross-sections adopted for numerical analysis (black line) [14].

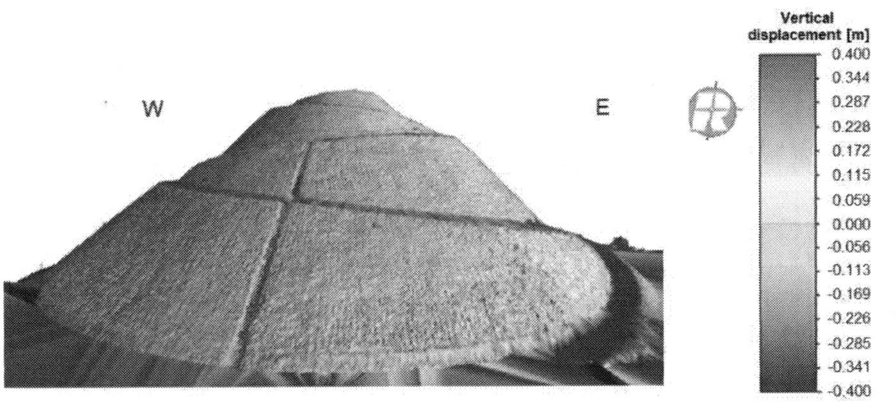


**Fig. 4.** Laser scanning differential model of the Kosciuszko Mound - deformations of the north slope [14].

## 4 Stability analysis of the Kosciuszko Mound

In order to determine the stability of the Kosciuszko Mound's embankment, numerical modelling was carried out [14]. A cross-section on the east-west line (Fig.3) was selected for the model. It was selected on the basis of the results of terrestrial laser scanner measurements, covering the most deformed areas (Fig. 3).

The stability of the Kosciuszko Mound's slope was checked using the Finite Element Method (FEM) in the MIDAS GTS NX program by making a two-dimensional model analysed in a plane strain conditions. The numerical analysis was carried out using shear strength reduction (SSR) technique. The Coulomb-Mohr model was adopted to describe the behaviour of the anthropogenic layers, and the elastic model for the limestone layer found in the base. The model was loaded with its own weight. The geometry of the model (Fig.5) and the types of soil that build the layers (Tab. 1) were selected on the basis of the results of geological surveys presented in [13].

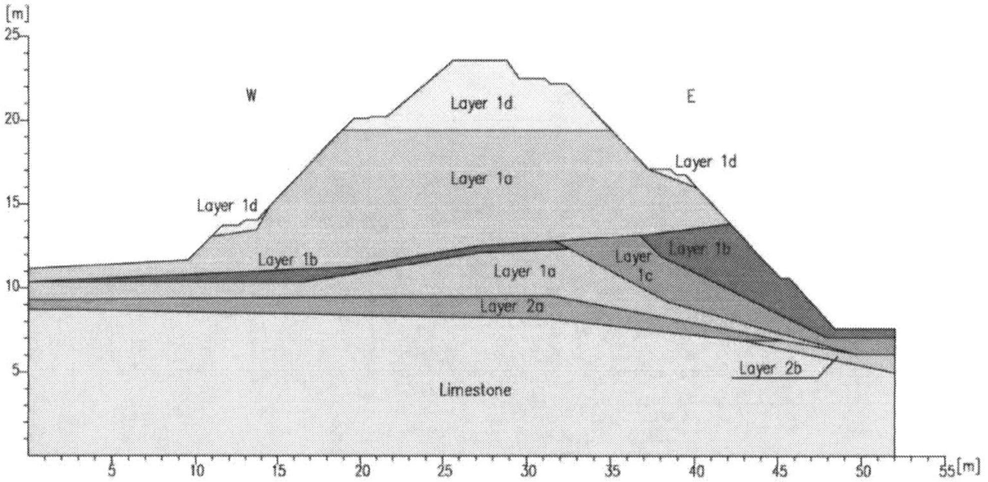


**Fig. 5.** Layout of W-E cross-sections adopted for the numerical model.

**Tab. 1.** Parameters adopted for the numerical analysis of soil parameters [14].

| Layer number | Volumetric density ρ [kg/m³] | Angle of internal friction φ [°] | Cohesion c [kPa] | Young modulus E [MPa] | Poisson ratio ν [-] |
|---|---|---|---|---|---|
| 1a | 2149 | 24 | 33 | 30.2 | 0.08 |
| 1b | 2094 | 20 | 29 | 20.7 | 0.12 |
| 1c | 2013 | 14 | 20 | 9.55 | 0.23 |
| 1d | 1960 | 30 | 0 | 28 | 0.25 |
| 2a | 2123 | 25 | 32 | 29.2 | 0.08 |
| 2b | 2088 | 18 | 30 | 20.7 | 0.12 |
| Limest-one | 2300 | - | - | 7445 | 0.25 |

The width of the model is 52 m, and the height of 23.5 m (at the highest point). The layers marked 1a-1d are anthropogenic soils and layers 2a and 2b are original soils. The upper part of the Mound is made of non-cohesive soils in the form of medium sand, medium sand with gravel, medium or fine sand with dust, sand with dust and gravel, gravel with sand and dust, gravel with sand. The lower part of the embankment is made of cohesive soils in the form of dusts, dusts with loams, clay with dust and sand, dust with clay and sand. The whole structure is based on limestone.

The results of the presented analysis refer to a situation with a critical stability. An analysis of slope stability with MIDAS GTS NX showed that the total factor of safety is 1.08. The factor of safety is slightly greater than 1, nevertheless this value is close to the critical value 1.00 (meaning the loss of stability) therefore it is necessary to control the development of landslide movements in this area.

The process of slope destruction starts in the lower part of the slope and then proceeds towards the top on the eastern side. The shear plane is clearly developed (fig. 6). The most serious damage to the slopes is in the lower part and decreases towards its top. The area covered by the occurrence of landslide movements and the direction of their development obtained on the basis of the analysis coincide with results of monitoring by a terrestrial laser scanner [16].

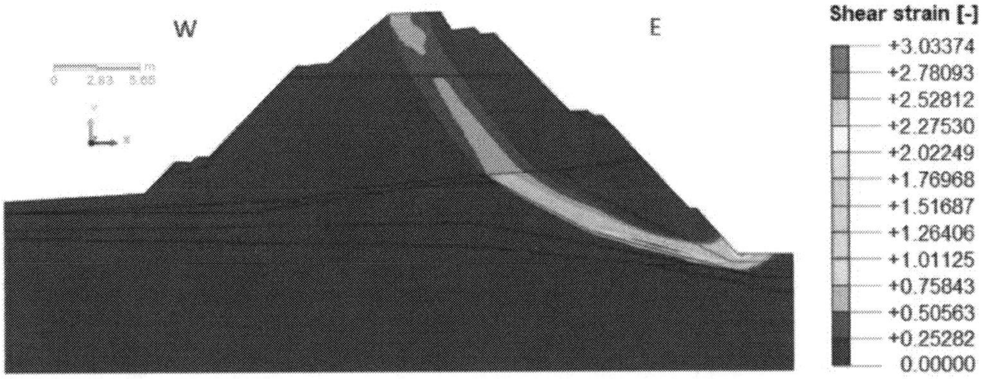


**Fig. 6.** Maximum shear strain for the WE cross section through the Kosciuszko Mound [14].

## 5 Conclusions

The example of landslides on the Kosciuszko Mound, which originated from anthropogenic soils, confirms how complicated are these kind of landslides. An important factor is the stability analysis based on detailed recognition of geological and engineering conditions occurring in the landslide area and the further terrestrial laser scanning of this area. On such a basis it is possible to analyse more effective the process of slope destruction.

Mound monitoring using terrestrial laser scanning has made it possible to create a differential terrain model. It is now possible to assess the development of the landslide process, determine the most deformed areas and the size of the displacements. On this basis, a representative cross section was selected, which was then subjected to numerical analysis.

Based on the study, it was found that:

- Results of numerical analysis confirmed measurements results made with a terrestrial laser scanner. It showed the most likely slide plane in the same place where the biggest deformations of the Kosciuszko Mound were observed.
- The Kosciuszko Mound priority is to temporarily secure or repair damage that threatens the stability of slopes. Among the works requiring immediate intervention, we can specify: protection of cracked walking paths, repair of the drainage system, filling the hollows, protection of balustrades surrounding the observation platform on the top of Kosciuszko Mound.
- The factor of safety determined by modelling is 1.08 and is close to the critical value 1.00. With further degradation of the ground structure, one can expect a reduction in the factor of safety, which means a loss of stability.

The presented study results may serve as a basis for a project to secure against these landslides.

## References


1. R. Kaczmarczyk, B. Olek, J. Stanisz, H. Woźniak, Z. Pilecki, *Wpływ gruntów nasypowych na powstanie i rozwój osuwiska (Influence of dump soil on the initatiing and development of landslide)*, Przeglad Geologiczny **62(2-10),** 594–600 (2014) (in Polish)
2. J. Stanisz, A. Borecka, Z. Pilecki, R. Kaczmarczyk, E3S Web Conf **24**, 03002 (2017) doi: 10.1051/e3sconf/20172403002
3. J. Kogut, E. Pilecka, and D. Szwarkowski, *Analysis of landslide effects along a road located in the Carpathian flysch*, Open Geoscience 2018 (in press)
4. E. Pilecka, D. Szwarkowski, E3S Web Conf **24**, 01006 (2017) doi: 10.1051/e3sconf/20172401006
5. R. Dikau, D. Brundsen, L. *Schrott*, M.-L. Ibsen, *Landslide recognition: identification, movement and causes*, Wiley, Chichester, UK (1996)
6. P. Harba, Z. Pilecki, Landslides **14**, 1225–1233 (2016) doi: 10.1007/s10346-016-0779-2
7. Z. Pilecki, E3S Web Conf **24**, 01001 (2017) doi: 10.1051/e3sconf/20172401001
8. Z. Pilecki, J. Ziętek, E. Pilecka, J. Karczewski, J. Kłosiński, *The effectiveness of recognizing of failure surface of the Carpathian flysch landslide using wave methods.* Proc. 13th European Meeting of Environmental and Engineering Geophysics "Near Surface 2007", EAGE, 3-5 Sept. 2007, Istanbul, Turkey, P48 (2007)
9. E. Pilecka, M. Moskal, *The influence of foundation for the initiation and growth of the landslidein the Carpathian Flysch*, Technical Transactions, **114**, 10, 113–121 (2018)
10. E. Pilecka, M. Moskal, *Numerical analysis of landslide next regional route No. 975 in Borowa before and after securing it*, Autobusy: technika, eksploatacja, systemy transportowe **19**, 6, 202–206 (2018)
11. A. Wójcik (ed.), *Kopiec Kościuszki i jego położenie geologiczne: atlas - przewodnik (Kosciuszko Mound and its geological location: atlas – guide)*, Urząd Miasta Krakowa. Wydział Kształtowania Środowiska, Krakow (2016) (in Polish)
12. Rozporządzenie Ministra Transportu, Budownictwa i Gospodarki Morskiej z dnia 27 kwietnia 2012 (Dz.U. 2011 nr 163, poz. 981, ze zmianami) w sprawie ustalania geotechnicznych warunków posadowienia obiektów budowlanych *(Regulation of the Minister of Transport, Construction and Maritime Economy of April 27, 2012 (Journal of Laws 2011 No. 163, item 981, as amended) on the determination of geotechnical conditions for the foundation of construction works)*(in Polish)
13. *Dokumentacja geologiczno-inżynierska dla rozpoznania podłoża, dla potrzeb opracowania zabezpieczenia obiektu stożka Kopca Kościuszki wraz z otoczeniem, zniszczonego przez aktywne osuwisko (Geological and engineering documentation for ground recognition, for the purpose of the protection of the Kosciuszko Mound's cone object with its surroundings, destroyed by an active landslide),* Przedsiębiorstwo Geologiczne S.A., Krakow (2013)
14. K. Tomaszkiewicz, *Analiza stateczności osuwiska okresowo czynnego na Kopcu Kościuszki w Krakowie wraz z propozycją zabezpieczenia (Stability analysis of periodically active landslide on the Kosciuszko Mound in Cracow with proposal for reinforcement)* engineering diploma thesis, Politechnika Krakowska, Krakow (2017) (in Polish)

15. L. Wysokiński, *Instrukcja ITB nr 424/2011. Ocena stateczności skarp i zboczy. Zasady doboru zabezpieczeń, (ITB instruction no. 424/2011. Eveluation of slope stability. Rules for selection of protection)* Instytut Techniki Budowlanej, Warszawa (2011) (in Polish)
16. B. Wrana, N. Pietrzak, *Slope stability estimation of the Kościuszko Mound in Cracow*, Studia Geotechnica et Mechanica **37**, 2, 93–101 (2015)

# Remote sensing of Carpathian flysch landslides with the use of terrestrial laser scanner

*Janusz P.* Kogut[1,*], and *Ievgen* Tymoshenko[2]

[1]Cracow University of Technology, Faculty of Civil Engineering, Cracow, Poland
[2]National Mining University, Dnipro, Ukrainep

**Abstract.** Terrestrial laser scanning helps us to detect unstable subsurface behaviour, assessing the slope stability and potential landslide failure modes. If the slopes are regularly observed, the risk of slope movement and subsequent consequences may be considerably reduced. This allows for optimum land use conditions that are economically justified. Landslides in the Carpathian flysch have a peculiar susceptibility to activation due to the region's geological structure. This work addresses the problem of monitoring and analysing the effects of landslides associated with the operation of routes (roads and railway lines) running through the slopes of the Carpathian flysch. The terrestrial laser scanner enables site remote sensing in a simple and automated manner. Regular measurements with multiple scanner positions may be used for long term slope monitoring. A detailed geological structural model allows for risk assessment with regards to failure modes, and it allows for a slope stability assessment. The model, along with the substrate parameters, introduced into the Finite Element Analysis package enables an analysis of the effects of landslide susceptibility and the displacements of the terrain surface in time, as well as due to different loading cases.

**Keywords:** laser scanning, numerical modeling, landslide, Carpathian flysch, slope stability

## 1 Introduction

The Carpathian Mountains, known as the Carpathians, do not actually form a continuous chain of mountains. Rather, they consist of a number of orographic groups that belong to the Alps. The Carpathians, which do not exceed an altitude of 2,700 m, do not have snowfields or large glaciers. The Carpathians are separated from the Alps and the Balkan Mountains by the Danube River. The Carpathians are surrounded by plains on all sides, namely the Pannonian Basin to the southwest, the Moesian Plate to the south, and the Eastern and Western European plates to the north. Figure 1 presents the Carpathians and its surroundings. The Carpathian Mountains were formed during the Alpine orogeny in the Mesozoic and Tertiary by moving the plates over subducting oceanic crust. The mountains take the form of

* Corresponding author: jkogut@pk.edu.pl

a fold and thrust belt with generally north vergence in the western part, northeast to east vergence in the eastern part and southeast vergence in the southern part. The external, generally northern, portion of the orogenic belt is a tertiary accretionary prism of the Carpathian Flysch Belt created by rocks scraped off from the sea bottom and thrust over the North-European plate [1]. The flysch is an alternating marine deposit of claystones, shales, and sandstones, which were detached from their substrate and transported several kilometres away. The northern slope of the Carpathians, located in Poland, is linked to the Baltic Sea by the drainage basins of the Vistula River. Larger rivers originating in the Carpathians include the Vistula, the Dniester, and several Danube tributaries. The Carpathian rivers are characterized by a rain–snow regime; high-water periods occur in the spring and in summer, the latter usually being more powerful. The floods sometimes have catastrophic dimensions caused by the poor ground retention of the rainfall. The rainfall may also trigger landslides.

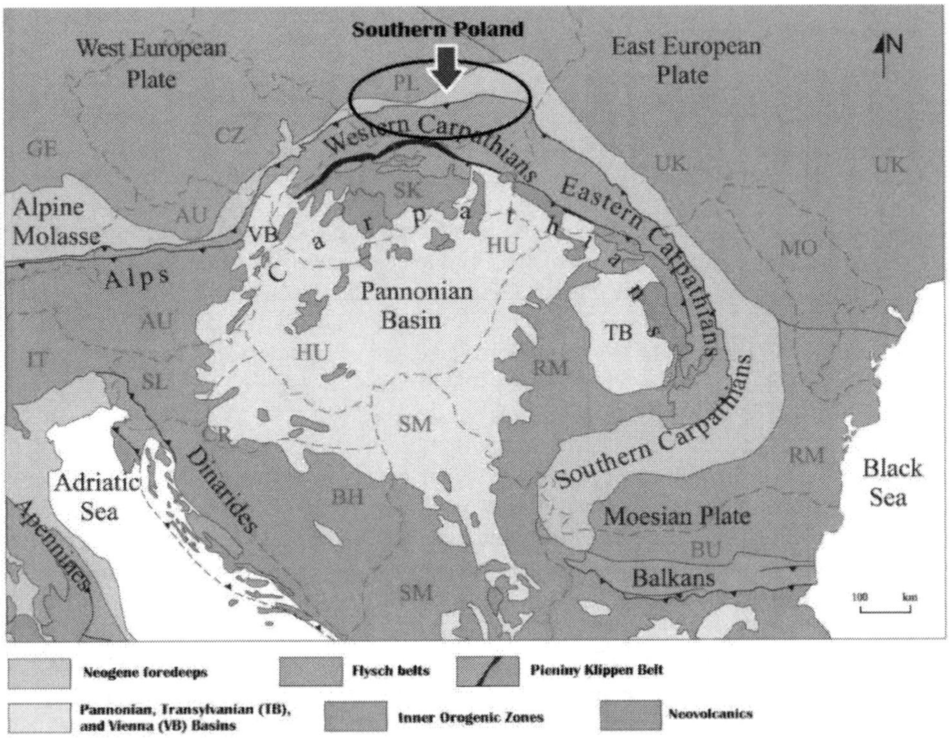


**Fig. 1.** The Carpathian Mountains and the surrounding plates [2].

In the Carpathians, which cover about 6% of Poland, about 95% of all landslides registered in Poland were identified and are currently being recorded. In the most susceptible regions of the Carpathians, about 40% of the area is covered by landslides or other forms of soil mass movements. It is estimated that, in this region of Poland, for every square kilometer of area and for every 5 km of the road infrastructure network, there is one landslide, on average [4]. The average slope of its mountains ranges from 15 to 35%, and the deep slots of the area with valleys and significant differences in height are conducive to the formation or the activation of landslides. Figure 2 presents the southern part of Poland, indicating the places where the soil mass movements and landslides are observed, including two landslides analysed in this paper.

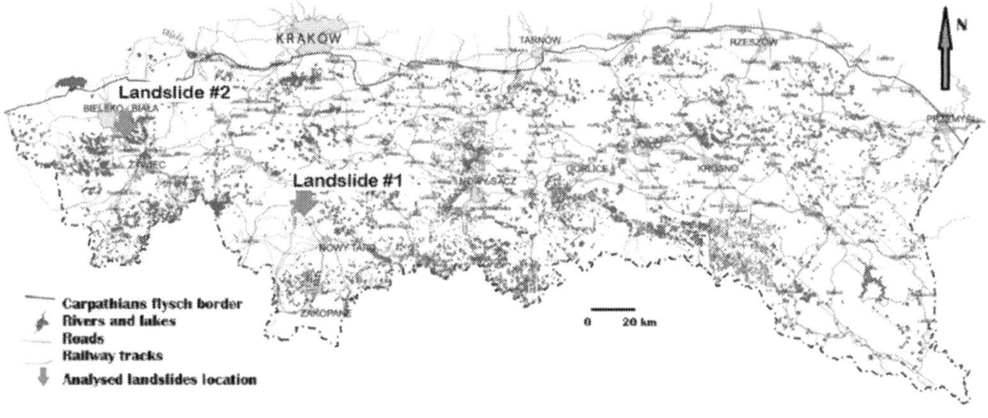


**Fig. 2.** Landslides in the southern part of Poland [3].

## 2 Survey of the retaining structure with the damaged section

The terrestrial laser scanning is a modern terrain mapping method. The measurements are based on recording a laser beam, which is reflected from different objects. The signal, registered by the unit, is converted into an electric one and then is subjected to further digital processing. Due to the waveform analysis, and when signal reaches the receiver, the calculations are performed for the signals reflected in time intervals, for the subsequent reflecting objects. For the known light beam velocity and its run time from the source to the receiver, the distance between the scanner and the reflecting body is determined. The high frequency wave emitted is by scanner is equal to f=122 kHz [5]. The mapping of the scanned area obtained is then transformed to the Cartesian coordinates (x,y,z). The data set as the form of a point cloud, creates an image of the scanned space. The measured space data is then archived for further processing. The difficulty in the laser recording processes is the vast amount of data. Therefore, the crucial parts of the area are combined data from various measurement stands. The removal of unwanted objects (particularly trees, shrubs, fences, etc.) is necessary and influences the simplicity of the scanned relief of the area. The measurements were taken in so called 'long scan mode', which is characterized by maximum resolution. As a part of the surveying work, a local situation and altitude map of the landslide was made using a Riegl Vz-400 terrestrial laser scanner [6]. The scanner has high speed data acquisition with up to 500,000 measurements/sec, with a wide field of view 100° over 360°. It has a range of measurements up to 800 m, with 5 mm accuracy and a precision of measurements about 3 mm. The ordinates were determined using a method of direct survey in the field in the local coordinate system.

The research area is located in the active part of the landslide #1 [7]. Therefore, the section of the provincial road located along this area is at risk. The as-built inventory shows that the sections of the retaining structure have been visibly dislocated. The displacement is clearly visible in the upper part of the plate and measures 8 cm horizontally and 3 cm vertically. Figure 3 presents a picture of the damaged slab section of the anchored retaining structure from different perspectives, taken during the summer and winter measurement campaigns.

**Fig. 3.** Retaining structure with the damaged slab section (photo by authors).

**Fig. 4.** 3D laser scan of the road and the retaining structure (own elaboration).

Figure 4 shows a terrestrial laser scan of the site. It displays an image of the road and the landslide stabilized by the retaining structure. The image was prepared using the RiScan PRO software package on a stretch of 75 meters [8]. The data set, in the form of a point cloud, creates an image of the scanned space. It illustrates the research area described here. The measured space data were then archived for further processing. The image obtained, in the form of a cloud of points, may be compared with subsequent scans in order to monitor landslide movements. The factors influencing measurement quality deterioration include precipitation resulting in disturbances in the cloud of points. Figure 5 displays the differential scan of the total displacements of the surface of the damaged section of the retaining wall. The intersection is still under the investigation and the results of two consecutive scans visible in figure 5 were performed within 6 months.

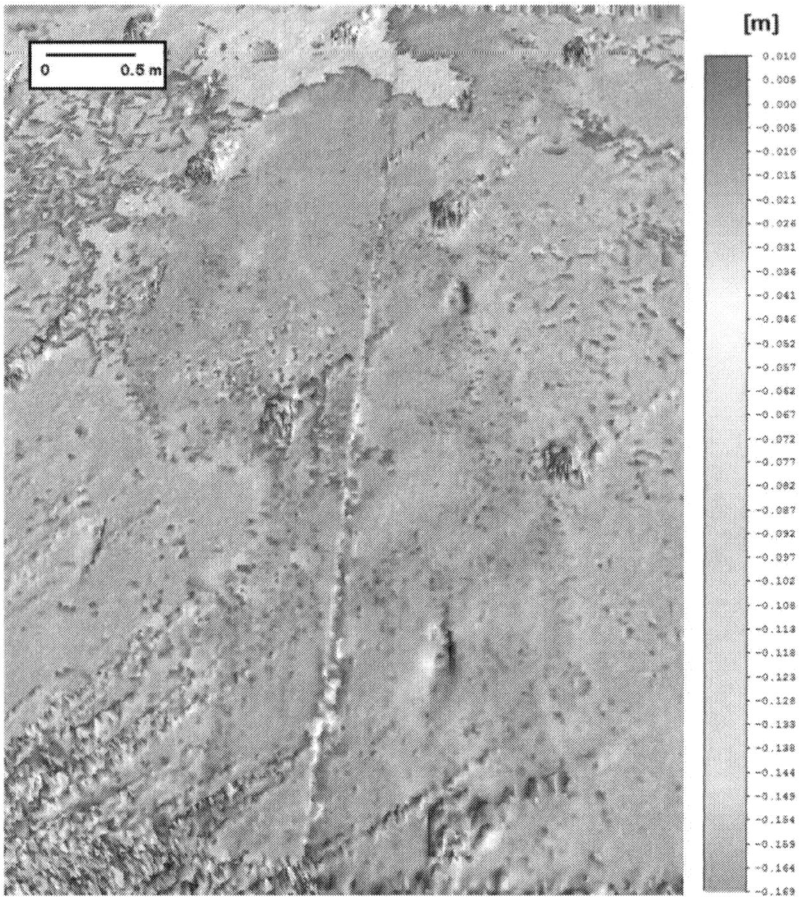


**Fig. 5.** Differential model of the total displacements of the surface of the analysed sections of the retaining wall visible in fig. 3 (own elaboration).

A terrestrial laser scanner survey allowed us to apply the cloud map of the terrain to the numerical 3D model. Figure 6 presents the surface of the analysed slope with the road and the retaining structure extracted from the scan. The surface was then introduced into the FEA package [9]. There are various slope stability analysis methods. Strength Reduction Method (SRM) is a non-linear method based on the limit equilibrium theory [10]. It analyses the minimum safety factor and failure behaviour using various shapes, loads, and boundary conditions. The SRM may be used to simulate the failure process without any previous assumptions and may be applied to 3D problems. The SRM method gradually decreases the shear strength and internal friction angle until the calculation does not converge; the obtained point is considered to be the failure point of the slope. The maximum shear strength reduction ratio, at that point, is used to calculate the minimum safety factor of the slope. Stability by SRM is achieved by weakening the soil in an elastic-plastic discrete analysis until the slope fails. The Factor of Safety is considered to be the factor by which the soil strength is to be reduced to reach failure.

https://doi.org/10.1051/e3sconf/20186601019

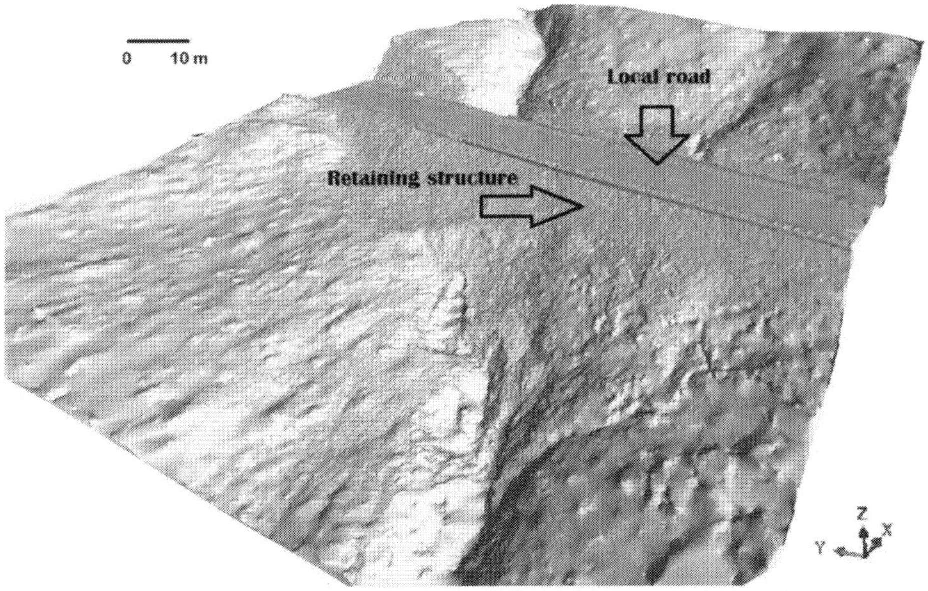


**Fig. 6.** 3D surface scan (own elaboration).

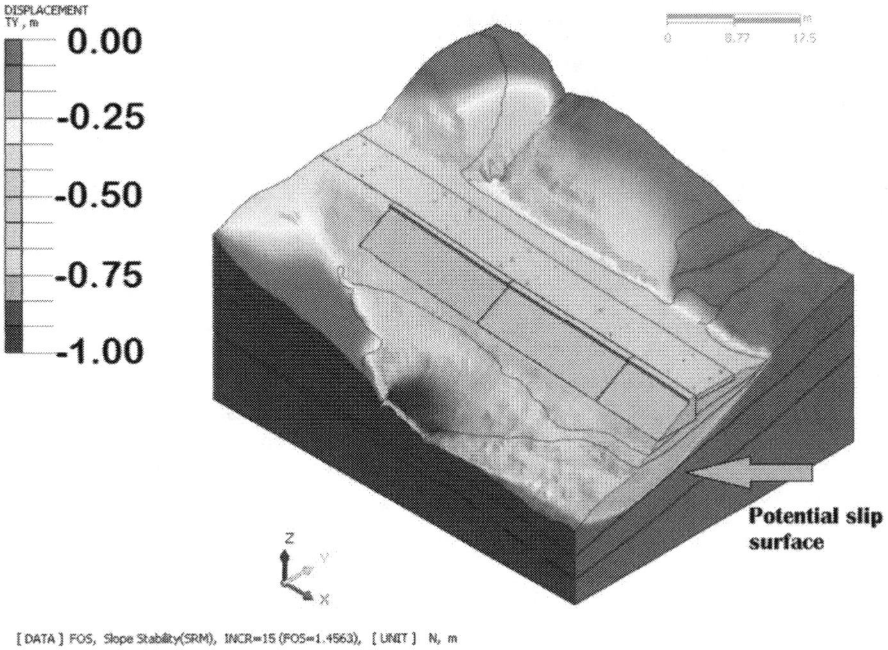


**Fig. 7.** 3D FEM model of the slope exhibiting horizontal displacements with a potential slip surface (own elaboration).

Figure 7 presents the FEM model of the slope with a slip surface computed using SRM. The FEM model included pre-stressed anchors mounted in the cross-section. Potential slip surface is indicated in figure 7, where the essential part of the terrain visible in figure 6 is used to create a real 3D map, while the subsurface data is taken from the geological prospection. More elaborated results of the FEM study and their discussion has already been done [7] and it is not presented here.

## 3 A route with the noise barrier and the retaining wall

The next example analyses the cross-section located in the area of the Carpathian flysch (landslide #2). The situation presented in this chapter focuses on forecasting an extreme threat, especially in current climate changes. It demands a detailed recognition of soil conditions to an appropriate level below the ground and diagnosing the occurrence of one or more potential slip surfaces.

Quaternary deposits recognized here are composed of silty soils - silty clays, and clays on the layer of debris. Below it, the Magura sandstone is present. The groundwater table is variable, and it is located 2 m below ground level. There is a single-family dwelling house located on the hill slope. It is considered to be an additional load on the site. Figure 8 displays the differential scan of total displacements of surface model in the natural conditions of the slope. The results of two consecutive scans visible here were performed within 6 months during the autumn and spring measurement campaigns.

**Fig. 8.** Differential model of the surface of the analysed case (own elaboration).

In the vicinity of the slope, a dual expressway was planned and built, along with a noise barrier located within the traffic lane. Therefore, earthworks were carried out involving the removal of the overburden soil and the construction of the escarpment. Figure 9 presents the overview of the slope with the retaining structure and noise barriers. The results of the analysis indicate the impact of the road excavations on reducing the stability of the built-up slope. Numerical calculations showed that the slope stability factor with the excavation made is about 15% smaller than the slope stability factor without excavation. The results concern only the dry slope. An additional destabilizing factor may be the groundwater table rise. The groundwater reduces the stability factor by about 20% [11]. This means that an excavation of the road and raising the groundwater level increases the threat to the road infrastructure.

**Fig. 9.** Overview of the slope with retaining structure and the noise barriers (photo by authors).

The Carpathian flysch is a difficult subgrade to design roads due to its complicated soil conditions and the landslide threat. Very often, the first slip surface is located on the border of the weathered zone of the flysch and the rock [12], as it turned out to be in this considered case. The built-up scarp presented here has failed, as a result of a road excavation in complicated soil conditions during the construction process. Hence, the construction of a retaining structure has become crucial. It was built during a period of unfavorable weather conditions, including heavy rains. The situation presented here concerns forecasting an extreme threat, especially in the current climate changes.

Figure 10 presents the differential model of the slope with the retaining structure and noise barriers. The displacements of the slope are visible only in the behaviour of the noise barriers. It seems to be stable with small vertical settlements around the retaining structure. The retaining structure measurements are not credible regarding the fact that the aggregate has been disintegrating due to weathering process.

In Figure 11 the FEM numerical model of the slope with the retaining structure and noise barriers is visible. The total displacements of the noise barriers located on the slope are similar to those registered by the terrestrial laser scanner. Also in this case the slope is still under the investigation.

**Fig. 10.** Differential model of the slope with retaining structure and the noise barriers (own elaboration).

**Fig. 11.** The total displacements of FEM model of the slope with retaining structure and the noise barriers (own elaboration).

## 4 Final remarks

This work presents the results of laser scanning in the Carpathian flysch conditions. It may help to evaluate the usefulness of the method for monitoring the deformations on the terrain surface over the potential slip surfaces. The measurement results, taken from two sites and carried out over several months, indicated possible existence of the local landslide threat even after stabilization works. However, the deformations found are rather small and cannot be dangerous for the nearby road infrastructure. The obtained results prove the usefulness of terrestrial laser scanning to recognize discontinuous deformations of the terrain surface in such harsh soil conditions. They are a starting point in the discrete analyses of slope stability as visible in two analysed cases, illustrated by the results of FEM analyses. They can also be used to assess the causes of damage to buildings associated with landslide activity.

## References

1. N. Oszczypko, *Late Jurassic-Miocene evolution of the Outer Carpathian fold-and-thrust belt and its foredeep basin (Western Carpathians, Poland)*, Geol. Quart., 50 (1): 169–194 (2006)
2. J. Golonka, F.J. Picha. The Carpathians and their foreland: Geology and hydrocarbon resources. AAPG Memoir 84, ISBN-10: 0891813659, ISBN-13: 978-0891813651 (2006)
3. D. Poprawa, W. Rączkowski, Landslides of the Carpathians, Przegląd Geologiczny 51, 8, 685-692 (2003) (in Polish)
4. L. Bober, L. Thiel, L. Zabuski, Landslide phenomena in Polish flysch Carpathians. Geological and engineering properties of selected landslides , IBW PAN Gdańsk (1997) (in Polish)
5. E. Pilecka, D. Szwarkowski, An application of the ground laser scanning to recognise terrain surface deformation over a shallowly located underground excavation, E3S Web of Conferences, vol. 24 (2017), DOI: 10.1051/e3sconf/20172401006
6. www.riegl.com
7. E. Pilecka, J. Kogut, D. Szwarkowski, Analysis of landslide effects along a road located in the Carpathian flysch. Open Geosciences, 10 (1): 517-531 (2018), DOI:10.1515/geo-2018-0041
8. RiScan PRO Manual. Mena3D GmbH, Stuttgart, Germany (2014)
9. MIDAS GTS NX, Manual specifications (2016)
10. D. Grifiths, P. Lane, Slope stability analysis by finite elements. Géotechnique, 49(3), 387–403 (1999), DOI:10.1680/geot.1999.49.3.387
11. E. Pilecka, J. Kogut, Soil mass movements as threats to the road transportation infrastructure, TTS, Technika Transportu Szynowego. Koleje, Tramwaje, Metro, 12, 2825-2830 (2015), ISSN 1232-3829, (in Polish)
12. P. Harba, Z. Pilecki, Assessment of time-spatial changes of shear wave velocities of flysch formation prone to mass movements by seismic interferometry with use of ambient noise. Landslide 14 (3): 1225-1233 (2017), DOI: 1007/s10346-016-0779-2

# Slope stability monitoring in open pit mines using 3D terrestrial laser scanning

*Miroslawa* Bazarnik*

Cracow University of Technology, Civil Engineering Department, Cracow, Poland

**Abstract.** In open pit mines the aspect of preventing and forecasting the threat of landslides and rock falls is crucial issue because of the significant consequences that instabilities may have. Systematic slope stability monitoring is necessary to ensure safe and continuous mining operations. The development of innovative technologies, such as 3D laser scanning, opens up new possibilities, especially in the case of large and hard-to-reach areas, such as open pit mines. Terrestrial laser scanners (TLS) provide fast, efficient, detailed, and accurate three-dimensional data. The article discusses the use of 3D terrestrial laser scanning method to monitor slope displacements and landslides in open pit mines. The first part of the article discusses the risk scale of gravitational displacement on the slopes, on examples of Polish open pit mines, and introduces the most common slope monitoring methods. Then, the principles of 3D terrestrial laser scanning were defined, and some examples of TLS applications in the open pit mines were presented.

**Keywords:** open pit mines, terrestrial laser scanner, displacement monitoring, landslide

## 1 Introduction

The mining of mineral raw materials using the opencast method is accompanied by a number of natural hazards, among which mass movements, that is the movement of rock masses due to the gravity, deserve special attention. Gravitational deformations of the rock mass are landslides and rock falls, sometimes covering the area beyond the upper edge of the excavation. During the mine exploitation period, numerous landslides of various sizes are usually recorded on the open-pit slopes. The volume of displaced earth masses for the largest landslides in the open pits reaches 6-8 million $m^3$, but most often they amount to several dozen to several hundred thousand $m^3$ [1]. The stability loss of excavations or dumping ground slopes, as a result of mining operations and natural geological processes taking place, can lead to serious changes in the rock mass causing landslides, thus posing a threat to the health and life of the mine crew. Such hazards occur on a large scale in brown coal mines, but they can also occur at exploitation of ceramic and rock raw material deposits, especially in karstic areas.

The issue of geodetic and geotechnical monitoring is of key importance in the aspect of preventing and forecasting the threat of landslides and rock outcrops. Monitoring is an important element of threat management, which includes both the identification of threats as well as their assessment and characteristics (supported by geographic information systems and related databases, including historical information). The threat extent depends on the geological structure of the deposit, including variability of the geotechnical and hydrogeological conditions. Moreover, it depends on the area and size of the excavation, character of the exploitation systems, as well as the machinery and equipment used. The development of innovative technologies, such as 3D laser scanning, opens up new possibilities, especially in the case of large and hard-to-reach areas, such as open pit mines. Terrestrial laser scanners (TLS) provide fast and efficient, detailed and highly accurate three-dimensional data. TLS cover a wide range of applications, including topographic mapping, as-built and architectural studies, and monitoring of engineering structures.

This article presents the use of 3D laser scanning technology in the mining industry. It is a review of existing solutions, aimed to discuss the use of the laser scanner to monitor the movement of slopes and landslides in open pit mines. In the case of slopes already unstable, monitoring is often the only chance to predict the degree of threat and direction of the mass movement development. 3D scanning methods can complement traditional geodetic methods.

## 2 The risk scale of the slope movement occurrence in Polish open pit mines

Open pit mining is an important element of Polish economy, providing raw materials for, among others, energy, road and construction industries, and gaining more and more interest among capital managers, both domestic and foreign. As a result of mining exploitation, the rock mass balance is disturbed, which may lead to risks for the mine and the adjacent environment. In case of violation of the natural equilibrium, geological and engineering changes may take place, resulting in loss of stability and, as a consequence, deformation of the rock masses [2].

The basic geotechnical threats include sliding ground masses and formation of rocky outcrops on permanent and operational slopes of outcrops and dumps, which may threaten people, machines, objects and the continuity of mining operations. Prevention of negative effects of phenomena occurring in the rock mass is based on detailed recognition of the geological structure and mechanical properties of rocks and soils, as well as continuous monitoring of the processes in due course. Moreover, the landslide hazard may be reduced by analysis of geological and engineering conditions and identification of threatened areas within individual mining facilities. The scale of threat of landslide movements is reflected in a number of examples of landslides quoted below, which took place in lignite mines, affecting the continuity and safety of mining operations. They include landslides, which occurred in the "Turów" Coal Mine, deformations on the western steep slope in the area of the Nysa River Pillar threatening the pillar's stability and the interruption of the Nysa riverbed in July 1989 [4]. In the years 1993 – 2007, only four landslides were detected on the northern slope of the "Turów" Coal Mine [5]. In 1994, on the slope of the external dump, the landslide was the largest in terms of the number of displaced masses, although not the most dangerous in terms of the potential impact on the mine [6]. In September 2016, the landslide encompassed the area of 6 square kilometers [7] on the internal dump, caused gigantic material losses, stopping coal mining and sale to individual recipients, as well as limiting the work capacity of the Turów Power Plant (Fig. 1). Some devices, including belt conveyors, have been destroyed. The spoil was also covered.

In the Brown Coal Mine "Bełchatów" since the beginning of earthworks, the formation of landslides was recorded on permanent and working slopes, with varying severity and of various sizes, often affecting mining works, drainage of the excavation, or power supply of the mine with electricity. Landslides which formed on the steep south and north slopes, each time affected the mining activity. In 2005, volume of landslide masses displaced on the southern slope was estimated for approx. 3.5 million $m^3$ [2, 6].

**Fig. 1.** Landslide in the open pit mine "Turów" – 2016 [3].

In 1996 in the Brown Coal Mine "Konin"the cut of organic land on the foreland of the "Jóźwin" and "Kazimierz Północ" open pits resulted in a rapid fall of the cold masses to the working level [2,8]. Displaced grounds covered the working level with a layer of up to 3–4 m, immobilizing the mechanical equipment. In 2002 in the Brown Coal Mines "Konin" and "Adamów" landslides were caused by insufficient discharge of residual waters as well [2,8]. Examples of landslides that posed a threat to the plant's operations and general safety could be found not only in brown coal mines, e.g. in 1999 in the "JARO" SA coal mine, in the "Stanisław" excavation, a landslide developed in the area immediately adjacent to the protective pillar established for the buildings of the village of Rusko, which forced evacuation of the population from the nearest buildings [2].

In 2004, two landslides were noted at the Diabase Mine "Niedźwiedzia Góra", on the northern slope and on the southern part of the eastern slope, which caused partial destruction and backfilling of technological roads, destruction of the pipeline and damage to LV and MV power lines [2]. In 2004, a landslide took place in the Basalt Mine "Gracze" on the western slope [2]. In rock salt mines, the threat posed by peeling rocks from solid and working exploitation walls is common. This threat is related to the formation of rock overhangs, cracks and fissures as a result of the blasting technique applief for the deposit exploitation. Formation of landslide hazard is also influenced by the geological structure, i.e. deposit retention, tectonics, karst phenomena and directions of exploitation applieed in mining plants [8].

Exploitation of brown coal deposits requires transfer and storage of earth and rock mass on the dumping ground. The most frequently occurring processes that may arise at formation of external and internal dump are landslides on boulder slopes as well as runoff and sludge scarp escaping. The process of displacement of the soil surface can also occur in front of the face of the dumping ground. Experience and practice acquired in the exploitation of lignite, in particular in the Brown Coal Mines: "Bełchatów" and "Turów", allow for more and more precise analysis of the state of geotechnical threats, including these due to natural conditions as well [2,8].

The most common factors affecting the formation and development of landslides include: inclination of rock layers, tectonics and geomechanical properties of rocks, such as cohesion, internal friction angle and pore pressure. Both exploitation pits and dumps are designed in such a way as to meet the geotechnical safety requirements of slopes, and thus the safety of operation and impact on nearby infrastructure. As a result of mine operations, terrain deformations may occur. Both continuous and discontinuous deformations can potentially affect housing developments within the limits of their impact. There may be discontinuous deformations, in the form of slopes, and deformations of the terrain surface. It is also possible that continuous deformations occur due to subsidence of the terrain surface brought about by dehydration of the rock mass. The risk of slope deformation can be diagnosed through comprehensive reconnaissance and multidirectional monitoring and appropriate mining and technological activities.

## 3 Slope monitoring methods

The fundamentals of effective monitoring consist in geological recognition and execution of precise geodetic and geotechnical measurements, as well as possibilities of fast data entry and processing, which the basis for undertaking the necessary preventive actions, and the actions aimed at eventual liquidation of the effects which did occur.

The selection of measurement methods used to determine the displacement of building objects depends on three basic factors [9]:

- rate of changes occurring on the tested object,
- type of determined displacements :
  - horizontal (two components in the horizontal plane and time),
  - vertical (one component in the vertical line and time),
  - spatial (three spatial components: x, y, z and time),

  or with another broken down:
  - relative (relative to points located on the tested object),
  - absolute (related to fixed points located outside the range of impacts),
- required accuracy of the determined displacement figures of the points examined.

The correct determination of displacements is the basis for determining the direction and speed of changes taking place, as well as the development of safeguards to avoid dangerous events.

The monitoring project should be constructed on the basis of predicted magnitudes and speed of changes taking place, closely related to the staging of mine work. If the changes are slow and related to reaching the appropriate excavation depths or excavation passage through successive sections, it can be assumed that during one measurement cycle there is no change on the tested object larger than the accuracy of the measurement. With this assumption, measurements of the geodetic network covering the tested object and its surroundings can be carried out. The construction of slope monitoring network, absolute (related to fixed points located outside the range of impacts) or relative (referred to points located on the tested object), is crucial to ensure compliance with the accuracy requirements [10]. It should be shaped upon the established type of potential threat, e.g. based on archival data taking account the appropriately tailored measurement methodology, as well as external factors and accuracy of measurement. At the development of the monitoring network consideration should be given to logistic factors, such as access to the site, power availability, etc. The slope monitoring design should be constructed on the basis of projected magnitudes and speed of changes, closely related to the work being phased in during the mine work. If the changes are slow and related to the achievement of appropriate excavation depths - running maintenance requires constant monitoring to avoid disruptions and failures. The literature [10,11] shows how one can monitor deformations based on the use of various geodetic and geotechnical technologies depending on numerous factors, including the required accuracy and frequency of monitoring, etc.

In the case of the slope monitoring, the main purpose of geodetic measurements is to describe geometric changes (e.g. point coordinates)of the surface topography by measuring geometrical elements (e.g. angles, distances or height differences). In the case of typical geodetic monitoring, it is necessary to establish a reference geodetic network [11] with fixed control points and several observation points. The control reference points must be outside the expected deformation area. Observation points in which geodetic instruments are set can also be placed directly on a moving slope. Their coordinates must be determined each time before examining the object. Measurements taken from one fixed point to another point are absolute measurements. An alternative to traditional measurement methods has become the 3D laser scanning technique. The advantage over manual data collection techniques is that the laser scanning recording methods allow automated, dense sampling of the object's surface in a short time [12].

## 4 Characteristics of the 3D laser scanning method

The technique of Terrestrial Laser Scanning (TLS) has become a commonly used method of acquiring 3D data. The first commercial systems became available in the 1990s [12]. The laser scanner uses laser light to measure the distance from the sensor to the object in a systematic manner. Most laser scanners are pulse scanners, in which the laser transmitter emits a laser pulse, which, in turn, reflected from the surface returns to the receiver. Based on the flight time, the distance to the reflected surfaces is determined. The measured distances are converted to 3D coordinates with the information on the origin of the range vector and the scanning angle. In order to convert the measurements into a better, georeferenced spatial coordinate system, the laser measurement must be connected to the GPS device [12]. Many instruments also record the intensity characteristics of the backscattered pulse. As intensities depend on numerous factors, such as surface roughness and reflection of objects, they could be used for visualization and classification purposes. Usually, one scan is not enough to collect data covering the entire object or location, therefore it is required to perform scans from several observation points. Scans from different locations are combined within the same coordinate system in post-processing. In order to ensure a full representation of the terrain surface, the obtained scans should be closely connected with each other and input in the geodetic coordinate system. This is important for the integration of TLS data with other spatial data (e.g. GPS measurements). As a result of data processing, a high resolution digital terrain surface model is obtained that can be exported to many geographic information system (GIS) programs, and used, for example, for modeling and visualization of the measured object. The main advantage of the TLS protocol in comparison with traditional geodesy techniques consists in direct, quick and detailed mapping of the object's geometry in 3D. TLS provides cost reduction and significantly faster data processing. Everything that you see on the stage is captured at the same time, which allows for the comprehensive use of data, both now and in the future [12]. The 3D laser scanning technique allows to record the state of the object or phenomenon independently from the availability of external data, whose spatial and temporal range often limits the observations. At the same time, it maintains high efficiency of measurements resulting from the frequency of sending a light beam (depending on the model, it is from 1 kHz (registration 1000 pts/sec) to 1 MHz (approx. 1 million pts/sec)). Thanks to the active emission source of the TLS laser beam, it is characterized by the independence of measurements from the lighting conditions and from the transparency of the air. As a result, the efficiency of data acquisition increases. On the basis of the point clouds obtained, a traditional model can be made in the form of an irregular grid of triangles, whose high density allows to generate numerical models with a very realistic appearance (Fig. 2). From generated 3D surfaces, it is possible to extract 2D profiles or contour lines of objects. The obtained models can then be processed and used in GIS and CAD systems for further data integration. The obtained three-dimensional models are made documentation (depending on the user's needs) reflecting the actual state of the object we are interested in and giving a graphic image of the planned investments. The digital form of data gives the possibility of constructing environmental models, designing drainage systems, etc.

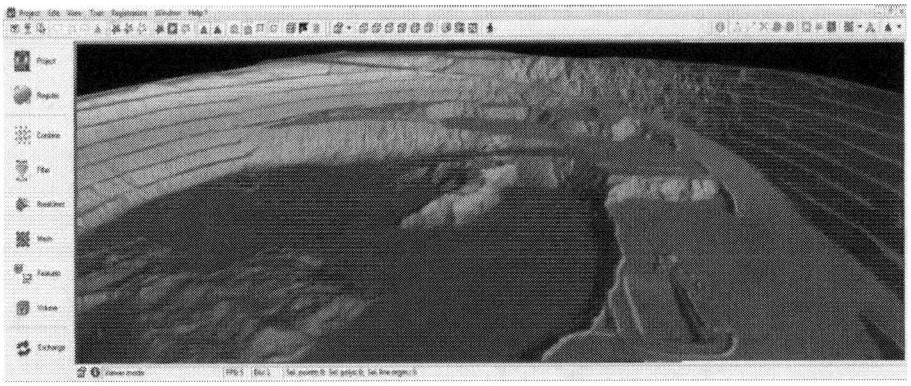

**Fig. 2.** DTM open pit mine model on the basis of a point cloud [13].

The concept of using terrestrial laser scanning to study deformation processes, which include landslide movements, is based on performing scanning of an area threatened by deformation processes at successive time points, especially during periods of intensification of linear erosion, e.g. after heavy rains or during spring thaws [14, 15]. As a result of superimposing two scans in a local geodetic coordinate system, a differential model is created. An example of such a model is presented in Figures 3 and 4.

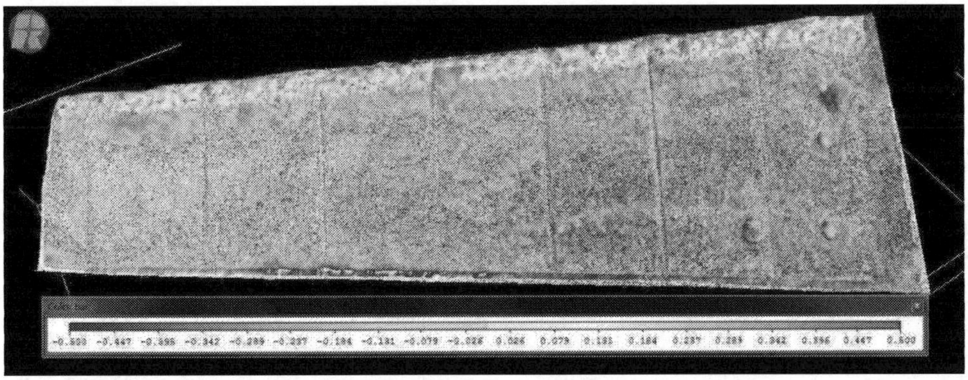

**Fig. 3.** The differential model of the area threatened by a landslide in Sadowie [16].

To ensure the best results of the tests, models made at particular time points should have the same resolution and accuracy. On the basis of the differential model, it is possible to observe changes in the terrain surface due to landslides [16]. It is possible to almost immediately formulate the prognosis of the development of the phenomenon and to determine the degree of threat. The obtained data in digital form can be used, for example, to simulate dynamic processes [17–20]. The use of 3D terrestrial laser scanning technology to analyze erosive phenomena on the slopes allows us to move away from traditional point measurements using geodetic methods. We are able to evaluate erosive phenomena (both in terms of process intensity as well as events initiating erosion) and also the area coverage of existing processes. The use of TLS involves the need to develop a research methodology, select a scanner, link to a general reference system or the object's own reference system, as well as determine the spatial resolution of the inventory being performed. The consequences of the adopted methodology and technical parameters of scanning affect not only the time and manner of conducting measurements, but also the reliability and accuracy of analyzes.

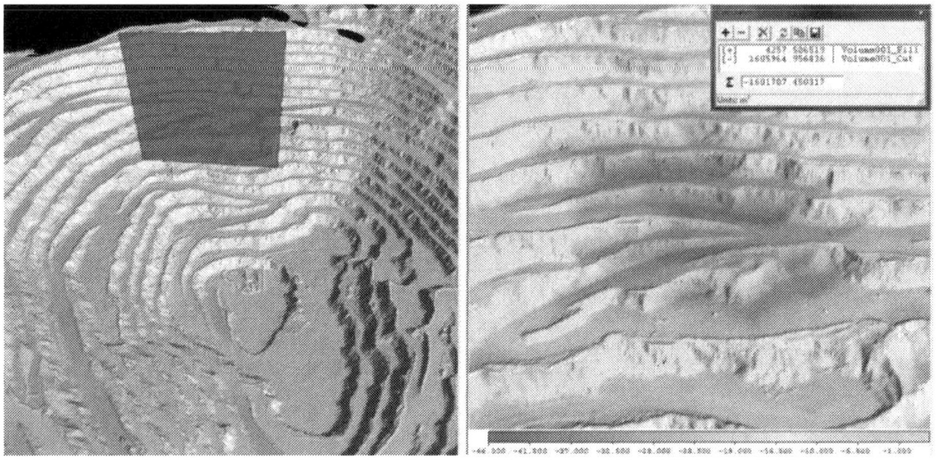

**Fig. 4.** A differential terrain model for open pit mine with visible surface changes [13].

## 5 Examples of using terrestrial laser scanning for slope monitoring in open pit mines

Terrestrial laser scanning 3D has become a widely used surveying method in the mining site [17-21]. TLS offer many possibilities in the field of terrain deformation monitoring, mainly due to the high spatial resolution and data capture speed, which makes it possible to repeat measurements to collect data in high resolution [6, 18, 20]. Laser scanners work even with reduced visibility caused by e.g. dust, fog, rain, snow, etc., which frequently occur in difficult environments, such as mining sites.

Open pit mines are very often characterized by their large size. They can include areas from 1000 meters to several square kilometers. In mines of such large expanse it is particularly important to prepare and organize the measurements. Usually, survey parameters have to be defined so that the scanning test will set-up properly, considering the mine's depth and range. Figure 5 shows an example of a differential model for the large open pit coal mine in Indonesia. Though large-sized the mine, it is possible to determine changes in the surface area and calculating the volume of rock masses in a relatively short time.

**Fig. 5.** The differential terrain model of coal open pit mine in Indonesia [22].

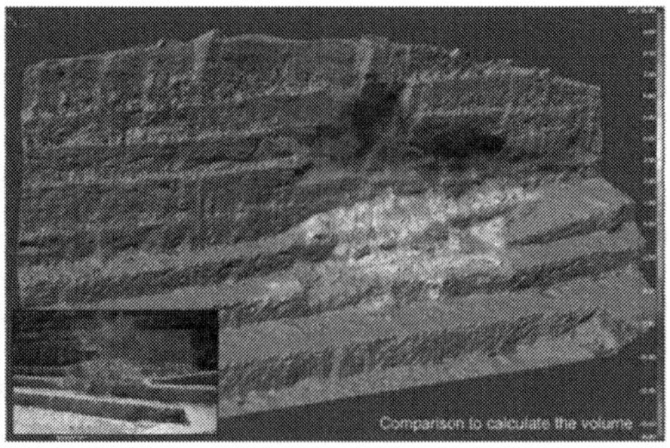


**Fig. 6.** Color-coded comparison before and after landslide activation [22].

There are examples in the literature of using 3D terrestrial laser scanners to provide information of a potential unstable subsoil detection, assessment of the grading structure performance, which includes identification of slope instabilities and/or evolving failure mechanisms [14-26]. Inter alia in platinum mines Mogalakwena and Potgietersrust Platinums Ltd (PPRust) in South Africa the TLS method, along with special software dedicated to open pit mines, is used to monitor pits and slope management [23, 24]. On the basis of point clouds volume calculations are made and the state of the pit slope is analysed. More importantly, progressive slope movement is calculated and plotted. TLS is used there for long term monitoring of trends and for identifying high risk areas. It provides significant information for determination geohazards and geotechnical risks, as well as is acting as a primary monitoring system providing long-term data to understand the history of the rock masses and its behaviour. The rockfall detection, structural mapping and displacement data aid in highlighting high-risk and hazardous areas. Based on the monitoring data in time, was identified movement locations and rock mass volumes disconnected and accumulated on the top north wall Magalakwena mine. The events analysis over time allowed to divided the mine area into zones from the risk exposure.

In Brazil gold mine, 3D laser scanners were used to analyse a landslide to calculating the volume of the material loss (Fig. 6). The volume was calculated for about 8,350 cubic meters in the upper part of the rock face [22]. To capture the entire extension, 1.5 km by 1km, with a depth of about 700 m, the mine was scanned from six positions, the coordinates of which were determined using a GPS receiver mounted on the top of the scanner. Accuracy was obtained at cm-level.

In Italian white marble open pit mine "Botticino", located in north Italy, a combination of digital photogrammetry by UAV and TLS was proposed. This resulted in an increase in local accuracy up to +/- 2 cm [25].In Poland, research is being carried out on the use of the 3D laser scanning method in the "Bełchatów" Brown Coal Mine [6, 26].

## 6 Conclusions

The laser scanning method becomes a competitive method for classical and photogrammetric measurements methods of slope monitoring in terms of accuracy. It is excellent for objects with a fan-shaped construction, which include open pit mines. The advantages and limitations of this method in the context of slope monitoring are shown in Table 1.

**Table 1.** Advantages and limitations of laser scanning in relation to conventional methods

| Advantages | Limitations |
|---|---|
| Capturing a large coverage of a rock surface with high resolution | Difficult to monitor small deformations, below few mm |
| A short time of capturing (a few minutes depending on the resolution selection) | Range of the laser scanner is limited by rock mass reflectivity. Maximum range decreases if the reflecting surface is wet. |
| All the data are digital and in 3D, both image and co-ordinates | |
| After reference surveying, all scanning data are registered in a global co-ordinate system, and the rock mass is referenced to its real position in space | |
| Possibility to determine geometrical features, e.g. length between points, orientation of a fracture, etc. | |
| Visual information, e.g. 2D and 3D digital image, 3D virtual model. | |

| | |
|---|---|
| Possibility to determine physical changes, e.g. the movement of rock masses, water leakage identification and different rock type identification by different reflective intensity | |

Terrestrial laser scanning provides very fast and easy access to data in real time. Nevertheless, terrestrial laser scanning should not be treated as the only measurement method and should be supplemented with more accurate geodetic measurements, established by engineering practice.

## References

1. P. Dobak, A. Drągowski, Z. Frankowski, A. Frolik, R. Kaczyński, A. Kotyrba, J. Pinińska, S. Rybicki, H. Woźniak., *Zasady dokumentowania warunków geologiczno-inżynierskich dla celów likwidacji kopalń [The principles for documenting geological and engineering conditions for the purposes of mine closure], Ministerstwo Środowiska*, ISBN 978-83-60117-86-6 Warszawa (2009)
2. M. Krawczyk, *Stan bezpieczeństwa w odkrywkowych zakładach górniczych oraz zmiany w przepisach obowiązujących przy eksploatacji kopalin metodą odkrywkową [The state of safety in open pit mining and changes in the regulations applicable to the exploitation of mineral resources opencast method] Proc.Conf. Mining Workshop*, 49-63; http://warsztatygornicze.pl/wp-content/uploads/2007_4.pdf (2007)
3. http://zzprckwb.org.pl/news/800
4. D. Milkowski, J. Nowak, *Zabezpieczenie i monitoring zagrożenia osuwiskowego na filarze rzeki Nysa Łużycka oraz osuwiska "Świniec", [Protection and monitoring of landslide hazard on the Nysa Łużycka river and "Świniec" landslide]*, Bezpieczeństwo Pracy i Ochrona Środowiska w Górnictwie **6**, 14–21 (2009)
5. D. Milkowski, A. Górecka, M. Wójcicka-Milewska, *Zabezpieczenie i monitoring osuwisk powstałych na zboczu północnym wyrobiska odkrywkowego BOT KWB TURÓW SA., [Protection and monitoring of landslides formed on the northern slope of the BOT KWB TURÓW SA open pit work]*, Górnictwo i Geoinżynieria **32 (2)** 247–256 (2008)
6. J. Maciaszek, P. Ćwiąkała, *Badania możliwości zastosowania skanowania laserowego do monitoringu osuwisk zboczy wyrobisk odkrywkowych na przykładzie KWB „Bełchatów" [Research on the possibilities of laser scanning method using for landslides monitoring on slopes in open pit mine, on the example of KWB "Bełchatów"]* Przegląd Górniczy **66 (6)** 52–57 (2010)
7. https://www.nj24.pl/article/gigantyczne-osuwisko-w-kopalni-turow
8. A. Mirek, L. Biały, *Stan zagrożeń naturalnych w kopalniach węgla brunatnego ze szczególnym uwzględnieniem zagrożeń występujących w KWB „Bełchatów" [The state of natural hazards in brown coal mines, with particular emphasis on hazards occurring in KWB "Bełchatów"]*, Górnictwo i Geoinżynieria **33 (2)** 327-341 (2009)
9. H. Bryś, S. Przewłocki, *Geodezyjne metody pomiarów przemieszczeń budowli [Geodetic methods of measurements of building displacements])*, Wydawnictwo Naukowe PWN SA, Warszawa (1998)
10. M. A., R. Cooper, *Control Surveys in Civil Engineering*, William Collins Sons & Co. Ltd., London (1987)

11. W. Welsch, O. Heunecke, H. Kuhlmann, *Auswertung geodätischer Überwachungsmessun-gen [Evaluation of geodetic monitoring measurements],* In: Möser et al. (edts.): Handbuch Ingenieurgeodäsie. Wichmann Verlag, Heidelberg, (2000)
12. N. Pfeifer, C. Briese, *Laser Scanning - Principles and Applications,* Austria (2007)
13. http://www.riegl.co.at/media-events/newsletter-archiv/0512-rimonitor-and-rimining/
14. J. Gance, J.P. Malet, T. Dewez, J. Travelletti, Eng. Geol. 172, 26–40, (2014) doi: 10.1016/j.enggeo.2014.01.003J.
15. Travelletti, J.P. Malet, Eng. Geo. **128**, 30–48, (2012) doi: 10.1016/j.enggeo.2011.05.003
16. E. Pilecka, M. Bazarnik, *Application of terrestrial laser scanner for monitoring the railway infrastructure threatened by landslides*, Monographie – recent Advances in civil Enginearing: Structural Mechanics edit. Stypuła K., 171–190, (2015) Wydawnictwo PK, ISSN 0860-097X
17. E. Pilecka, D. Szwarkowski, E3S Web Conf. **24**, 01006, (2017) doi : 10.1051/e3sconf/20172401006
18. M. Bazarnik, Bulletin of the Mineral and Energy Economy Research Institute of the Polish Academy of Sciences **94**, 149–160 (2016)
19. E. Pilecka, D. Szwarkowski, Budownictwo Górnicze i Tunelowe **4**, 21 (2015) (in Polish)
20. Z. Muszyński, W. Milczarek, *Modelowanie przemieszczeń pionowych na podstawie danych z naziemnego skanera laserowego [Modeling of vertical displacements based on data from terrestrial laser scanner]*, Geoinformacja jako metoda ochrony przed geozagrożeniami, Warszawa, 42–63 (2014)
21. V. Kajzar, R. Kukutsch, N. Heroldová, Acta Geodyn. Geomater. **12/1** (177), 51–58, (2015) doi: 10.13168/AGG.2015.0004
22. D. Conforti, *Using Static and Mobile Laser Scanners to Measure and Manage Open Pit Mines,* Australian Institute of Mine Surveyors Conference 11 -13 August 2010 *http://www.minesurveyors.com.au/files/members/2010Convention/Papers/Dario%20Conforti%20-%20Optech/Da rioConforti-FinalPaper.pdf*
23. M. Radzilani, *3D Laser Mapping,* Position IT, EE Publisher (2015) http://www.ee.co.za/article/slope-monitoring-solution-open-pit-mines.html
24. M. J. Little, *Slope monitoring strategy at PPRust open pit operation,* The South African Institute of Mining and Metallurgy International Symposium on Stability of Rock Slopes in Open Pit Mining and Civil Engineering, 211–230, (2006)
25. G. Vassena, A. Clerici, *Open pit mine 3D mapping by TLS and digital photogrammetry: 3D model update thanks to a slam based approach,* The International Archives of the Photogrammetry, Remote Sensing and Spatial Information Sciences**XLII-2**, 1145–1148 (2018) doi : 10.5194/isprs-archives-XLII-2-1145-2018
26. J. Wajs, Mining Science **22**, 75–83, (2015) doi :10.5277/msc152207

# The comparative analysis of the seismic test results performed using the cross-hole and down-hole methods in terms of test reliability and efficiency

*Wojciech* Tschuschke[1], *Sławomir* Gogolik[1,*], *Maciej* Kroll[1] and *Krzysztof* Janicki[2]

[1]Institute of Construction and Geoengineering, Poznań University of Life Sciences, Poland
[2]KGHM Polska Miedź („Polish Copper") SA, Division Hydrotechnical Department, Poland

**Abstract.** Seismic tests are becoming more and more meaningful in modern geoengineering since they allow precise assessment of changes in the analyzed profile of soil deformation parameters at very small strain. The engineering practice generally uses two methods for such kinds of tests: reference cross-hole test and commercial down-hole method. The costs of implementation of this kind of research and the reliability of the results obtained along with any additional benefits from the type of the selected testing technique are not without significance. Seismic measurements for which comparative analysis was performed were carried out in a strongly anisotropic materials base of a tailings. The results of tests conducted with a standard cross-hole method were compared with those conducted using the down-hole method, in which the seismic piezocone (SCPTU) and seismic dilatometer (SDMT) were used. Test results have shown that in terms of assessment of their credibility, there were no fundamental differences in the quality of the recorded signals. It has been pointed out, however, that there are many advantages to penetration in situ tests, which complement identification of subsoil structure with many crucial elements that cannot be achieved using the cross-hole method.

**Keywords:** CHT, SCPTU, SDMT, comparative analysis, tailings

## 1 Introduction

The non-destructive seismic testing is one of the types of in situ geophysical tests used. In general, it relies on registration of the velocity of a particular type of seismic wave passing through the subsoil. Seismic test methods distinguish surface test techniques, which implement the phenomenon of seismic refraction (SR), as well as the in-hole testing techniques conducted between cross-holes (CHT) or down-holes (DHT) [1, 2]. Seismic measurements may be conducted as stand-alone test or in combination with another kind of in situ tests as the so-called combined test (invasive and non-destructive). According to the

* Corresponding author: s.gogolik@wp.pl

elasticity theory, the seismic waves velocity can be used to determine the deformation modulus of the subsoil within the range of very small strain. The results of seismic surveys can be also used to analyze the distribution of the initial void ratio in issues related to the assessment of the liquefaction potential of the soil, as well as the identification of full saturation zones in the subsoil [2, 3]. Due to the fact that the seismic signal in the subsoil propagates very quickly, which causes the measurement period to be very short, the precision of the measurement of the in situ seismic signal and the accuracy of the accepted interpretation method, i.e. time interval, are extremely important for the reliability of the test results. In the above context, the choice of an appropriate testing method may result in more or less precise identification of the subsoil parameters, which are determined based on the measurements of the seismic wave velocity. The primary aim of the study presented in the paper is comparison in terms of quality of seismic wave velocity test results.

## 2 Investigation site

Specialist seismic tests with the tested cross-hole (CHT) and down-hole (DHT) methods were carried out in the tailings massif deposited within the Mining Refuse Neutralisation Facility (MRNF) Żelazny Most. This facility has been operating for over 40 years and is used to accumulate tailings from processing copper ore while serving a key role in the management of process water and water from mine drainage. The scale of the facility, due to the amount of the accumulated waste, classifies it into the group of the world's largest hydraulic facilities (Fig. 1).

**Fig. 1.** Aerial view of the Żelazny Most depository, (https://polska-org.pl/783674,foto.html).

The storage facility area amounting to approximately 1400 ha is closed by the surrounding dams of a total length of about 14.3 km. The pipelines distributing tailings with outlets which spigot the tailings that form the beach are located on the crown of the dams. The central part of the storage facility is the pond formed from the drained spigotted tailings. After over 40 years of the storage facility operation, as a result of progressive accumulation, it built up approximately 630 million $m^3$ of tailings with continuously maintained overlying water of volume ranging between 5 million $m^3$ and 7 million $m^3$.

The waste material deposited in the facility, as a result of the process of spigotting accompanied by the tailings segregation process, can be distinguished by spatial variability of grain size and physical properties. The coarsest sandy material depositing on the beaches in the area where it is discarded is used to superstructure dams as filling of the bowl of the facility with tailings is taking place. Finer material flows along with the process water into the beach forming a number of silty laminations to dominate the construction profile of tailings in the pond area. The way the facility is superstructured, by the 'upstream' method, causes constantly higher dam embankments to be formed on the deposited material of increasingly inferior physical properties and lower bearing capacity [4, 5].

# 3 Methodology

The researches of the seismic wave propagation velocity in the tailings massif were carried out using three measurement methods: cross-hole seismic testing (CHT) and cone penetration testing SCPTU, seismic dilatometer testing SDMT, classified as the down-hole methods (DHT). The CHT used an alternative method according to the standard ASTM 4428 with two boreholes (Fig. 2). The boreholes were made by drilling in a bentonite slurry and secured with a PVC casing pipe. The quality of the prepared boreholes have investigated by sonic logging method. This technique consist of the measurements of the velocity and propagation modes of the elastic waves in the materials surrounding the boreholes, with a penetration of few tens of centimetre around the hole.

In order to identify the geological-engineering conditions of the tailings massif, specialized seismic tests were performed within 18 research nodes. The nodes were positioned on the line of 6 cross-sections located within three dams: northern, eastern and western in the edge zone of the beach and of the embankments in the area of active impact of the circumferential beach drainage. Within each node a grid of research points of a projected rectangle was positioned by locating two drilled holes piped on the diagonal of the grid (CHT) and by setting in the other two corners of the grid space research for the down-hole method using the seismic cone penetration testing (SCPTU) and seismic dilatometer testing (SDMT) (Fig. 2). Distances between particular test points within the node did not exceed 8 m (Fig. 3). The research was carried out at the bottom layer of the natural subsoil, which supports the tailings massif occurring at a depth of 40 m to 60 m depending on the location.

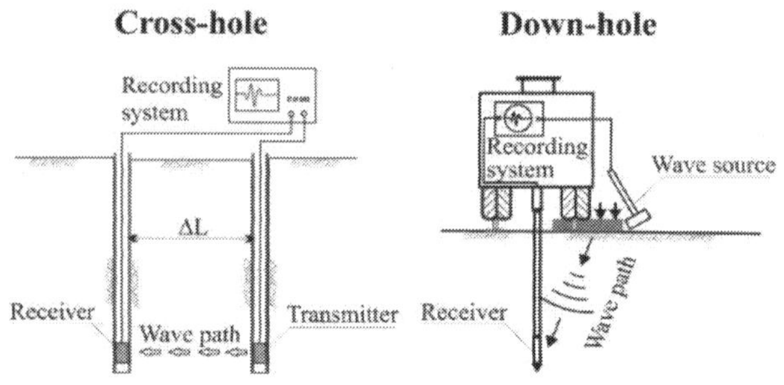


**Fig. 2.** Scheme of cross-hole and down-hole geophysical tests.

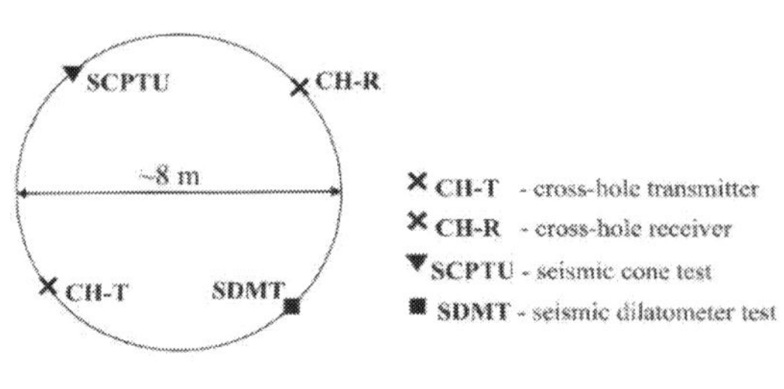


**Fig. 3.** Location of research points within the research node.

In the main studies, the source of the shear wave was the GEOS electrodynamic energizer. enabling to obtain vertical polarity waves ($S_{hv}$) and horizontal polarity waves ($S_{hh}$), whereas to generation the compression wave (P) the Sparker source was put into practice. Various types of converters suitable for recording the dominant wave type were used as receivers.

Wave velocity measurements were made at intervals of 1 m of the borehole depth increase. Each depth recorded 10 signals of the compression wave and 20 impulses of the vertically polarized shear wave signala (10 for positive and negative polarization). Signal acquisition was managed using the Win_CrossHole software developed by the SOLGEO company [6]. The system allowed the visualization and storage of all signals. In addition to estimating the period of real-time wave propagation, the system enabled averaging or stacking of the collected trace to obtain an signal quality. For precise determination of the velocity of the waves it was necessary to determine the exact distance between the source and the receiver. Such information was obtained by inclinometric measurements. The applied INCLIS-DH probe equipped with a three-axis magnetometer and a set of three accelerometers allowed to determine spatial deviations from the vertical axis of boreholes with depth. The propagation time of the compression wave from the source to the receiver determined the moment when the wave front arrived (first peak), whereas the shear wave crossing time was determined by the analysis of waves of the opposite polarity (Fig. 4).

The SCPTU tests employed the seismic cone, which in addition to the measurement of basic penetration parameters (cone resistance - $q_c$, friction sleeve - $f_s$ and pore water pressure - $u_c$), enable to determine the compression wave velocity ($V_p$) and shear wave velocity in the subsoil ($V_s$) [7, 8]. The cone used in the tests was equipped with two receivers of shear and compression wave, 0.5 m away. The wave source was a steel beam perpendicular to the axis of the penetrometer. In order to ensure good contact with the subsoil, the beam was pressed into the ground with the penetrometer leveling cylinders.

**Fig. 4.** Measurement system of CHT [6].

The SCPTU test was based on the pressing the piezocone into the subsoil at a constant speed and recording main penetration characteristics. At intervals of 1 m, the penetration was stopped and the program module responsible for the seismic measurements was started. Hammer striking the beam caused the electric circuit to be closed and the wave to be registered allowing the determination of the time of wave propagation from the source to the geophone (Fig. 5).

**Fig. 5.** Penetrometer and seismic piezocone.

At the given depth, two measurements of the shear wave propagation time by horizontal impact to the left and right end of the beam and one measurement of the compression wave generated by the vertical hammer blow to the beam were performed. After confirming the obtained results, the seismic piezocone was pressed into the next set depth and the previously described procedure was repeated. The wave velocity in a given layer was determined based on the difference in the time of arrival of the wave and the distance traveled by the wave between adjacent depths of the conducted seismic tests (equation 1).

$$V = (Z_2 - Z_1)/\Delta T \quad (1)$$

where: $\Delta T$ – travel time, $Z_2$-$Z_1$ – distance of wave propagation

The time interval was determined by the cross-correlation method (Fig. 6).

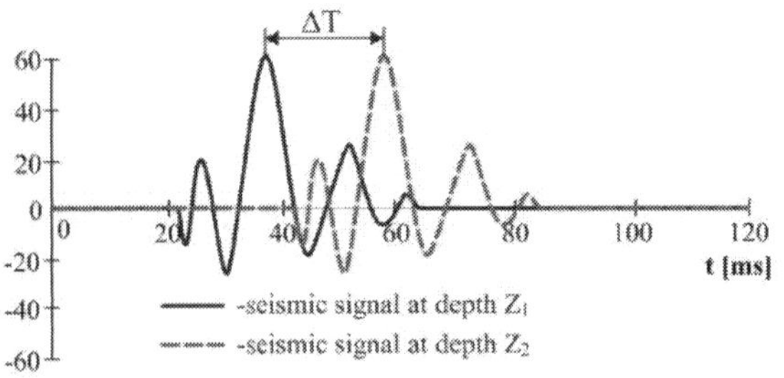


**Fig. 6.** Scheme of time interval determining, (t-time).

The SDMT test implemented the Marchetti seismic dilatometer which, in addition to the standard A and B pressure measurements, made it possible to determine the shear wave velocity ($V_s$) [9, 10]. The device used two geophones placed at a distance of 0.5 m from each other, whereas the wave was generated using a pendulum hammer. The seismic tests were performed at 1 m interval and at every depth 3 to 5 repetitions of the wave velocity measurement were recorded (Fig. 7).

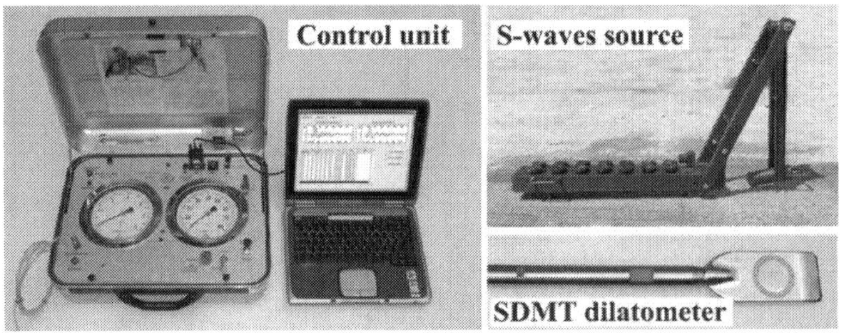


**Fig. 7.** SDMT measurement system [9].

## 4 Results and analysis

Regarding the fundamental questions of the reliability analysis of the seismic tests conducted with competing methods, the wave velocity distributions in the same test profiles were compared with depth. In the replication test, the cross-hole testing is limited only to the registration of seismic signals enabling determination of the shear and compression wave velocity (Fig. 8).

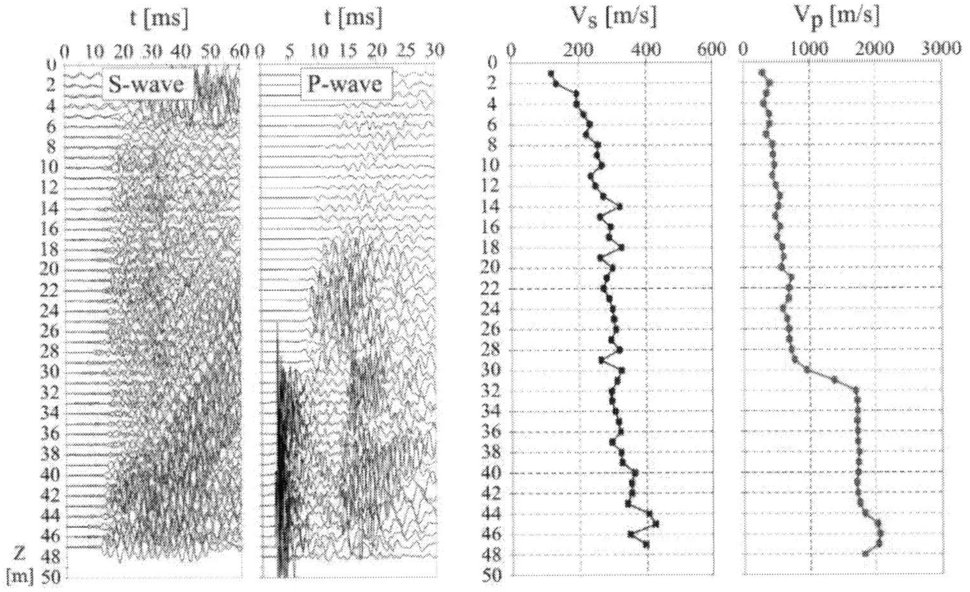


**Fig. 8.** Example of CHT results [6], (Z - depth).

The quality of the test results is controlled based on the analysis of the repeatability of the seismic signal expressed in a statistical measure of the variability of this survey. The ability to repeat the survey and select signals of the guaranteed quality minimizes measurement uncertainties associated with determining a reliable seismic wave velocity. The seismic measurement in the SDMT is also based on a replication test, though limited to the possibility of measuring only the shear wave velocity (Fig. 9).

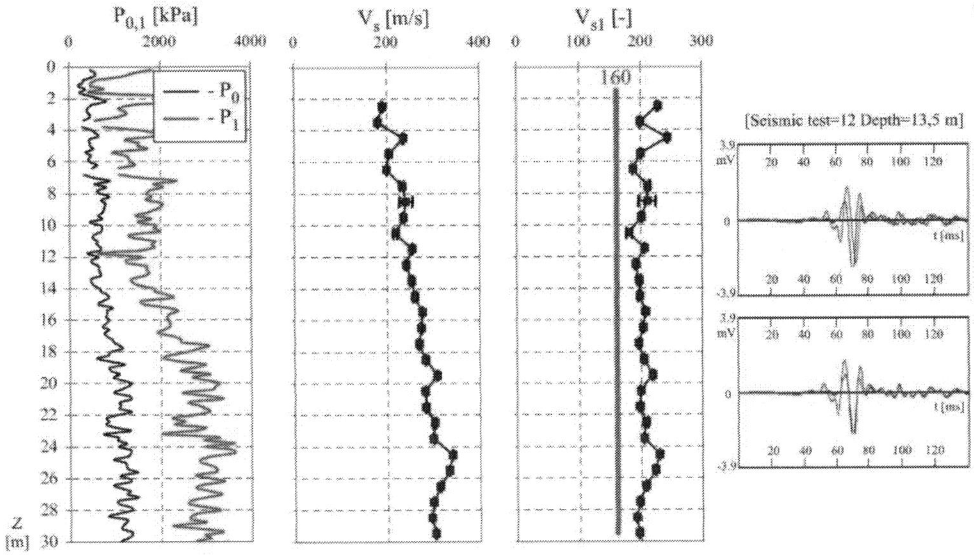


**Fig. 9.** Example of SDMT results [11], ($P_0$, $P_1$-characteristics from DMT test).

The SCPTU allows to establish the velocity of two waves, shear and compression, on the basis of single measurement of the seismic signals and thus without the possibility of the statistical analysis (Fig. 10).

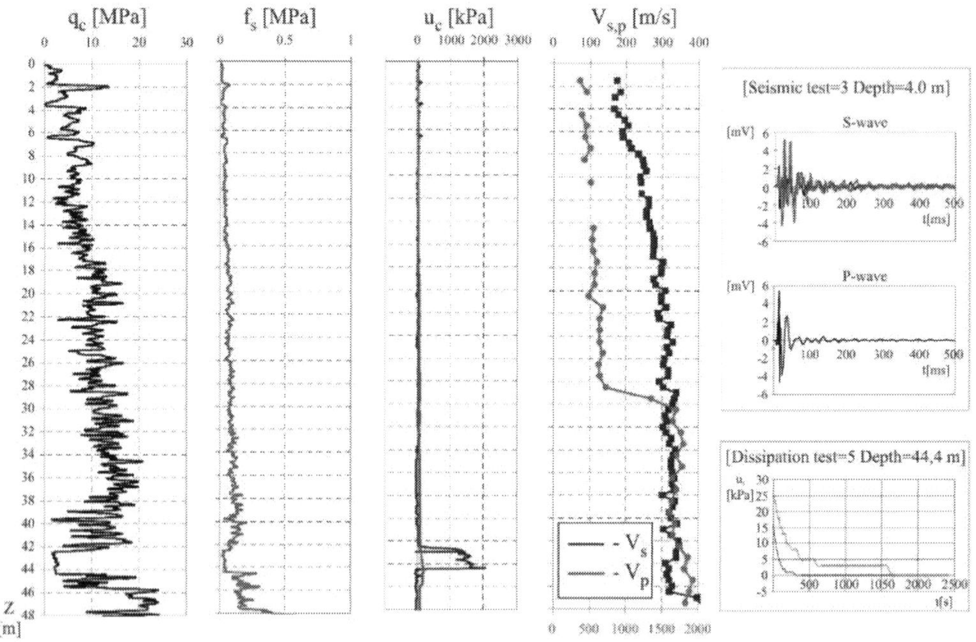


**Fig. 10.** Example of SCPTU results [11].

The results of the comparative analysis indicate that the velocity distributions of the seismic waves with the profile depth determined on the basis of different methods are strongly convergent (Fig. 11). This is confirmed by the results of the statistical analysis documenting the lack of statistical difference between the velocity distributions of the seismic waves determined by three independent methods. The biggest fluctuations in the velocity distribution of the shear wave with the profile depth with the repeatability of the seismic signal measurement expressed by the variation coefficient with values that do not exceed 1.5% are characteristic to the CHT. More even and overlapping distributions of the shear wave velocity with depth, with the quality of the seismic signal recorded with a dilatometer of an average variability coefficient of 2.0%, were obtained from the cone penetration tests SCPTU and SDMT. An important factor affecting the results of the comparative analysis is the profile of the subsoil in which the test is conducted. In the case of a highly layered subsoil, which is constituted by deposited tailings, the penetration test (DHT), in which the propagation of seismic waves is measured vertically, generates a result that averages the velocity of the seismic wave passing through various layers of the soil on a small section limited by the location of geophones (usually 0,5 m). Conversely, in the CHT, where the propagation of seismic waves is measured horizontally, the average measurement of the seismic wave velocity is determined by the local heterogeneity of the soil in a much larger space included in the conducted measurement. The above-mentioned impact of the subsoil heterogeneity on the seismic test methodology can be explained by local differences in the evaluation of the seismic wave velocities determined by the CHT and DHT.

Regarding the second question, the effectiveness of the test, it should be emphasized that in the CHT, the seismic test is a basic measurement while the DHT rather uses seismic tests as an additional testing method. In addition, the CHT requires much more financial and time expenditure compared to the DHT penetration testing.

According to the principles of the theory of elasticity, the results of the seismic wave velocity measurements can be used to estimate the initial shear and constrained modulus (equation 2 and 3):

- initial shear modulus

$$G_0 = \rho (V_s)^2 \quad (2)$$

- initial constrained modulus

$$M_0 = \rho (V_p)^2 \quad (3)$$

where: $\rho$ - soil density, $V_s$ - velocity of shear wave, $V_p$ - velocity of compression wave.

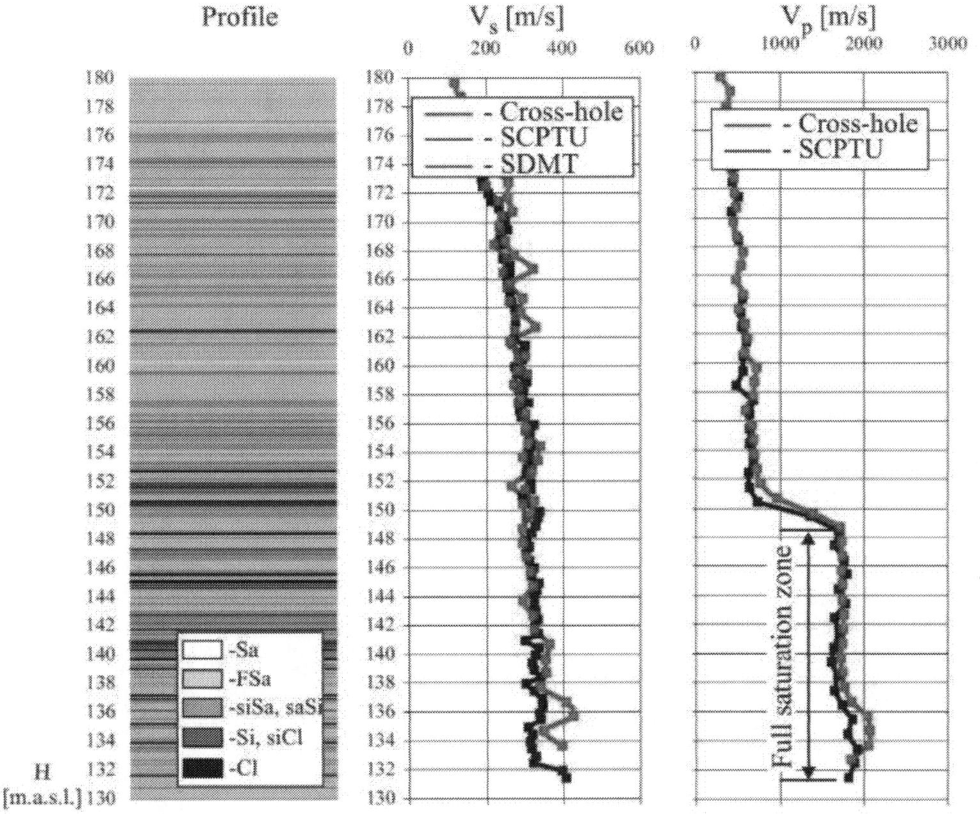


**Fig. 11.** Comparative analysis of tested methods [11].

Sample distributions of shear and constrained modulus calculated on the basis of the seismic wave velocities determined by the SCPTU are presented in Figure 12.

The soil density required to estimate the modulus values, which requires additional and independent testing in the case of the CHT. In competing DHT, the conventional classification systems based on penetration parameters (Fig. 9, Fig. 10), which are recorded in the standard CPTU and DMT, are used to determine the density of soils. On the basis of the same penetration parameters, using the original interpretation procedures and classification systems, continuous penetration characteristics of the subsoil are obtained in penetration studies including identification of physical, state, strength and deformation parameters, and, in the case of the CPTU testing, additional subsoil filtration parameters. The latter are determined on the basis of dissipation tests of excess pore pressure, the test carried out at the moment when cone penetration stops and when seismic measurements can be simultaneously carried out.

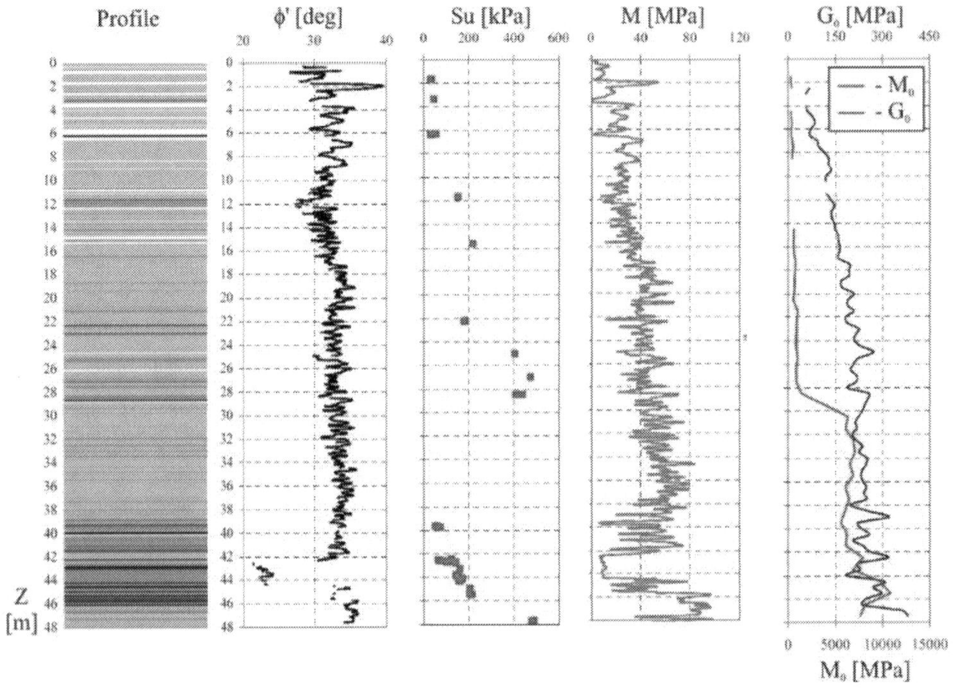


**Fig. 12.** Example of distribution of mechanical parameters from SCPTU, (ϕ' - effective friction angle, $s_u$ - undrained shear strength, M - constrained modulus).

Regardless of the implemented method of seismic tests, their result interpretation, tailings in particular, allows obtaining crucial information about the conditions in which the material was deposited. The first information is applied to resolve the scope of the full saturation zone of tailings, usually a zone of the least favorable strength characteristic. The full saturation zone is determined by the compression wave velocity of over 1550 m/s (Fig. 11). The second important information is the identification of potential static liquefaction zones in the tailings massif. For this purpose, the distribution of the normalized shear wave with depth (equation 4) is analyzed and the potential hazard zone is identified by the normalized wave velocity less than 160 m/s (Fig. 9).

$$V_{s1} = V_s * (\sigma'_v / Pa)^{-0,25} \tag{4}$$

where: $V_s$ – shear wave velocity, $\sigma'_v$ – effective vertical stress, $Pa$ – atmospheric pressure.

## 5 Summary

The calibration tests carried out using three methodologically different geophysical methods of seismic surveys and the analysis of their results showed that the statistical validity of the results of the tested methods is comparable. The method of choice, regardless of economic issues, should take into account preferences as well as limitations resulting from the type indicated to conduct the research. In the above context, the competitive methods are distinguished by the following characteristics presented in table 1.

**Table 1.** The advantages and limitations of CHT, SDMT and SCPTU.

| Method | Advantages | Limitations |
|---|---|---|
| CHT | - full statistical assessment of measurements quality,<br>- significant quality of preparatory works | - significantly more expensive tests as compared to the down-hole method |
| SDMT | - possibility of statistical assessment of measurements quality,<br>- characteristics from the penetration test ($p_0$, $p_1$),<br>- identification of soil deformation and shear strength parameters, | - no possibility to measu compression wave velocity ($V_p$) |
| SCPTU (preferred method) | - possibility to determine velocity of 2 waves ($V_s$, $V_p$),<br>- additional penetration characteristics ($q_c$, $f_s$, $u_c$),<br>- full stratigraphic and parametric identification of subsoil,<br>- possibility to assess ground filtration capabilities,<br>- possibility to create correlations between penetration parameters -$V_s$, $V_p = f(q_c, f_s, u_c)$ | - no possibility of statistical analysis of recorded signals |

## References

1. J. S. Lee, *Geo–characterization using waves – Principle to application*, Proc. of the 19th Int. Conf. on Soil Mechanics and Geotechnical Engineering, 245–264 (2017)
2. M. Jamiolkowski, *Role of geophysical testing in geotechnical site characterization*, Proc. of the 3rd De Mello Lecture, Portuguese-Brazilian Geotechnical Congress, Soils and Rocks **35**, 2, 117–140 (2012)
3. H.L. Giacheti, R.A. Rodrigues, *The seismic DMT in an unsaturated tropical soil, Proc. of the 19th Int. Conf. on Soil Mechanics and Geotechnical Engineering* **1**, 599–602 (2018)
4. M. Jamiolkowski, *Soil mechanics and the observational method: challenges at the Zelazny Most copper tailings disposal facility, Geotechnique **64**, 8, 590–619 (2014)*
5. W. Tschuschke, *Sondowania statyczne w odpadach poflotacyjnych (Cone penetration testing in mine tailings)*, Zeszyty Naukowe. Budownictwo, Politechnika Śląska, Gliwice (2006) (in Polish)
6. D. Milani, S. Priano, *Zelazny Most Poland – Cross–Hole Tests*, Report S16SN11–2. SolGeo Geophysical Solutions, Seriate, Italy (2017)
7. P. Mayne, Mc. Gillivray, *Seismic piezocone and seismic flat dilatometer at Treporti*, Proc. of the 2nd Int. Conf. on Site Characterization **2**, 1695–1700 (2002)
8. W. Tschuschke, G. Kmiecik, A. Mikos, *$G_0$ profiling in mine tailings based on SCPTU and SDMT*, Proc. of the 4th Int. Conf. on Geotechnical and Geophysical Site Characterization **2**, 1701–1706 (2013)
9. S. Marchetti, P. Monaco, G. Totani, D. Marchetti, *In situ tests by seismic dilatometer (SDMT)*, Proc. from Research to Practice in Geotechnical Engineering, ASCE Geotech. Spec. Publ. **180**, 292–311 (2008)
10. S. Gogolik, M. Waliński, *Applicability of SDMT test in evaluating the modulus $G_0$ in tailings*, Studia Geotechnica et Mechanica **XXIX**, 1–2, 39–46 (2007)
11. W. Tschuschke, S. Gogolik, *Interpretation of Results of SCPTU and SDMT at the Mining Refuse Neutralization Facility Zelazny Most*, Report 0281–2015/5, Poznan, Studio Geotechnika, Poland (2017)

# Microgravimetric and GPR surveys for detection of unconsolidated zones in a levee

*Monika* Łój[1,*], *Sławomir* Porzucek[1], *Tomisław* Gołębiowski[2] and *Mark E.* Everett[3]

[1]AGH University of Science and Technology, Cracow, Poland
[2]Cracow University of Technology, Cracow, Poland
[3]Texas A&M University, College Station, USA

**Abstract.** Selected results of complex geophysical surveys carried out on the Vistula river flood levee in Cracow are herein presented. Two complementary geophysical methods were applied for detection of potential unconsolidated zones in the body of the levee, i.e. microgravimetry and ground-penetrating radar (GPR). The surveys were carried out in 2D mode, along a profile at the crown of the flood levee. Microgravimetric data reveal anomalies showing zones of decreased bulk density. These zones provide information about poor quality of the levee. The main anomaly was interpreted in a quantitative manner using gravity modelling. Non-standard GPR processing and visualization of radargrams were employed to better extract information concerning the distribution of unconsolidated zones. High resolution GPR surveys allow to outline such zones which was the basis for construction of the 2D model used in the gravimetric modelling. Integration of these two geophysical methods provided important information about the spatial variations of mass density in several unconsolidated zones within the body of the flood levee.

**Keywords:** microgravimetry, GPR, levee, gravity modelling, unconsolidated zones

## 1 Introduction

Levees are critical hydraulic structures in terms of protection of human life and security of real property. For this reason, it is important to keep them in the best possible geotechnical condition. The performance integrity of levees is usually determined by standard geotechnical surveys. However, such surveys provide only point wise information about the condition of levees, therefore it is essential to apply methods which can reveal the geotechnical condition in a spatially continuous manner. Geophysical techniques such as electrical resistivity tomography (ERT), induced polarization (IP), electromagnetic induction (EM) [2, 4, 10] and ground-penetrating radar (GPR) [3, 5, 6, 11], are frequently used for this purpose. In addition, MASW method [12] and microgravimetry are increasingly being used to recognize the geotechnical condition of flood levee. Microgravimetry delivers information

* Corresponding author: mloj@agh.edu.pl

about the bulk density distribution within a study area [7]. In the case of a levee, locations of reduced mass density can be identified. Additionally, this method allows to define the near-surface geology and potentially find water seepage zones. In this paper we have applied microgravity together with the GPR method for assessment of geotechnical condition of selected flood levee.

## 2 Geophysical surveys

The geophysical surveys were carried out on the Vistula flood levee, between boreholes L89+782 and L90+090, as shown in Fig. 1. In 2014, this section of levee was designated for renovation due to its poor geotechnical condition. Our measurements were made along a 350 m long profile, located on the crown of the levee-crest axis (Fig. 1).

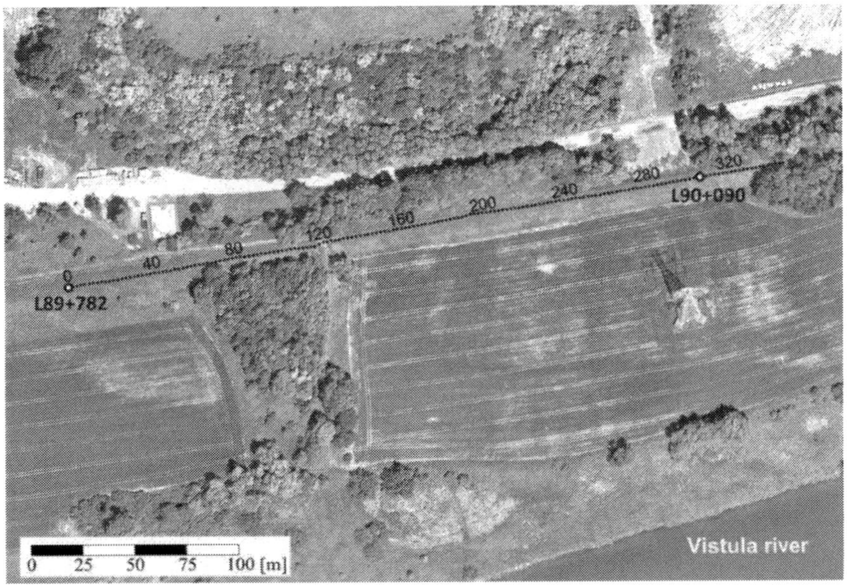


**Fig. 1.** Location of the profile for geophysical surveys.

In terms of geology, the study area is located within the Carpathian foredeep, which is filled with Miocene sediments. Within these sediments, the Vistula River has channelized its bed. The Vistula flood levees were constructed beginning at the end of the 19th century. These are the open levees which protect the surrounding private and industrial land and buildings against flood waters of the Vistula River.

Local raw materials i.e. sands, clays and gravels, as well as anthropogenic materials (ash, slags) were usually used during the original construction of the levee. These materials were not well sorted which means that there could be many heterogeneities within the levee structure.

As mentioned earlier, the geophysical profile passes between the two geotechnical boreholes L89+782 and L90+090 (Fig. 1). Geotechnical testing produced cross-sections (Fig. 2) which verify that the levee is made of various materials. In borehole L89+782, under a thin layer of fine sand with humus, there exists silt with cobble to depth 1.5 m. Below them is clay to depth 2.9 m. In the cross-section (Fig. 2, left), it appears that this part of the levee and its basement are composed of horizontal layers. In the other borehole L90+090 (Fig. 2 - right), mainly silty clay to depth 3.8 m is found along with compact silty clay deeper to a depth 5.0 m. In this cross-section (Fig. 2, right), it appears that the central part of the levee lies in a basin - the levee either fills a natural depression or this part of the levee subsided.

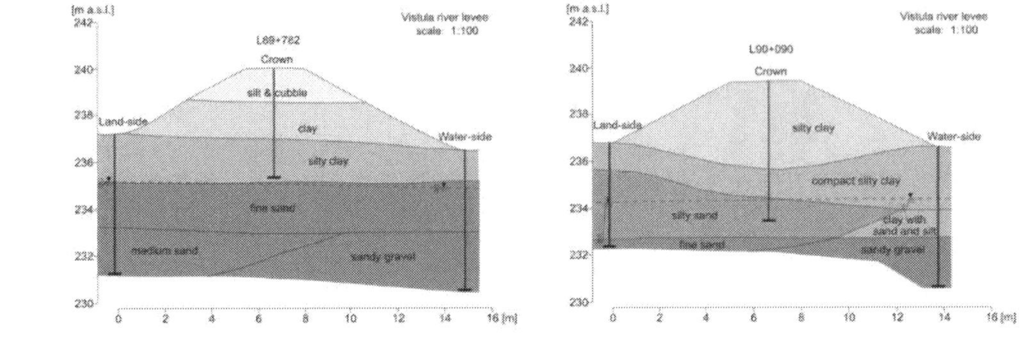


**Fig. 2.** Geotechnical cross-sections across boreholes: L89+782 (left) and L90+090 (right).

## 2.1 Microgravimetric surveys

Microgravimetric measurements were carried out at station spacing 2 m using the Scintrex CG-5 high-grade quartz gravimeter, which has automatic recording capabilities. The microgravimetric observations were repeated at many stations to calculate the mean square error of measurements, which amounted to ±0.004 mGal. This value is in correspondence with the technical specifications of the gravimeter and confirmed that our observations were carried out to high precision. The data were used to calculate the Bouguer anomaly at each of the measurement stations, based on a presumed average bulk density 2 g·cm$^{-3}$. In view of small changes in the station elevation along the profile, any error in the presumed density does not have a significant impact on the Bouguer anomaly.

The resulting Bouguer anomaly profile (Fig. 3) contains substantial small-scale variability. To try to reduce this variability, despite the small elevation changes from station to station, a terrain correction was calculated. The terrain-corrected Bouguer anomaly retains however much of the small-scale variability.

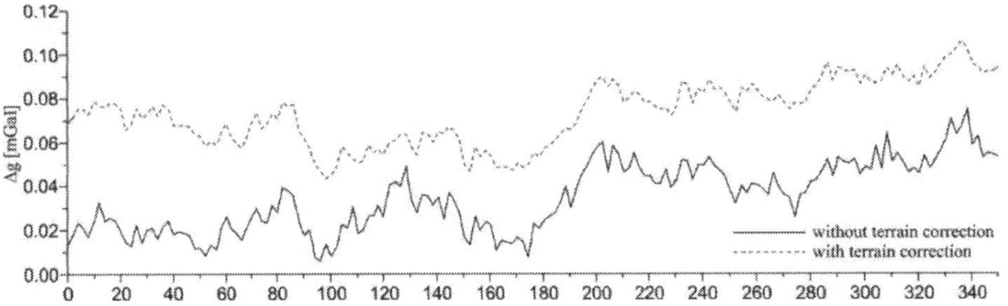


**Fig. 3.** Bouguer anomaly before and after terrain correction.

On this basis it could be assumed that the remaining small-scale variability comes from measurement errors and/or near-surface density changes within the uppermost layer, 1 m deep. In order to further reduce the variability, a low-pass Butterworth filter in the wavenumber domain with a cut-off wavelength of 20 m was applied. Comparison of the results before and after filter application confirmed that the cut-off parameter was chosen appropriately (Fig. 4).

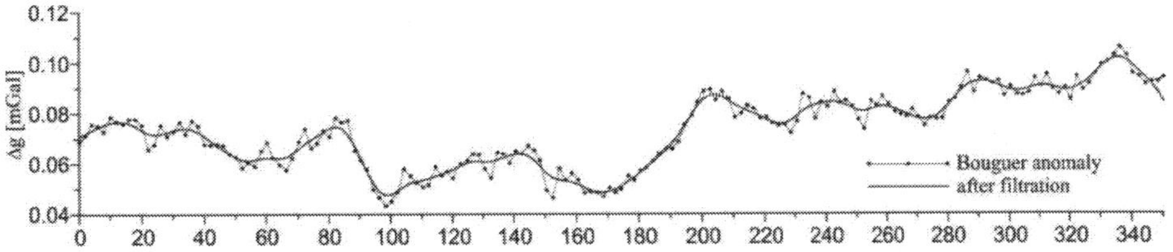


**Fig. 4.** Low-pass filter applied to Bouguer anomaly.

The filtered Bouguer anomaly shows a regional trend, reflected in the increase of the anomaly value in the direction of the profile. This trend was eliminated by subtracting a second degree polynomial approximation (Fig. 5).

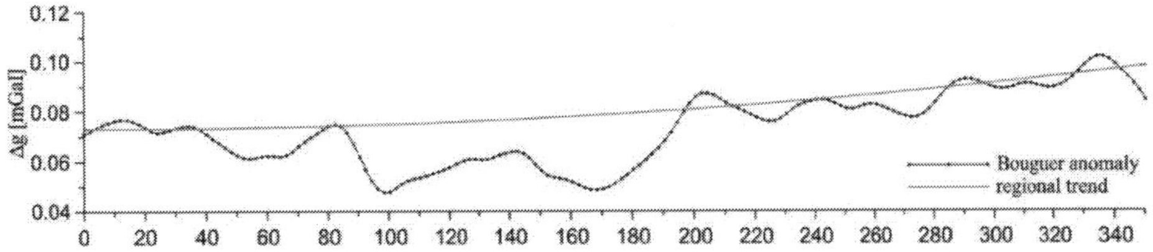


**Fig. 5.** Bouguer anomaly with the regional trend.

The residual anomaly, after the trend removal, is shown in Fig. 6. There are two significant negative residual anomalies; these are marked with the "A" and "B" symbols. Anomaly "B" is located between 90 and 190 m and has small amplitude ~0.03 mGal. The shape of the anomaly shows that it is generated by an laterally extensive, shallow-buried zone of relatively low bulk density. It can also be seen that anomaly "B" is bimodal, with a central maximum flanking two minima, which indicates a denser area in the centre of the anomaly. The above-mentioned area is at the point of a local road crossing the levee.

A smaller region of decreased density located within the profile section between 45 and 75 m (anomaly "A") is also observed. The amplitude of this anomaly is ~0.015 mGal. Other smaller residual anomalies with amplitudes below 0.01 mGal correspond to small local changes in a bulk density. As shown below, some of these are partially visible in the GPR sections.

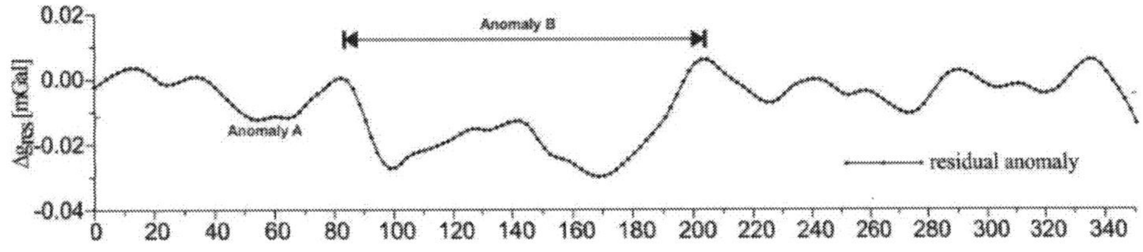


**Fig. 6.** Residual anomaly after trend removal.

## 2.2 GPR surveys

The GPR surveys were primarily designed for site reconnaissance and accordingly were carried out in a fast 2D mode, specifically as short-offset reflection profiling. The two-channel Swedish-made ProEx georadar (www.guidelinegeo.se) was used. The system enables the user to conduct simultaneous measurements using two different antennae. Taking into account the overall height of the levee and the necessity of resolving small-scale unconsolidated zones, a 500 MHz antenna (Table 1) was connected to the first channel of GPR system. However, the body of the levee is composed of clay and silt, such that high attenuation and consequently low depth penetration is expected at this frequency Thus, an antenna with lower frequency 250 MHz (Table 1) was connected to the second channel. Both antennae are bistatic and shielded and the standard orientation of Tx and Rx dipoles (i.e. co-pole orientation) was adopted during the measurements. Traces were recorded along the profile at spacing Δx=0.02 m (for the 500 MHz antenna) and Δx=0.05 m (for the 250 MHz antenna). During the GPR surveys the stacking number varied between 4x and 32x in order to improve the signal/noise ratio.

The radargrams presented in the paper are shown in normalized form, i.e. with amplitudes (energies) normalized to the maximum amplitude (energy) of the direct air wave. For time-depth conversion a constant mean velocity 0.09 m/ns is assumed; such velocity is appropriate for the mixture of silt and clay which comprises the body of the levee [1, 8, 9].

In the first stage of data processing, radargrams were digitally processed with the use of the ReflexW program (www.sandeiergeo.de). The following standard procedures [12] were applied: time-zero correction, de-wow, DC correction, Butterworth filtering, 1D median filtering, application of gain function, background removal, stacking, 2D average procedure. The processed radargrams for 250 MHz and 500 MHz antennae, for different stacking numbers, are shown in Figs. 7-10.

**Table 1.** Penetration/resolution parameters of applied antennae [6].

| Central frequency | Max. depth penetration | Mean resolution |
|---|---|---|
| 250 MHz | 10m | 0.1m |
| 500 MHz | 5m | 0.05m |

In Figs. 7-10, the reduction of the depth penetration of the electromagnetic wave in comparison with the maximum penetration (Table 1) of a given antenna, is caused by the high attenuation of clay (Fig. 2). In such situations, only those unconsolidated zones that are located near the top of the levee might be detected; thus a joint interpretation of GPR and microgravimetric data is possible only for this near-surface region.

Along the entire length of the profile (Figs. 7-10), a boundary at depth ~1.0 m was recorded by the 250 MHz and 500 MHz antennae; this anomaly may be interpreted as the boundary between silt and clay (Fig. 2) or as an internal boundary in silt. In the near-surface zone (i.e. in silt), two main laterally disrupted regions may be distinguished in Figs. 7-10, i.e.: between x=0 m and x=125 m, and from x=210 m to x=225 m. It is interesting also to note that there are sometimes large differences in radargrams recorded for different stacking numbers, e.g.:

- in Fig. 7B, a continuous, horizontal and thin group of signals between x=137 m and x=196 m is visible but in Fig. 8B this group of signals is wide and interrupted,
- in Fig. 7C, a high-amplitude wide anomaly between x=247 m and x=283 m is recorded but in Fig. 8C, a thin anomaly is visible in this place,
- in Fig. 9B it is difficult to depict an anomaly between x=210 m and x=225 m but in Fig. 10B anomaly at this location is easily distinguished,
- in Figs. 7B and 8B reflections from trees were recorded, between x=168 m and x=175 m and x=312 m and x=322 m, but in Fig. 10C these reflections are not visible.

There are many more such examples in Figs. 7-10 but only a few selected differences are discussed here in order to show how the stacking number influences GPR recordings.

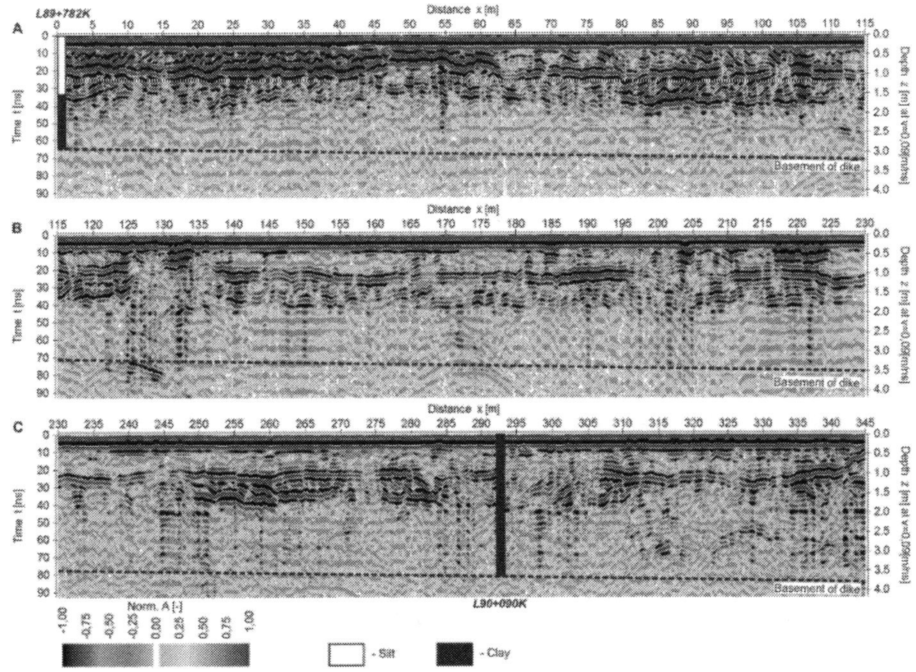


**Fig. 7.** Radargram for 250 MHz antenna, stacking 4 times: A) profile from x=0 m to 115 m; B) from x=115 m to 230 m; C) from x=230 m to 345 m.

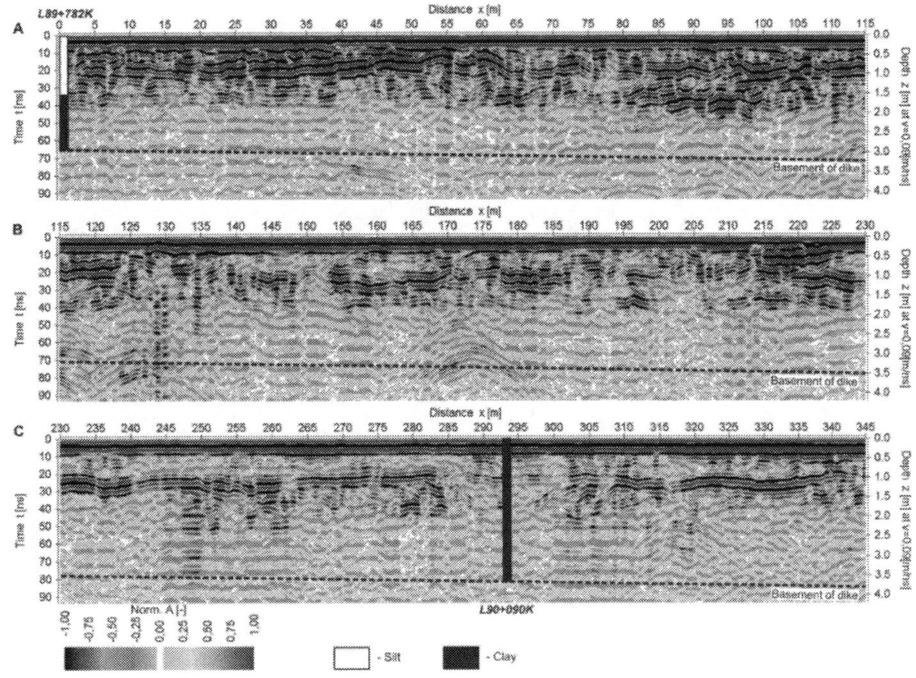


**Fig. 8.** Radargram for 250 MHz antenna, stacking 32 times: A) profile from x=0 m to 115 m; B) from x=115 m to 230 m; C) from x=230 m to 345 m.

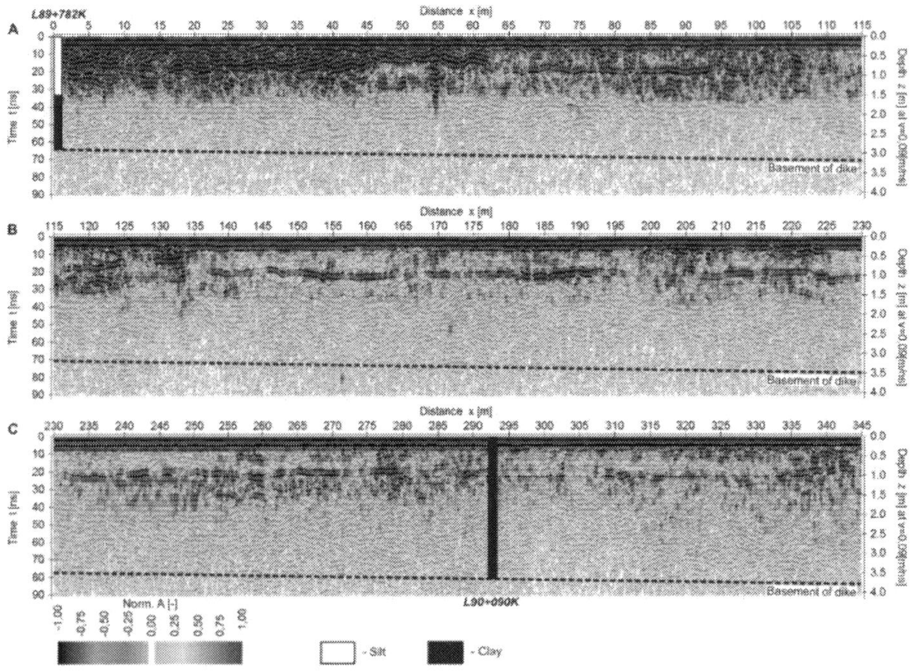


**Fig. 9.** Radargram for 500 MHz antenna, stacking 4 times: A) profile from x=0 m to 115 m; B) from x=115 m to 230 m; C) from x=230 m to 345 m.

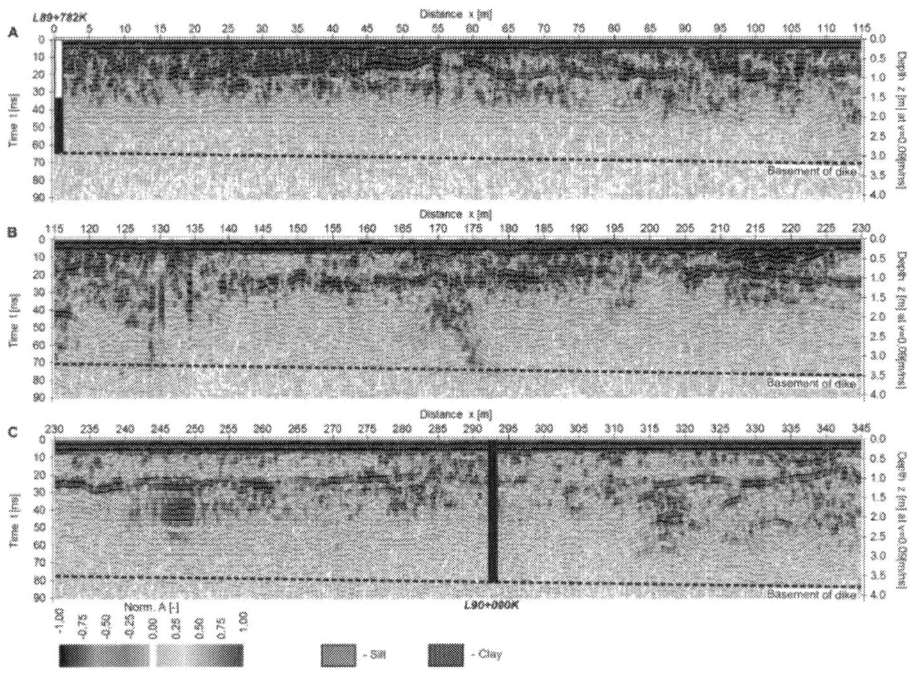


**Fig. 10.** Radargram for 500 MHz antenna, stacking 32 times: A) profile from x=0 m to 115 m; B) from x=115 m to 230 m; C) from x=230 m to 345 m.

In the second stage of data processing, we combined radargrams, first for 250 MHz antenna and second for 500 MHz antenna. This was done to better distinguish the main disintegrated regions. The procedure involves overlaying two radargrams recorded for

250 MHz antenna (Figs. 7 and 8) and two radargrams for 500 MHz antenna (Figs. 9 and 10). Afterwards, more advanced processing was applied [12] to the composite radargrams, specifically the following processing procedures were used: spectral whitening, a gain function in the form of an energy decay procedure, bandpass time-dependent frequency instead of Butterworth filtering, morphologic filtering, muting of direct air and ground waves and, finally, conversion of radargrams to envelope distributions (instantaneous amplitudes were evaluated using a Hilbert transform). The resulting radargrams in the form of an envelope distribution (which is a measure of the signal energy distribution) are shown in Fig. 11B (for 250 MHz antenna) an in Fig. 11C (for 500 MHz antenna); additionally, in Fig.11, information about the microgravimetric anomalies is presented (Fig. 11A). In Figs. 11 B, C depth axis was reduced to a depth 2 m, because no important anomalies were recorded at greater depths (Figs. 7-10).

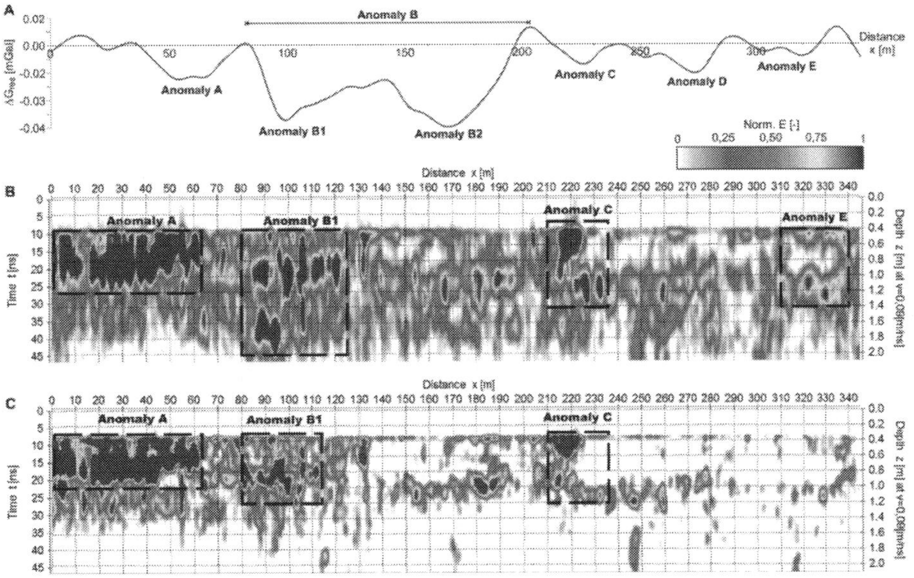


**Fig. 11.** A) Microgravimetric anomalies; B) Envelopes distribution for 250 MHz antenna; C) Envelopes distribution for 500 MHz antenna.

In Fig. 11, it is evident that the GPR anomalies only partially correlate with the microgravimetric anomalies. A shifting of the anomalies along the profile direction is observed; this may be caused by inaccuracy of the distance measurement while using the wheel odometer in the GPR system on the slippery surface (mud and grass) of the crown; it is also possible that microgravimetric anomalies are shifted against GPR anomalies because the gravity meter records the total effect of unsymmetrical unconsolidated zones that may not be recorded in the 2D radargrams.

GPR anomaly B1 (Figs. 11 B, C) correlates well with microgravimetric anomaly B1 (Fig. 11A) so two geophysical methods allow us to depict first, big unconsolidated zone.

It is very strange that huge microgravimetric anomaly B2 (Fig. 11A) was not confirmed by the results of GPR surveys (Figs. 11 B, C); effect observed in the radargrams in gravimetric anomaly B2 should be similar to this recorded in anomaly B1 (Figs. 11 B, C)

Although, a general correlation is observed in the results of the two geophysical methods. Especially, anomalies A, B1, C and E are well-correlated (Fig. 11). The small amplitude of microgravimetric anomaly A (Fig. 11A) is puzzling since GPR recorded the strong, horizontal anomaly in this part of the examined levee (Figs. 11 B, C).

## 3 Gravity modelling

In gravimetry, quantitative interpretation is often performed using gravity forward modelling [3]. 2¾D gravity modelling was applied to interpret the main anomaly "B". A model was created based on the available knowledge about the construction of the levee and the road crossing.

The modeling shows that the bulk density in the anomaly "B" area must have lower density relative to the rest of the levee by about 0.3 g·cm$^{-3}$ (Fig. 12). These areas are located on both sides of the road and may be associated with the retention of rainwater snow and flood waters at the corner of the road and the levee. The "zero" density body is the remnant after the material has been compacted during road construction. The left side of the body corresponds to the right side of the GPR anomaly "B1". Additionally on both sides of the levee, there are other bodies that have been added to the gravity model at the location of the road crossing. Their densities were higher, at ~0.15 g·cm$^{-3}$, relative to the mean levee bulk density. It should be noted that during the gravity modelling it was assumed that the near-surface layer (depths <0.5 m) has a constant mean bulk density along the entire length of the levee.

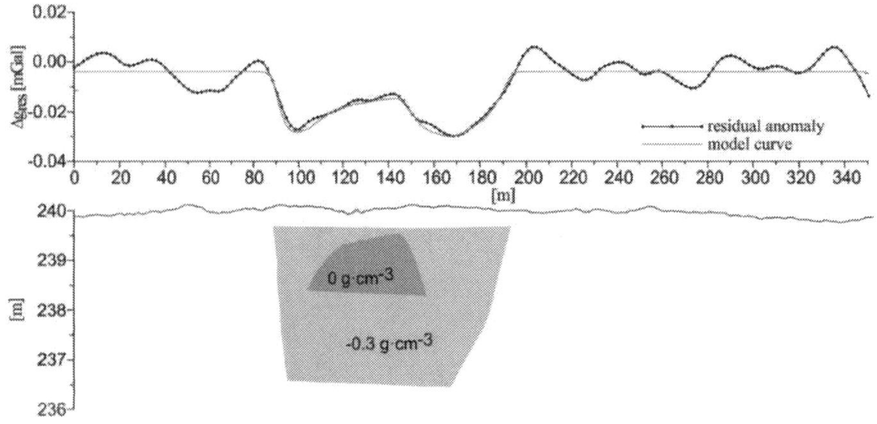


**Fig. 12.** The result of the gravity modelling of anomaly "B".

## 4 Conclusions

The results of the microgravimetric survey clearly demonstrate that this method can be applied to identify the potential weak zones in the levee. In described surveys several negative gravity anomalies with different amplitudes and horizontal extents are recorded. They can correspond to unconsolidated zones in levee structure. Nevertheless, they can be generated by a relatively higher content of non-cohesive soil (silt) in the levee.

The largest of the gravity anomalies was submitted to gravity forward modelling. The obtained density model unambiguously shows that within the profile section between 90 and 190 m the levee bulk density is significantly lower than at other places. This means that the levee may be susceptible there to failure during a flood.

As the levee consists of clay and silt, the GPR method allows to examine the body of levee only to depth c.a. 2 m; therefore, correlation of the results obtained from the two geophysical methods was possible only in the near-surface zone. Hardly part of GPR anomalies were confirmed by microgravimetric surveys and vice versa. GPR method allowed to distinguish a boundary inside the levee, between either silt and clay or internal boundary in silt; this anomaly was not observed in microgravimetric data. Microgravimetric curve allowed to depict 6 anomalies, but only 4 was confirmed by GPR surveys.

An extensive suite of radarogram data processing has been performed, which allowed the better extraction of information about anomalies located in the near-surface area; some of GPR anomalies might be correlated with unconsolidated zones.

Short analysis, concerning the influence of stacking number to recorded data was also presented in the paper; higher stacking number not always results with better recordings.

*The works were financed by a grant of AGH University of Science and Technology no. 11.11.140.645 and Grant of Cracow University of Technology, no. Ś-2/371/2018/DS.*

## References

1. A. Annan, *Ground Penetrating Radar*, Workshop notes, Sensor and Software, Canada (2001)
2. A. Mecke, I. Lee, J.R. Baker jr., M.M. Banaszak Holl, B.G. Orr, Eur. Phys. J. E **14**, 7 (2004)
3. A. Cygal, M. Stefaniuk, A. Kret, M. Kurowska, Geology, Geophysics and Environment, **42**, 3, 279–287 (2016), doi: 10.7494/geol.2016.42.3.279
4. T. Golebiowski, S. Porzucek, B. Pasierb, Near Surf. Geophys. **14**, 4, 371–384 (2016), doi: 10.3997/1873-0604.2016017
5. G. Jonesa, P.Sentenaca, M. Zielinski, J Appl Geophys **106**, 196–211 (2014), doi: 10.1016/j.jappgeo.2014.04.018
6. J. Karczewski, A. Lesniak, J. Zietek, T. Danek, Pol. J. Environ. Stud. **18**, 3A, 142–148 (2009)
7. M. Loj, S. Porzucek, Science and Technologies in Geology, Exploration and Mining, SGEM **1**, 743–749 (2015)
8. MALA 2004, *Rules of thumb*, MALA Geoscience AB, Mala, Sweden
9. MALA 2003, *Basic and Advanced Radar Theory*, MALA Geoscience AB, Mala, Sweden
10. E. Niederleithinger, A. Weller, R. Lewis, R. Stötzner, Conference: 4th International Symposium on Flood Defense, Toronto, Canada (2008)
11. S. Oryński, M. Okoń, W. Klityński, Acta Geophys. **64**, 6, 2322–2336 (2016), doi:10.1515/acgeo-2016-0092
12. M. Di Prinzio, M. Bittelli, A. Castellarin, P. Rossi Pisa, J. Appl. Geophys. **71**, 2-3, 53–61 (2010) doi: 10.1016/j.jappgeo.2010.04.002

# Preliminary results of pore pressure profiling on theTęgoborze-Just landslide

*Jacek* Stanisz[1,*] and *Zenon* Pilecki[2]

[1]AGH University of Science and Technology, Krakow, Poland
[2]Mineral and Energy Economy Research Institute, Polish Academy of Sciences, Krakow, Poland

**Abstract.** In this study, we present the preliminary results of deep profiling of pore pressure in the near surface formations of the Carpathian flysch, on the Tęgoborze-Just landslide near Nowy Sącz in Southern Poland. The aim of the study is to identify zones with significant changes in pore pressure influenced landslide processes. These pore pressure changes correspond to a greater change in the degree of saturation or water flow paths. Profiling was performed in four series using a CPTU static probe with a NOVA Acoustic cone. The measurement of pore pressure in flysch formations is very complicated due to the strong heterogeneity of the medium properties and variable water conditions, which are strongly influenced by the intensity of precipitation. The tests were carried out in a colluvium to a depth of approx. 5.0 m, i.e. to the border of a less weathered bedrock, under varying water conditions. The results obtained indicate the presence of four zones of significant changes in pore pressure. The location of these zone is consistent with the location of the greater displacement measured with the inclinometer. There is a greater probability that these weak zones may form a rupture surface.

**Key words**: pore presure, CPTU sounding, inclinometer, landslide, Carpathian Flysch

## 1 Introduction

Pore pressure profiling provides important information about structure, geological-engineering properties and physical-mechanical processes occurring in the geological medium [1-4]. Investigations concerning changes in water conditions, including pore pressure, may be related to the development of the landslide process (e.g. [5-8]).

The aim of the study is to identify zones with significant changes in pore pressure near the surface formation of Carpathian flysch influenced landslide processes. These pore pressure changes correspond to a greater change in the degree of saturation or water flow paths. We used static probing of CPTU with a NOVA Acoustic cone. The study was carried out in the colluvium on the Tęgoborze-Just landslide in the south-eastern part of Poland (Fig. 1).

* Corresponding author: jstanisz@agh.edu.pl

Three typical parts may be distinguished in the geological-engineering profile of the Carpathian flysch: near surface soils, weathering zone and bedrock [9-10]. Near surface soils, generally regarded as quaternary, are related to flysch bedrock, as they have generally formed as a result of its weathering. Their physical and mechanical properties are very variable. In these parts of the flysch medium, landslide processes most often develop [9-11]. The properties of near-surface soils depend on many factors, such as: the type of bedrock, the sort of weathering and the morphology of the slopes. The bedrock is characterized by alternating layers of medium and coarse sandstone and conglomerates as well as fine-grained siltstones and claystones with other kinds of rocks (marls, hornstones) [12-13].

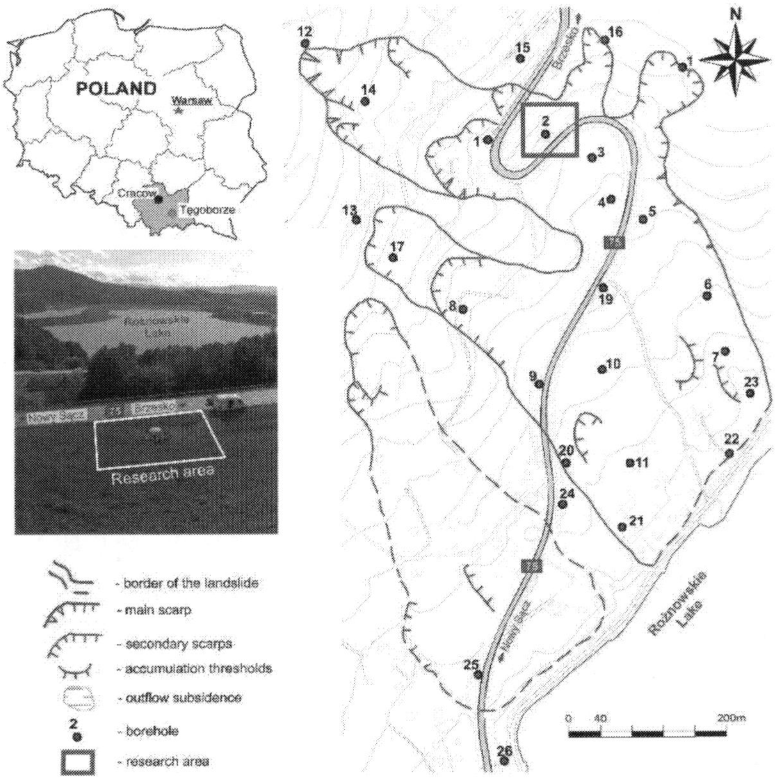


**Fig. 1.** A situation map showing the border of the Tęgoborze-Just landslide and a general view of the research area.

Measurement of pore pressure in the highly inhomogeneous Carpathian flysch medium is difficult from a technological point of view. The relationship between changes in pore pressure in the geological medium and the development of landslide processes is also complicated. Bednarczyk [6] conducted research into pore pressure changes in order to identify the slip surface in landslides occurring during flysch formation near Gorlice. He developed a monitoring system consisting of pneumatic and vibrating wire sensors of pore pressure, inclinometers, piezometers and an atmospheric station. Bednarczyk [6] confirmed that the development of deformations was associated with an increase in pore pressure in the range from 50 to 100 kPa. The increase in pore pressure recorded was particularly noteworthy after the occurrence of intense precipitation above 200-250 mm.

Zabuski [5] studied the changes in water conditions depending on the intensity and duration of precipitation for several landslides occurring during flysch formation in the region of Szymbark. He used 25 open piezometers and systems of inclinometers. Zabuski [5] observed that displacements and changes in water conditions were related to the mechanical and hydraulic properties of the geological medium.

Similar studies concerning landslides occurring during flysch formation in the Szymbark region were also conducted by Gil et al. [14]. They observed that the groundwater level decreased for less intense precipitation below 30 mm. For more intense precipitation above 30 mm, the groundwater level increased from 40cm to 80cm. The intense precipitation was also related to the displacement increase recorded on inclinometers.

In the study, geological-engineering conditions in the near-surface part of the Carpathian flysch in the Tęgoborze-Just landslide are presented. The methodology of pore pressure profiling and its limitations are discussed. Finally, we present the results of pore pressure measurements together with inclinometer measurements and their analysis with regard to the identification of a possible slip surface.

## 2 Geological - engineering conditions

The research area is a portion of an active landslide located in Tęgoborze-Just in the south of Poland, about 90 km south-east of Kraków. Mass movements develop on the south-eastern slopes of the Kamienna and Jaworze mountain range, including the peaks: Jodłowiec Wielki (482 m a.s.l.) and Saint Just. Geographically, the area is located on the border between two mesoregions: the Beskid Wyspowy and Pogórze Rożnowskie being a part of the Western Outer Carpathians [15] (Fig. 1).

From a tectonic point of view, the research area is located on the border of two thrusting tectonic units - Magura and sub-Magura (grybowska-michalczowej) (Fig. 2). They contain folds and transverse dislocations.

The landslide area is approx. 21.33 ha, about 330 m wide and about 780 m long. The azimuth of the movement direction is close to 130°, and the average inclination is 11.8°. The landslide is periodically active. In the research area, the bedrock lies at a depth of approx. 24-28 m. Over this layer, a weathering zone is situated which has a thickness of 5-24 m (Fig. 2C). In table 1, we present the parameters of geotechnical layers identified on the Tegoborze-Just landslide. These layers are also shown in the geological profile on figure 4.

The main landslide movement takes place in the near-surface zone to a depth of 15.5 m (Fig. 2C). Within the colluvium, different soil and rock fragments move each other [16]. The effects of the landslide process are visible in the form of the folding and cracking of the road surface. In addition, cracks in the foundations and walls of buildings in the vicinity are also registered.

There is no continuous water level within the landslide colluvium. Its location is variable and related to both the intensity of precipitation and permeability of flysch formation. In the area of the landslide, water leaks are observed.

The measurements were taken in the northern part of the Tęgoborze-Just landslide along national road no. 75 in the area of the 2I/P measuring point (Fig. 2). This point consists of two boreholes used for piezometric and inclinometer tests. At a distance of about 5 m from the measuring point 2I/P to NE, the borehole J-1 was hand drilled to a depth of 5.5 m (Fig. 2). In the immediate vicinity of one another, four static sounding CPTU tests were conducted together with the profiling of pore pressure to a depth of about 5.0 m (Fig. 2A and B).

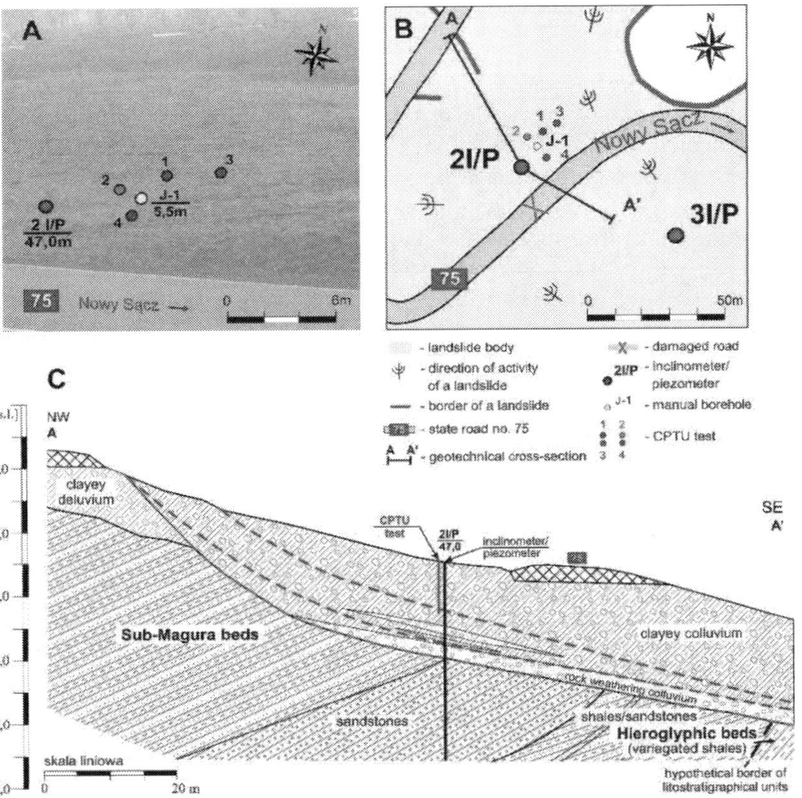


**Fig. 2.** General view of the research area of the Tegoborze-Just landslide (A), with the scheme of research (B) and the longitudinal geological-engineering crosssection of the landslide (on the basis of [16]) (C).

**Table 1.** Physical and mechanical parameters of the geotechnical layers (Fig. 4) in the area of the Tęgoborze-Just landslide (on the basis of [16]).

| Number of geotechnical layer | Type of geotechnical layer | Plasticity index | Bulk density [Mg/m³] | Cohesion [kPa] | Angle of internal friction [deg] | Uniaxial compressive strength [MPa] |
|---|---|---|---|---|---|---|
| Ia | clayey colluvium: (saclSi, clSi)+ shale/sandstone | semi-solid | 1.90–2.15 | 23 | 15 | – |
| Ib | clayey colluvium: (saclSi, clSi)+ shale/sandstone | plastic | 1.90–2.10 | 18 | 12 | – |
| IIa | clayey colluvium: (sagrSi)+ shale/sandstone | plastic | 2.00–2.14 | 23–30 | 14–21 | – |
| IIb | clayey colluvium: (sagrSi)+ shale/sandstone | plastic/liquid | 1.90–2.10 | 16 | 8 | – |
| III | rock weathering colluvium: sandstones rubble, breccia | – | 2.30–2.44 | 0–5 | 32 | 3.5–19.8 |
| IVa | shales+mudstones | – | 2.31–2.43 | – | – | 5.5–15.0 |
| IVb | sandstones +mudstones | – | 2.14–2.90 | – | – | 19.0–91.0 |

saclSi - sandy,clayey silt clSi - clayey silt sagrSi - sandy, gravelly silt

## 3 Methodology

Pore pressure profiling was carried out using the static sounding CPTU test of AB Geotech (Sweden). A NOVA Acoustic type electric cone was used (Fig. 3).

The measurement procedure consisted of [17]:

- The installation of the measuring set - anchoring and leveling.
- Preparation of a vented filter - it is protected in a container filled with glycerol.
- Venting the measuring chamber and filter assembly.
- Connecting the transmitter with the cone and protecting the cone tip with a thin insulation cover.
- Performing a control measurement.
- Obtaining a profile of the pore pressure.
- Performing a control measurement.

**Fig. 3.** A view of the CPTU measurement set during a test on the landslide Tegoborze-Just (A), a view of the components of the CPTU measurement set (B) (based on [17]).

The measurement procedure was repeated each time that a new filter was applied. The measurement of the pore pressure was carried out at intervals of 2 cm.

The measurement of pore pressure during static soundings – the CPTU tests, depend on several factors: soil type, the degree of humidity, its stiffness, stress state, preconsolidation, water flow conditions and the degree of geological medium destruction as a result of mass movements [18]. During soundings, the penetration of the cone in the geological medium causes a change in the effective stress and displacement of soil particles and/or rocks. The changes in effective stress are associated with positive or negative pore pressure (suction pressure) [19, 20]. Below the groundwater table, the soil is fully saturated, and the pore pressure is greater than the atmospheric pressure and it is recorded as a positive value. Above the groundwater table where the soil is partially saturated, the pore pressure is less than the atmospheric pressure and hence is negative in value. The magnitude of the negative pore pressure is determined by the surface tension at the air-water boundaries within the pores and is governed by grain size. In general, the finer the soil particles, the larger the saturation capillary head, and hence the greater the negative pore pressure. Precipitation water infiltration may rapidly reduce the magnitude of negative pore pressure [21]. However, the magnitude of negative pore water pressure may vary widely. Any change in this pore pressure alters the shear strength of the soil and therefore has a considerable effect on the slope stability.

A negative pressure may also arise behind the cone tip during the CPTU sounding. This process mainly concerns soils with a solid and semi-solid consistency. The negative pressure in soils with a plastic consistency often leads to the silting-up of the filter.

Bearing in mind the specifics of the measurement of pore pressure, its deep profiling may provide information concerning the location of greater saturation zones or water flow paths. The identified zones may be related with the development of slip surfaces in the geological medium threatened by the landslide process.

In the study, four measurements were realized in 2017. Two measurements designated 1 and 2 were carried out in dry conditions on the 10th and 18th of August 2017. Measurement no. 3 was carried out on the 8th September 2017 after a heavy precipitation of 41 mm. In turn, measurement no. 4 was made immediately after a slight precipitation of 2 mm on the 11th of October 2017.

## 4 Results and analysis

The results of the research are shown in Figure 4. Figure 4A presents four graphs of pore pressure changes for each deep profile and a graph of their arithmetic mean changes. Figure 4B presents changes in the average value of pore pressure and inclinometer displacement in two perpendicular directions. The displacement was measured using an inclinometer at the 2I/P measuring point. In general, the pore pressure in all four measurements varied from -86 kPa to +47 kPa. The maximal positive pore pressure value was measured in dry conditions. The maximal negative value was measured in wet conditions after precipitation. In dry conditions in clayey colluvium, the value of pore pressure was equal 0 kPa to a depth of 1.5 m.

The results of pore pressure measurements have been analyzed for significant positive or negative changes. Such changes may be associated with significant changes in the water content or flow water path. Therefore, such changes may indicate the presence of weak zones that accompany slip surfaces in a geological medium threatened by the landslide process.

On both figures 4A and 4B, four main zones of significant changes in pore pressure may be distinguished, they are numbered from 1 to 4:

- Zone 1 occurs at a depth of 0.4–0.7 m (Fig. 4A). In wet conditions, the negative pressure is indicated from -50 to -10 kPa. This is particularly visible on the graph in wet conditions after precipitation of up to 41 mm. In this zone there are silty clays with a plastic consistency. In dry conditions, the pore pressure is equal to 0 kPa.
- Zone 2 occurs at a depth of 1.5 to 2.0 m. It is divided into parts a and b, due to contact with silty clays with various consistencies. Pore pressure varies from -82 to 48 kPa. A greater value of pore pressure was measured at the bottom of the silty clay layer (b). An increase in pore pressure up to about 48 kPa may be related to the occurrence of shale and sandstone breccia and different water flow paths. In zone 2 a displacement increase of approx. 7 mm in the Y direction perpendicular to the longitudinal axis of the landslide was measured.

 https://doi.org/10.1051/e3sconf/20186602001

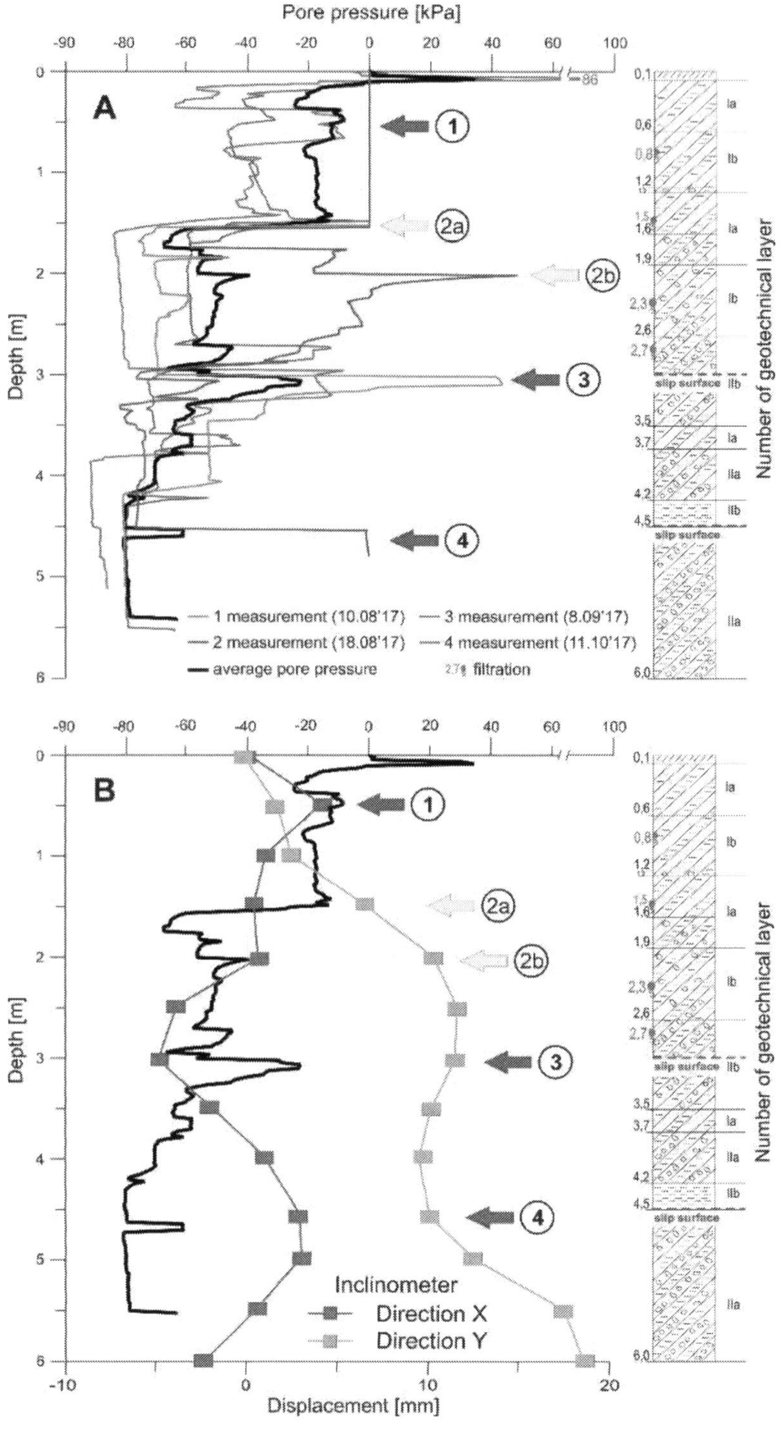


**Fig. 4.** Graphs of deep profiling of pore pressure changes for four field measurements and changes in the average value of the pore pressure (A); graphs of the average value of the pore pressure and inclinometer displacement (B). On the right hand side, a geological-engineering profile based on the J-1 borehole is shown. (Explanations in the text).

- Zone 3 is characterized by great increases in pore pressure in dry conditions to a value of about 45 kPa. An exception occurred in the case of just one measurement, for which no increase in the pore pressure occurred. In wet conditions, the value of the pore pressure varied over a small range from -5 to -2 kPa. The displacements measured with the inclinometer considerably increased by approximately 11 mm, especially in the X direction along the longitudinal axis of the landslide. The rotational movement was small and amounted to 4 mm. Presumably, zone 3 includes an active slip surface at a depth of approx. 3.0 m.
- In zone 4, at a depth of approx. 4.5 m, the increase in pore pressure in dry conditions was measured. The wet conditions due to precipitation did not cause significant changes in pore pressure. In this zone a great increase of displacement in the X direction of about 10 mm was observed. At a depth of 4.5 m, the slip surface was determined based on archival research [16].

## 5 Conclusions

The measurement of pore pressure in the highly heterogeneous colluvial of the Carpathian flysch is difficult to realize and interpret. These difficulties result from the lack of a continuous aquifer in the colluvium of the landslide. Water conditions are variable and closely related to the occurrence and intensity of precipitation as well as the permeability of flysch formation. The relationships between the changes in pore pressure and the development of landslide processes are also complicated.

Based on the four deep profiles of pore pressure and inclinometer measurements in the near-surface colluvium of the Tęgoborze-Just landslide, the following conclusions may be formulated:

1. The static sounding CPTU of AB Geotech with NOVA type electric cones allows for the deep profiling of the pore pressure in the near-surface medium up to about 5m. In dry conditions, the pore pressure varied from -82 kPa to 48 kPa. In wet conditions with a varied degree of saturation, the pore pressure changed from -85 to 2 kPa. Significant changes in the pore pressure values were measured in wet conditions as well as in dry conditions.
2. Four zones of distinct pore pressure changes to a depth of approx. 5.0 m have been determined. In these zones displacements of different value were registered with the inclinometer. This information allows for the determination of weak zones in which the slip surface of a landslide may develop.
3. The distinct changes in pore pressure are mainly influenced by the type of geological medium (rock/soil), its degree of saturation and degree of medium deterioration by landslide processes.
4. The deep profiling of the pore pressure may be useful in determining weak zones in the near-surface geological medium, even those as complicated as the Carpathian flysch. This information may contribute to a more reliable assessment of the stability of the slope of landslides in the Carpathian flysch.

*The research was supported by the Faculty Geology, Geophysics and Environmental Protection of AGH University of Science and Technology, Cracow, Poland, No. 11.11.140.649.*

## References

1. J. Stanisz, *Możliwości rozpoznania zagrożenia osuwiskowego na podstawie obserwacji zmian ciśnienia porowego w ośrodku geologicznym (Possibilities of recognition of landslide hazard based on observation of pore pressure changes in a geological medium)*, Zeszyty Naukowo-Techniczne SITK RP, o. w Krakowie **3** (102), 1–8 (2013) (in Polish)
2. J. Stanisz, *Czujniki do pomiaru ciśnienia porowego dla potrzeb rozpoznania położenia powierzchni poślizgu osuwiska (Sensors for pore pressure measuring for identification landslide slip surface)*, Zeszyty Naukowe IGSMiE PAN **89**, 77–91 (2015) (in Polish)
3. Z. Skutnik, M. Biliniak, *Suction controlled triaxial testing of unsaturated medium sand and sandy clay*, Ann. Warsaw Univ. Life Sci. - SGGW, Land Reclam. **46**, 301–307 (2014) doi:10.1515/sggw-2015-0004
4. P. Zawrzykraj, *Ocena parametrów filtracyjnych iłów warwowych ze stanowiska Plecewice k. Sochaczewa w warunkach in situ (Assessment of permeability parameters of in situ tested varved clays from Plecewice near Sochaczew)*, Przegląd Geologiczny **65**, 587–596 (2017) (in Polish)
5. L. Zabuski, E. Gil, W. Bochenek, *Interdependece between groundwater level and displacement of the landslide slope*, Polish Geol. Inst. Spec. Pap. **15**, 39–42 (2004)
6. Z. Bednarczyk, *Monitoring of Rainfall Induced Landslides in Relation to Weather Conditions at Selected Locations in Polish Carpathians*, In: Lollino G. et al. (eds.) Engineering Geology for Society and Territory - Volume 2 (2015) doi: 10.1007/978-3-319-09057-3_207
7. P. Harba, Z. Pilecki, *Assessment of time-spatial changes of shear wave velocities of flysch formation prone to mass movements by seismic interferometry with use of ambient noise*, Landslide **14**, 1225–1233 (2017) doi: 10.1007/s10346-016-0779-2
8. J. Stanisz, P. Krokoszyński, R. Kaczmarczyk, *Impact of precipitation on dissipation of pore pressure in colluvium of the Carpathian Flysch landslide*, Proc.of China-Europe Conference on Geotechnical Engineering (2018) doi: 10.1007/978-3-319-97115-5_138
9. K. Thiel, *Kształtowanie fliszowych stoków karpackich przez ruchy masowe (Formation of flysch Carpathian slopes by mass movements)*, Prace Instytutu Budownictwa Wodnego PAN Gdańsk (1989) (in Polish)
10. J. Dziewański, Z. Pilecki, W. Sroczyński, *Zagadnienia badań geologiczno-inżynierskich w projektowaniu tuneli komunikacyjnych w utworach fliszu karpackiego na przykładzie tunelu w Lalikach (Geological-engineering research in the design of transportation tunnels in the Carpathian flysch on the example of a tunnel in Laliki)*, Studia, Rozprawy, Monografie **96**, Wyd. IGSMiE PAN, Kraków, 64 (2001) (in Polish)
11. Z. Pilecki, *Basic principles for the identification of landslides using geophysical methods* E3S Web Conf. **24**, 01001 (2017) doi: 10.1051/e3sconf/20172401001
12. H. Marcak, Z. Pilecki (eds.),*Wyznaczanie właściwości utworów fliszu karpackiego metodą sejsmiczną dla potrzeb budownictwa tunelowego (Determination of properties of the Carpathian flysch rocks using the seismic method for the needs of tunnel construction)*, Wyd. IGSMiE PAN, Kraków (2003) (in Polish)
13. T. Majcherczyk, Z. Pilecki, Z. Niedbalski, E. Pilecka, M. Blajer, J. Pszonka: *Wpływ warunków geologiczno-inżynierskich i geotechnicznych na dobór parametrów obudowy wstępnej tunelu drogowego w Lalikach (Influence of geological engineering and*

*geotechnical conditions on the selection of parameters of preliminary support for a road tunnel in Laliki),* Gospodarka Surowcami Mineralnymi **28/1**, 103–124 (2012) (in Polish)

14. E. Gil, L. Zabuski, T. Mrozek, *Hydrometeorological conditions and their relation to landslide processes in the Polish Flysch Carpathians (an example of Szymbark area),* Studia Geomorphologica Carpatho-Balcanica **43**, 127–143 (2009)
15. T. Ziętara, *Obszary osuwiskowe w dolinie Dunajca nad Jeziorem Rożnowskim (Landslide areas in the Dunajec Valley on the Rożnowskie Lake),* Rocz. Sądecki **14** (1974) (in Polish)
16. G. Czudec, J. Bulanda, I. Andrasik, W. Kawa, Z. Paul, Z. Bestyński, J. Bulanda, D. Mach, P. Gawron, J. Ryznar, C. Biskup, P. Jurczyk, A. Gałus, M. Gruca, *Dokumentacja geologiczno-inżynierska dla opracowania systemu zabezpieczenia strefy osuwiskowej nr MPL0051, w ciągu drogi krajowej nr 75, km od 51+900 do 52+700 w miejscowości Tęgoborze-Just (Geological and engineering documentation for the development of the landslide protection system No. MPL0051, along the national road No. 75, km from 51 900 to 52 700 in the town of Tęgoborze-Just)* (2012) (in Polish)
17. Z. Sikora, *Sondowanie statyczne. Metody i zastosowanie w geoinżynierii (Static sounding. Methods and application in geoengineering* (2006) (in Polish)
18. R. G. Campanella, P. K. Robertson, D. Gillespie, *Factors affecting the pore water pressure and its measurement around a penetrating cone*, 39th Canadian Geotechnical Conference, Ottawa (1986)
19. K. Terzaghi, *Die Berechnung der Durchlassigkeitsziffer des Tones aus dem Verlauf der hydrodynamische Spannungserscheinungen*, Abt. Iia Sitzungsber. Akad. Wiss, Wien (1923)
20. J. E. Jennings, J. B. Burland, *Limitations to the use of effective stresses in partly saturated soils*, Geotechnique **12**, 125–144 (1962)
21. N. Lu, W. Likos, *Unsaturated soil mechanics*, John Wiley and Sons Inc., New York (2004)

# Application of the mineral liberation analysis (MLA) for extraction of grain size and shape measurements in siliciclastic sedimentary rocks

*Joanna* Pszonka[1,*], *Dariusz* Sala[2]

[1]Polish Academy of Sciences, Mineral and Energy Economy Research Institute, Cracow, Poland
[2]Polish Academy of Sciences, Institute of Geological Sciences, Krakow Research Center, Cracow, Poland

**Abstract.** The mineral liberation analysis setup (MLA) consists of a scanning electron microscope (SEM) – based backscattered electron (BSE) image with an energy dispersive X-ray system (EDX) for elemental analysis and a computer software that integrates images and X-ray identification of minerals and maps their distribution. Thereby, various quantitative and qualitative data sets are collected including grain size distribution and shape parameters such as aspect ratio, shape factor and angularity. Other techniques, e.g. the Gazzi-Dickinson point counting method or frequently questionable image analysis software to extract data for textural analysis are time consuming, strenuous and with limitations that need to be addressed. Significant productivity of the mineral liberation analysis provides statistical representation and thereby stringent arguments to detect and suggest some potential solving in uncertainty and complexity of the submarine gravity flows phenomenon that is extremely difficult to monitor, however volumetrically the most significant processes moving sediments on Earth. The mineral liberation analysis seems to be one of the most suitable method to acquire such data set.

**Keywords:** scanning electron microscope (SEM), mineral liberation analysis (MLA), grain size distribution, textural parameters

## 1 Introduction

The mineral liberation analysis (MLA) is done by an automated measurement system, created to provide quantitative analyses of mineral species and material fragments. This system is based on a scanning electron microscope (SEM) with an energy dispersive X-ray (EDX) spectrometer and a software:

1. backscattered electron (BSE) image analysis allows to determine grain boundaries and sites for X-ray spectral acquisition,
2. X-ray spectra allow to classify samples by comparison to a library of reference spectra,

* Corresponding author: jpszonka@min-pan.krakow.pl

3. software automates microscope operations and data acquisition.

The main advantages of using the MLA system are:

1. systematic, computer-automated and unattended operations reducing human errors and bias, and tedious manual analyses,
2. analysis of multiple samples increasing productivity and providing statistical representation,
3. cost effective,
4. ability to analyse very fine material at the scale of micrometres.

The main MLA–disadvantage is problematic distinguishing samples with similar compositions and polymorphs.

Creation and development of the MLA system was focused on applied mineralogy and metallurgical processes [1], such as ore body evaluation, geological exploration, geometallurgical mapping. However, the MLA-application has been extended for other studies including sedimentology [2, 3, 4, 5]. The aim of this paper is to display how to obtain size- and shape-grain parameters of submarine gravity flow sediments. These textural characteristics are important indicators for submarine flow hydrodynamics.

## 2 Material and Methods

### 2.1 Material

The Cergowa sandstones from the Dukla and Fore-Dukla Units in the Outer Carpathians were deposited by high-density turbidites and hyperpycnal flows fed from a directly connected delta in the proximal part of the basin, and surge-type turbidity currents in the more distal localities [6, 7]. These sandstones are fine- and very fine-grained and represent lithic wackes and arenites [8].

10 samples of the Cergowa sandstones were crushed to the gravel size in the Retsch Jaw Crusher BB 200, sieved to extract fraction <315 μm and ferromagnetic minerals were removed by a neodymium hand magnet. Further removal of ferromagnetic fraction was carried out using the Carpco MIH-13-111-5 and next the Frantz LB-1 magnetic separator. Subsequently the non-magnetic fraction was subjected to a heavy liquid separation by the tetrabromoethane (density of 2.97 g/cm$^3$) and the methylene iodide (density of 3.32 g/cm$^3$). The separation allowed to segregate light and heavy minerals. In this experiment heavy minerals were chosen for further proceeding and homogenized. Then, separated and homogenized minerals for each sample were prepared as standard petrographic thin sections, coated with carbon and labelled as 1-10CS.

### 2.2 Methods

The SEM generates a beam of incident electrons in a vacuum that is formed by a series of electromagnetic lenses separated by apertures in an electron column. The energy of incident electrons can be as high as 30 keV. Incident electrons react with the sample and produce three basic types of signals:

1. backscattered electrons (BSE) – high energy electrons that are ejected by an elastic collisions of incident electrons with nuclei of atoms in the sample and originated from a wide region within it; the energy of BSE is comparable to that of the incident electrons,
2. secondary electrons (SE) – lower energy electrons emitted by an inelastic interactions of incident electrons with valence electrons of atoms in the sample and

scattering from a surface or a near surface region of the sample; the energy of SE is 50 eV and less.

3. characteristic X-rays (EDX) – X-ray is emitted when the electron beam removes an inner shell electron and is replaced by an outer shell electron.

To create SEM images the electron beam is scanned in a raster pattern over a sample area, named a frame. Electrons are sensed for each position in the frame by electron detectors. BSE images provide information about distribution of elements, topography, crystallography and the magnetic field of samples. SE imaging can produce a very high resolution image of a sample surface. The intensity of the emitted electron signal is displayed as brightness in images. The magnification of an image is controlled by size of the frame, thereby reducing size of the frame results in an increase in frame resolution. Otherwise incident electrons that interact with the sample generate characteristic X-rays with measured energy levels that identifies present elements.

The MLA software integrates BSE images and X-ray data to identify materials and maps their distribution in the scanned sample area and then moves the electron beam to the next frame where measurements are retaken. Two MLA measurement methods are the most common:

1. the extended BSE (XBSE) for particle mounts, where the single X-ray analysis is used for identification of minerals and the BSE imaging for discrimination of their boundaries,
2. the grain X-ray mapping (GXMAP) for thin sections, where detailed X-ray analysis on a grid pattern is used for both identification of minerals and discrimination of their boundaries because BSE imaging is insufficient to distinguish them.

The backscattering coefficient is the ratio of the number of electrons backscattered by a mineral to the number of incident electrons from the SEM and is a function of the atomic number Z of the mineral. Minerals composed of lighter elements backscatter fewer electrons and appear darker in images, whereas minerals composed of heavier elements backscatter more electrons and appear brighter. The MLA system receives BSE data as a 256 greyscale image, from 0 that is black to 255 that is white, and relates the backscattering coefficient to the greyscale value. These greyscale variations provide distinguishing boundaries between minerals by the image separation function. It outlines regions with more or less homogeneous grey level in a particle image. Each BSE grey value of every region is related to minerals with unique Z. However, in the case when minerals have similar or the same atomic number Z and grey level, it is impossible to distinguish them by BSE imaging alone and X-ray analysis is needed for mineral classification.

The X-ray analysis links the spectrum of energy peaks from an unknown mineral with a library of X-ray spectra for known reference minerals collected by the same instrument parameters, except the acquisition time. The MLA software collates the spectrum of the unknown mineral to each known reference considering peak position, peak shape and intensity ratio. The program MLA Image Processing verifies the BSE images and X-ray data. The analysis products false colour images, assigned to minerals and linked with BSE images and quantitative data. Subsequently the program MLA Dataview evaluates results for all particles/grains as tables, graphs and diagrams or exports them to the Microsoft Excel. Before an extraction data do the Dataview, the set can undergo clean-up processes such as: de-agglomeration, particle removal, frame removal, touch-up and mineral grouping in order to obtain proper liberation results.

Mineral liberation analysis measurements were performed at the Geometallurgy Laboratory at the TU Bergakademie Freiberg, on the scanning electron microscope FEI Quanta 600 FEG equipped with the Bruker Quantax Dual X-Flash Detector 5010 and the MLA–software Suite 3.1.4 for automated data acquisition. The imaging was carried out in the backscattered electron (BSE) with following parameters: accelerating voltage of 15 keV, electron beam current of 10 μA, and working distance 12-14 mm. Thin sections were measured in the GXMAP method, where each X-ray analysis was made for 8 msec with a step size of 10 pixels on a 1.5 x 1.5 mm frame with resolution of 500 x 500.

## 3 Results

The data set obtained from the MLA analysis with GXMAP technique from 10 samples of the Cergowa sandstones after clean-up processes are extracted to database and depending on study aims examined, processed, presented and stored by the Dataview software. The aim of this experiment is to obtain textural characteristics of turbiditic sandstones: size and shape measurement of grains, necessary for the comparative study of textural differences, such as four moment measures (mean, standard deviation, skewness, kurtosis), as well as supplemental median value. All values are here calculated with respect to grain's digital image (Fig. 1) with a basic measurement of shape:

1. particle/grain polygon – is determined from the boundary pixels of each shape,
2. convex hull polygon – is determined using the particle/grain polygon,
3. minimum bounding rectangle (MBR) is determined by finding the maximum vector onto which the projection width of the polygon points is at its maximum; then using the vector at 90° to get the projection width for the other dimension for the minimum bounding rectangle.

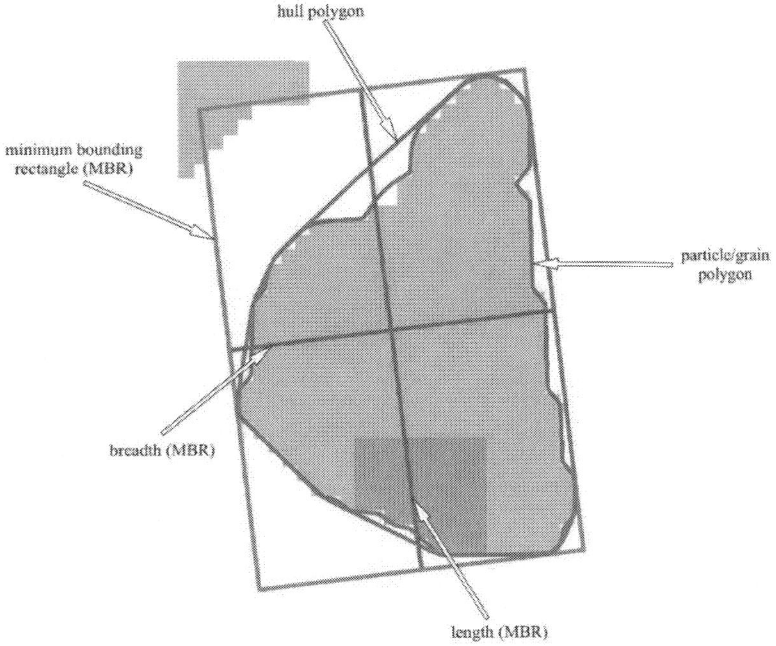


**Fig. 1.** Digital image of an exemplary particle made in GXMAP mode, with shape (particle/grain polygon, convex hull polygon, minimum bounding rectangle) and size (length and breadth MBR) measurements of particle.

Fig. 2 presents exemplary MLA results processed by the Dataview software for the sample 1CS, where following size measurements of grains are collected:

1. area – the area of the particle/grain in micron squared (μm²),
2. perimeter – the surface perimeter of the polygon perimeter (μm),
3. max span – the diameter of the minimum bounding circle (μm),
4. length (MBR) – the length of the minimum bounding rectangle (μm),

5. breadth (MBR) – the breadth of the minimum bounding rectangle (μm),
6. diameter – the equivalent circle (EC)diameter based on polygon area (μm),

in order to compute the shape calculations [3]:

7. aspect ratio (AR) = $L_{MBR}/B_{MBR}$, where $L_{MBR}$ and $B_{MBR}$ are the length and breadth of the minimum bounding rectangle of the particle/grain,
8. angularity (ANG) = $\Sigma(R_P - R_{EE})^2/(R_{EE})^2$, where $R_P$ is the radius of the particle measured from its geometric center and $R_{EE}$ is the radius of an equivalent area ellipse,
9. shape factor (SF) = $P/[2\pi \sqrt{(A/\pi)}]$, where P and A are the surface perimeter and area of the particle.

Thereby for the sample 1CS obtained 118012 results, for the sample 2CS – 77605, the 3CS – 75328, the 4CS – 71144, the 5CS – 273, the 6CS – 16861, the 7CS – 19093, the 8CS – 1793, the 9CS – 15162 and for the sample 10CS – 16997. Statistical results viz: mean, median, standard deviation, skewness and kurtosis for all samples are summarized in the Table 1. Additionally, the difference in the number of results in each sample informs about possible divergence in the content of minerals with density > 2.97 g/cm$^3$.

The Dataview software offers many possibilities to compile and compare obtained results using tables, graphs and diagrams, that facilitate imaging and interpretation of sediment flow dynamics. As an example, in the Fig 3 is presented a comparison of automatically generated grain size distributions of quartz grains (those with density > 2.97 g/cm$^3$) for all analyzed samples 1-10CS.

**Table 1.** Statistical results (mean, median, standard deviation, skewness, kurtosis) for 10 samples of the Cergowa sandstones (1-10CS) for indicated textural parameters (area, perimeter, max span, length, breadth, diameter).

| PARAMETERS → DESCRIPTIVE STATISTICS ↓ | **Area [μm²]** | **Perimeter [μm]** | **Max span [μm]** | **Length (MBR) [μm]** | **Breadth (MBR) [μm]** | **Diameter [μm]** |
|---|---|---|---|---|---|---|
| Sample 1CS = 118012 results | | | | | | |
| Mean | 1964.88 | 209.83 | 55.61 | 56.56 | 36.32 | 41.08 |
| Median | 1025.93 | 173.99 | 47.87 | 48.93 | 30.93 | 36.14 |
| Standard deviation | 2953.29 | 173.73 | 42.02 | 41.21 | 27.05 | 28.54 |
| Skewness | 4.23 | 2.14 | 1.38 | 1.38 | 1.49 | 1.40 |
| Kurtosis | 31.78 | 8.61 | 2.91 | 2.93 | 3.40 | 2.96 |
| Sample 2CS = 77605 results | | | | | | |
| Mean | 2872.11 | 251.85 | 64.02 | 64.56 | 41.42 | 46.07 |
| Median | 1007.93 | 173.99 | 48.05 | 48.83 | 30.51 | 35.82 |
| Standard deviation | 5320.66 | 254.84 | 57.62 | 56.35 | 37.42 | 39.17 |
| Skewness | 4.13 | 2.27 | 1.58 | 1.58 | 1.79 | 1.75 |
| Kurtosis | 24.79 | 7.80 | 2.87 | 2.90 | 3.88 | 3.72 |
| Sample 3CS =75328 results | | | | | | |
| Mean | 2427.16 | 229.61 | 60.02 | 60.66 | 38.11 | 43.05 |
| Median | 962.94 | 161.99 | 45.45 | 47.38 | 29.19 | 35.01 |
| Standard deviation | 4608.07 | 232.85 | 53.77 | 52.74 | 33.32 | 35.17 |
| Skewness | 5.10 | 2.78 | 1.87 | 1.88 | 2.02 | 1.92 |
| Kurtosis | 44.67 | 12.91 | 5.03 | 5.13 | 5.63 | 5.11 |
| Sample 4CS = 71144 results | | | | | | |
| Mean | 2856.19 | 263.02 | 66.14 | 66.70 | 42.51 | 47.05 |

| | | | | | | |
|---|---|---|---|---|---|---|
| Median | 1115.93 | 191.99 | 52.08 | 53.52 | 33.00 | 37.69 |
| Standard deviation | 4889.90 | 254.54 | 56.67 | 55.42 | 36.31 | 37.72 |
| Skewness | 3.79 | 2.17 | 1.46 | 1.46 | 1.67 | 1.59 |
| Kurtosis | 20.53 | 7.03 | 2.38 | 2.40 | 3.37 | 2.99 |
| Sample 5CS = 273 results | | | | | | |
| Mean | 2602.28 | 240.85 | 65.25 | 66.10 | 42.66 | 48.22 |
| Median | 1502.90 | 215.99 | 59.01 | 60.22 | 38.24 | 43.74 |
| Standard deviation | 3314.95 | 171.08 | 43.62 | 42.80 | 29.89 | 31.49 |
| Skewness | 2.63 | 1.12 | 0.72 | 0.68 | 1.13 | 0.96 |
| Kurtosis | 9.03 | 1.66 | 0.30 | 0.17 | 1.70 | 0.98 |
| Sample 6CS = 16861 results | | | | | | |
| Mean | 3159.60 | 263.14 | 74.86 | 75.42 | 49.79 | 58.39 |
| Median | 2816.81 | 269.99 | 76.30 | 76.67 | 50.78 | 59.89 |
| Standard deviation | 2389.38 | 116.09 | 33.13 | 32.59 | 22.72 | 24.76 |
| Skewness | 2.00 | 0.43 | 0.11 | 0.13 | 0.09 | -0.03 |
| Kurtosis | 11.24 | 4.67 | 1.18 | 1.25 | 0.59 | 0.64 |
| Sample 7CS = 19093 results | | | | | | |
| Mean | 2804.76 | 235.58 | 66.96 | 67.69 | 44.18 | 52.25 |
| Median | 2195.85 | 233.99 | 66.51 | 66.85 | 44.92 | 52.88 |
| Standard deviation | 3269.54 | 140.37 | 40.99 | 40.26 | 25.57 | 29.01 |
| Skewness | 5.53 | 1.59 | 1.63 | 1.65 | 0.67 | 0.84 |
| Kurtosis | 63.19 | 9.84 | 9.40 | 9.65 | 2.17 | 3.43 |
| Sample 8CS = 1793 results | | | | | | |
| Mean | 2185.37 | 214.90 | 59.02 | 59.94 | 39.96 | 46.38 |
| Median | 1709.89 | 209.99 | 58.70 | 60.00 | 40.40 | 46.66 |
| Standard deviation | 2595.68 | 130.46 | 34.51 | 33.96 | 22.81 | 25.14 |
| Skewness | 10.29 | 1.57 | 1.17 | 1.14 | 0.60 | 0.85 |
| Kurtosis | 231.70 | 9.00 | 8.41 | 7.86 | 1.75 | 5.47 |
| Sample 9CS = 15162 results | | | | | | |
| Mean | 4527.05 | 305.69 | 86.42 | 86.42 | 57.41 | 66.79 |
| Median | 3730.25 | 311.99 | 89.09 | 88.47 | 58.37 | 68.92 |
| Standard deviation | 4161.76 | 167.44 | 47.12 | 46.09 | 32.90 | 36.09 |
| Skewness | 1.54 | 0.24 | 0.16 | 0.18 | 0.26 | 0.17 |
| Kurtosis | 3.98 | 0.06 | -0.13 | -0.07 | -0.30 | -0.38 |
| Sample 10CS = 16997 results | | | | | | |
| Mean | 3147.48 | 261.95 | 74.56 | 75.08 | 49.70 | 58.24 |
| Median | 2807.81 | 269.99 | 76.20 | 76.23 | 50.83 | 59.79 |
| Standard deviation | 2373.16 | 115.40 | 33.07 | 32.51 | 22.79 | 24.82 |
| Skewness | 1.95 | 0.27 | 0.04 | 0.07 | 0.07 | -0.06 |
| Kurtosis | 11.35 | 2.79 | 0.79 | 0.87 | 0.59 | 0.56 |

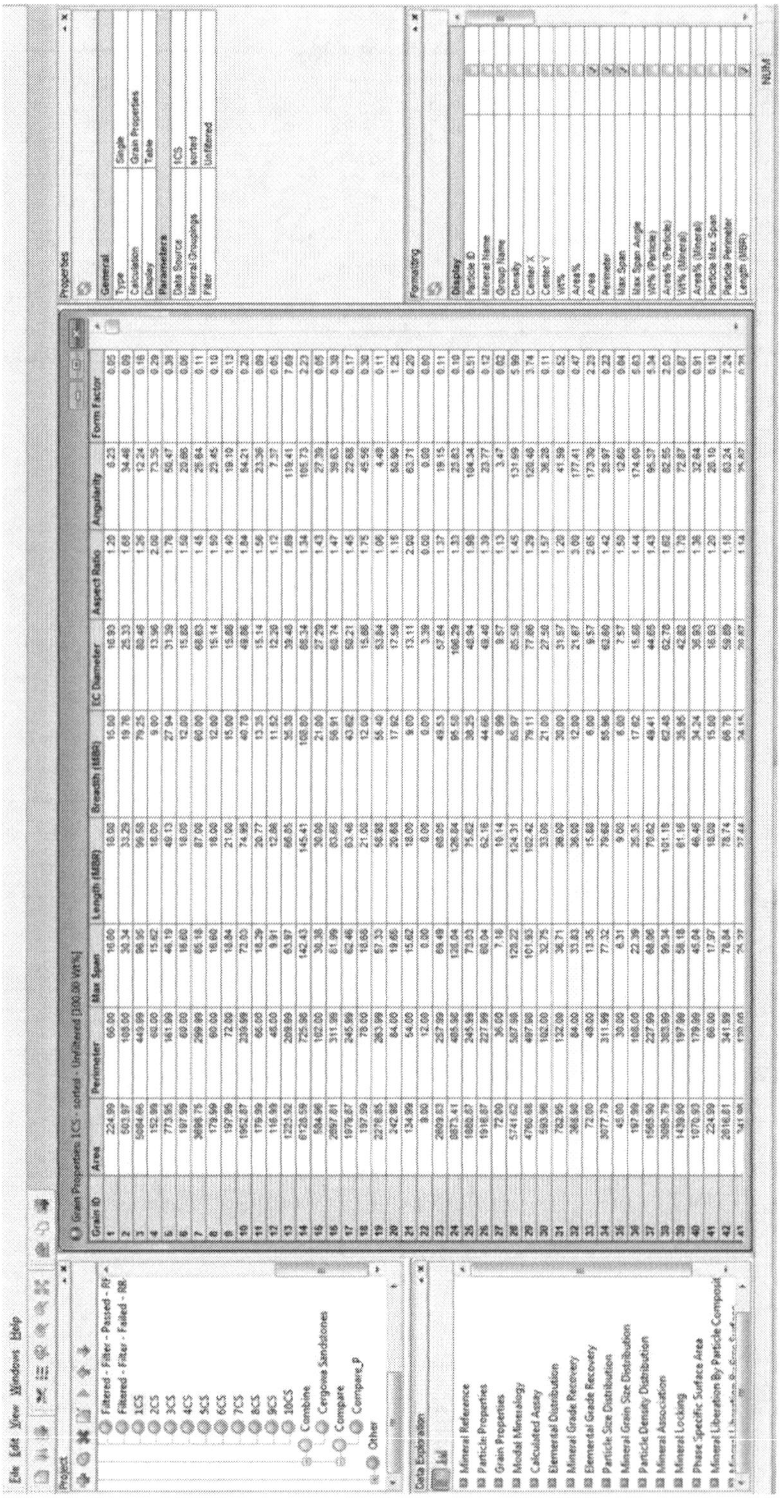

**Fig. 2.** The exemplary MLA results for the sample 1CS with the following size and shape measurements of grain: area, perimeter, max span, length (MBR), breadth (MBR), diameter, aspect ratio, angularity, form factor.

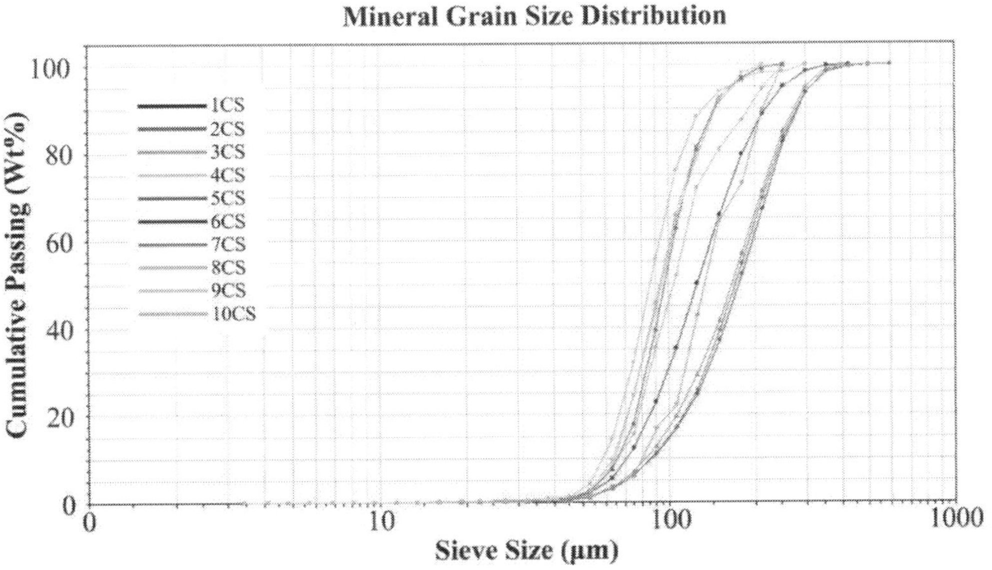


**Fig. 3.** Grain size distributions of quartz grains in samples 1-10CS.

## 4 Discussion

Textural characteristics of siliciclastic sediments are important indicators of flow hydrodynamics. Their differences shown by the size-shape sorting reflects depositional conditions. Sorting takes place over the entire distance travelled with the least transportable grains coming to rest first and the more transportable carried farther. The depositional sorting of grains is distinctly different for various sedimentary environments and even for various flow-regimes of transport in the same environment. Besides, a single flow-event may comprise several different flow types, and transformations can occur between these flow types. Understanding of sediment flow dynamics is considered as extremely complex, especially for enormously difficult to monitor submarine sediment flows.

Kneller and Buckee [9] underlined difficulties in understanding the dynamics of submarine suspended sediments and determined them immensely composite by virtue of turbulence. The phenomenon is non-linear, non-uniform (variation in space) and unsteady (variation in time). In addition, flow state, deposit type and flow transformation there are strongly dependent on the volume fraction of cohesive fine mud within a flow [10]. Their complexity expands even more with an increase of loads of cohesive sediments in suspension. However, in cases where mud content is volumetrically less significant, as in arenites, size and shape parameters of grains are crucial.

In this experiment, where the Cergowa sandstones represents mainly arenites [8], was shown the possibility of quick obtaining a huge amount of data, including statistical results, necessary for further analysis to define flow types and boundaries between them that form a continuum [9, 10]. These data display both quantitative findings, such as size and shape measurements of grains, as well as qualitative findings, as a differentiation of mineral phases. The textural characteristics complete with sedimentary structures are significantly usable to interpret the last-stage evolution of density flows and depositional processes. Furthermore, textural properties next to mud content in deep water facies impact on the porosity and permeability distribution in a hydrocarbon reservoir, which are the most important petrographic parameters of any reservoir rocks, due to the ability to transmit and store fluids.

## 5 Conclusions

On the basis of the study we can conclude:

1. The mineral liberation analysis is an automated measurement system that provides quantitative analysis, including size and shape measurements of grains in sedimentary rocks, such as: area, perimeter, max span, length (MBR), breadth (MBR), diameter, aspect ratio, angularity, form factor, necessary to determine depositional flow types and their transformations. Qualitative MLA–findings, such as mineral identification, complete these data.
2. The mineral liberation analysis seems to be one of the most suitable methods to extract data set for textural analysis in a large amount and relatively quick time.
3. The data set obtained from the mineral liberation analysis has great and strong potential to detect and suggest some solving in uncertainty and complexity of the difficult to monitor submarine gravity flows phenomenon because of large amount of data. Submarine gravity flows are especially noteworthy to track due to their importance in transporting sediments on Earth and industrial meaning.

*This work was funded by the National Science Centre (NCN) Poland grant Miniatura 1 no. 2017/01/X/ST10/00048 for Joanna Pszonka. Acknowledgement to Professor Bernhard Schulz and Sabine Gilbricht for technical and reliable introduction to the MLA system.*

## References

1. R. Fandrich, Y. Gu, D. Burrows, K. Moeller, Int. J. Miner. Process. **84**, 310–320 (2007) doi: 10.1016/j.minpro.2006.07.018
2. D.G. Lowe, P.J. Sylvester, M.E. Enachescu, AAPG Bull. **95**, 1295–1320 (2011) doi: 10.1306/12081010005
3. V. Tsikouras, G. Pe-Piper, D.J.W. Piper, M. Shaffer, Sediment. Geol. **237**, 150–165 (2011) doi: 10.1016/j.sedgeo.2011.02.011
4. M. Redwan, D. Rammlmair, J.A. Meima, Sci. Total Environ. **414**, 480–493 (2012) doi: 10.1016/j.scitotenv.2011.10.038
5. H. von Eynatten, R. Tolosana-Delgado, V. Karius, K. Bachmann, L. Caracciolo, Sediment. Geol. **336**, 68-80 (2016) doi: 10.1016/j.sedgeo.2015.10.008
6. P. Plink-Björklund, R.J. Steel, Sediment. Geol. **165**, 29–52 (2004) doi: 10.1016/j.sedgeo.2003.10.013
7. J. Pszonka, *Sedimentological study of the Cergowa Beds in the Dukla and Fore-Dukla Units of the Flysch Carpathians (in Polish, with English abstract)* (Studia, Rozprawy, Monografie 196, IGSMiE PAN, Kraków, 2015)
8. J. Pszonka, J. Götze, J. Petrol. Sci. Eng. **161**, 582–589 (2018) doi: 10.1016/j.petrol.2017.11.069
9. B.C. Kneller, C. Buckee, Sedimentology **47/s1**, 62–94 (2000) doi: 10.1046/j.1365-3091.2000.047s1062.x
10. P.J. Talling, D.G. Masson, E.J. Sumner, G. Malgesini, Sedimentology **59**, 1937-2003 (2012) doi: 10.1111/j.1365-3091.2012.01353.x

E3S Web of Conferences **66**, 02003 (2018) https://doi.org/10.1051/e3sconf/20186602003
*AG 2018 – 4th International Conference on Applied Geophysics*

# Validation of Rock Testing Methods in Determination of Apparent Density

*Joanna* Hydzik-Wiśniewska*, and *Sebastian* Olesiak

AGH University of Science and Technology, Cracow, Poland

**Abstract.** Research on the physical properties of rock materials and aggregates used for construction should be based on standardized methods. Whereas scientific research is focused more on the development of new methods, testing and evaluation of new properties, etc. In case of own testing procedures and standardized but modified methods, they should be validated before being put into use. Validation is the confirmation of the ability to designate that method and tests its usefulness. In order to investigate the method's possibilities, the following assessment methods can be used: calibration or precision evaluation using reference standards or reference materials, systematic evaluation of factors affecting the result, resistance of the test method to variability of controlled parameters, comparison of test results obtained by various methods, inter-laboratory comparisons, and uncertainty of measurement. The paper presents mathematical formulas allowing to evaluate the precision of research methods and the consistency of results, which are the basis for validation of research methods. In the practical part of the article, own method of apparent density testing, was validated based on the analysis of repeatability, internal laboratory reproducibility and between laboratory reproducibility.

**Keywords:** validation, rock testing methods, physical properties

## 1 Introduction

Testing of physical properties of rock material used in construction should be based on standardized methods. The assessment of results obtained during such tests makes it easier to determine whether individual raw materials or products comply with declared requirements. In addition to normalised test methods, the literature mentions other ways to determine the properties of rock, stone products or aggregate [1, 2, 3]. Proprietary research procedures, even those based on known rules, also standardized but modified methods, widened or applied out of range, should be validated prior to use [4]. The validation is a process that allows a verification and confirmation and provides an objective proof that requirements pertaining to a research method have been complied with and that the obtained result will be reliable, dependable and consistent [4, 5, 6]. This means that it has to be proven that it is possible to determine the parameters characterising a given method, and then use

*Corresponding author: hydzik@agh.edu.pl

such a determination to conclude if the method is useful and meets the requirements. The validation is always a balance between costs, risk and technical capabilities of a laboratory. The validation of chemical methods is widely described in the references [7, 8, 9]. In the case of validation of physical properties testing, particularly of rock materials, the literature does not give too much detail. It is difficult to find specific examples of compatibility assessment of test results, accounting for dispersion and the necessity to perform tests on a few or even a few dozen specimens from the same batch. The unique character of testing the physical and mechanical properties of rock lies in the fact that there is a large variability of rock parameters within a single formation or even a deposit. The standards require that individual tests be made on a specific number of specimens representing a material, and the results be given as an arithmetic mean or a higher/lower expected value, accounting for the dispersion of results and the specimen size.

The paper presents mathematical formulas allowing for an evaluation of the precision of research methods and the consistency of results which are a basis of the testing method validation. The practical part uses an example of apparent density determination in a validation of a proprietary method based on the analysis of repeatability, intralaboratory reproducibility and interlaboratory reproducibility.

## 2 Testing Method Design and Assessment Criteria

The development of a testing procedure should particularly account for a correct identification and scope of a given property and description of the object to be tested. The requirements for the measuring instruments should be defined in terms of the equipment and environment conditions, the instruments need to be calibrated, and standards and reference materials should be used, if applicable. The research procedure description must also indicate the data to be recorded and the data analysis method, criteria for the evaluation of results, and the algorithm to estimate the uncertainty of results [4, 10].

The following assessment methods can be used to verify the test method applicability:

- calibration or precision assessment using standards or reference materials;
- regular assessment of factors affecting the result;
- research method resistance to variability of controlled parameters;
- comparison of results obtained with different methods;
- interlaboratory comparison;
- results of uncertainty evaluation.

According to the requirements of PN-EN ISO/IEC 17025:2018-02, it is possible to use one assessment method or a combination of methods. The value of parameters characterising the research method accuracy generally depends on the type of tested objects or materials and on the value level of the measured property. The parameters characterising the method accuracy include the dispersion of results and conformity of results with the measured value [6, 7, 9, 10 - 12].

In terms of measures of the dispersion of results, one can mention precision and resistance. The test result precision is evaluated based on repeatability and reproducibility. The repeatability is a degree of conformity of the results of successive measurements of the same magnitude, conducted in the same measurement conditions. The repeatability conditions include the same measurement procedure, the same observer, the same measuring instrument used in the same conditions, the same place and repetition at short time intervals. Reproducibility, on the other hand, is a degree of conformity of the results of measurements of the same magnitude performed at different measurement conditions, including e.g. measurement methods, personnel, measuring equipment, reference standards, place, application conditions and time. There is an interlaboratory reproducibility (results obtained in different laboratories) and intralaboratory reproducibility (results obtained in one laboratory). The resistance is a measure of the test method capability of providing identical results despite slight changes of the method parameters. The resistance testing involves identification of method parameters which can significantly affect the measurement results, and examination of the impact of small, intentional changes of these parameters on the obtained results. The tested parameters can include environment parameters (e.g. temperature) and/or test method parameters (e.g. a reagent volume).

The parameters characteristic for the assessment of results conformity with the measured magnitude include the method bias (correctness) and selectivity. The test method bias is a difference between the expected test result and the accepted reference value. The selectivity means the method's ability to respond to a correct measured value in the presence of factors that prevent a correct determination. This is a qualitative parameter, and knowing it is particularly important in testing chemical properties. Other method parameters are also important in chemical tests, such as limit of quantification, limit of detection and linearity. The last, very important parameter used to assess the test method is the measurement uncertainty which describes the measured value dispersion.

## 3 Determination Methods of Apparent (Bulk) Density of Rock Materials

The apparent (volumetric) density is the ratio of the mass of a dried specimen to the specimen total volume including pores.

$$\rho_b = \frac{m_d}{V_b} \tag{1}$$

where:

$m_d$ - dried specimen mass, kg,

$V_b$-specimen total volume including pores, $m^3$.

Knowing the rock volumetric density is very important and used, for example, in the determination of other physical and mechanical properties or horizontal and vertical pressures in the rock mass. References [1, 2, 3] mention many apparent density determination methods. Depending on the rock type and condition (firm, cohesive or loose), the following rock volume determination methods can be used:

- displacement of a non-wetting liquid (e.g. mercury) by the specimen:
- direct method on regular specimens,
- covering the specimen with paraffin and immersion in water,
- previous specimen saturation with water,
- testing the specimens of intact cohesive rock in a ring,
- loose rock testing in a cylinder.

The paper analyses the volumetric density determination results obtained with the use of the proprietary procedure and the following standardized methods:

- for natural stone according to PN-EN 1936:2010 [13],
- for stone materials according to PN-B-04100:1966 [14],
- for aggregates according to PN-EN 1097-6:2013 [15].

According to PN-EN 1936:2010 [13], the volumetric density is determined on regular specimens cut with a diamond saw (cylinders, cubes or prisms) of a minimum volume of 60 $cm^3$. In addition, the area-to-volume ratio should be from 0.08 $mm^{-1}$ to 0.20 $mm^{-1}$.The test is performed using the hydrostatic method on specimens previously saturated with water. Air is removed from the specimens prior to saturation by maintaining the 2.0±0.7 kPa pressure in the vacuum vessel. The method from PN-B-04100:1966 [14] also refers to cubic or cylindrical specimens of the side (diameter) equal to 50 mm. The specimen volume is calculated from averaged side (diameter) values. PN-EN 1097-6:2013 [15] is for aggregate testing, however aggregates with grain size above 31.5 mm can be treated as small rock fragments. The method recommended in this standard - the so-called wire-basket method - is based on the hydrostatic method, but all weighing takes place for the entire volume of a given aggregate.

Regardless of the shape and type of the stone material, the volume in the hydrostatic method is determined according to the Archimedean principle as the volume of displaced liquid and i.e. equal to:

$$V_b = \frac{m_s - m_h}{\rho_{rh}} \quad (2)$$

where:

$m_s$ - mass of material totally saturated with water in air, kg,

$m_h$ - mass of material totally saturated with water in water, kg,

$\rho_{rh}$ - water density at test temperature, kg/m$^3$.

The proprietary procedure is also based on the hydrostatic method, but it differs from the method referenced in the standard [13] in that it is used for testing specimens of irregular (any) shape fragments chipped away from an undisturbed rock. In addition, prior to weighing, the specimens are saturated to obtain constant mass under the atmospheric pressure (the difference in two successive weighings24-hours apart must not exceed 0.1% of the initial mass). Regular specimens from standardized methods will always have the area-to-volume ratio as small as possible, and a smooth surface created by cutting with a diamond saw. On chipped specimens, on the other hand, the surface area will be highly diversified, and the surface will be very rough, sometimes not free of cracks. This results in the collection of a much larger amount of water on the surface and consequently can distort the specimens' volume. The specimens' saturation with water (vacuum in the standard method, and pressure saturation in the proprietary method) does not affect the volume because in both cases the surface pores are closed and do not allow a further specimen saturation during the weighing on a hydrostatic scales. The vacuum specimen saturation is relevant only in the assessment of open porosity.

## 4 Measures of test method compatibility assessment

The following methods have been applied to check the test results: repeatability and intralaboratory reproducibility, and interlaboratory comparisons taking into account the measurement uncertainty [6, 7, 11, 12].

The repeatability standard deviation is one of the measures of the method accuracy and is determined based on the results of independent repeated measurements of the same object. The repeatability standard deviation is calculated based on many series of results. *J* series with *K* independent repetitions of measurements of the same magnitude were conducted. The measurements in each individual series were performed under the repeatability conditions, but they related to different objects with similar properties (many specimens from the same rock type). The tests were performed using the PN-EN 1936:2010 method on regular specimens, and on irregular specimens according to the test method to be validated. The standard deviation evaluation $s_{rj}$ was calculated for each series based on the obtained $x_{jk}$ results.

$$S_{rj} = \sqrt{\frac{1}{K-1}\sum_{k=1}^{K}\left(x_{jk} - \bar{x}_j\right)^2} \quad (3)$$

where:

$\bar{x}_j$- arithmetic mean from the *j*-th series results under the repeatability conditions.

Then, the repeatability standard deviation (4) and the repeatability limit *r* (5) were calculated:

$$S_r = \sqrt{\frac{1}{J}\sum_{j=1}^{J} S_{rj}^2} \tag{4}$$

$$r = 2.8 \cdot S_r \tag{5}$$

The absolute difference between two results obtained under the repeatability conditions should not be greater than the repeatability limit *r* determined with the 95% probability.

The reproducibility standard deviation was determined for variable conditions but within the same laboratory (intralaboratory reproducibility). Again, *J* series (2 test methods) with *K* independent repetitions of measurements of the same magnitude were conducted. The assessment was performed for the volumetric density testing with the use of various instruments, i.e. the volumetric density test results according to PN-B-04100:1966 [14]) were compared with the results obtained with the use of hydrostatic (proprietary) method. Similar to the repeatability, the standard deviation evaluation $S_{Rk}$ (6) and the reproducibility standard deviation $S_R$ (7) were calculated:

$$S_{Rj} = \sqrt{\frac{1}{K-1}\sum_{k=1}^{K}\left(x_{jk} - \bar{x}_j\right)^2} \tag{6}$$

where:

$\bar{x}_j$ - arithmetic mean from the *j*-th series results under the reproducibility conditions

$$S_R = \sqrt{\frac{1}{J}\sum_{j=1}^{J} S_{Rj}^2} \tag{7}$$

Again, the absolute difference between two results obtained under the reproducibility conditions should not be greater than the reproducibility limit *R* (8):

$$R = 2.8 \cdot S_R \tag{8}$$

The t-Student's test was also used to evaluate the method correctness under the intralaboratory reproducibility conditions. The obtained results were used to calculate mean volumetric density values using the proprietary method $\bar{x}_A$ and the standardized method $\bar{x}_B$, and then the difference $\Delta\bar{x}$. The *t* statistics were calculated to evaluate the statistical significance of this difference (9).

$$t = \frac{|\Delta\bar{x}|}{S_{\Delta\bar{x}}} \tag{9}$$

where:

$S_{\Delta\bar{x}}$ - standard deviation of the difference of the means, equal to (10):

$$S_{\Delta\bar{x}} = \sqrt{\left(S_{RA}^2 - \frac{K_A-1}{K_A}\cdot S_{rA}^2\right) + \left(S_{RB}^2 - \frac{K_B-1}{K_B}\cdot S_{rB}^2\right)} \tag{10}$$

where:

$S_{rA}$ and $S_{rB}$ - repeatability standard deviation for methods A and B, respectively,

$S_{RA}$ and $S_{RB}$ - interlaboratory reproducibility standard deviation for methods A and B,

$K_A$, $K_B$ - number of repetitions in each series.

If the $t$ statistics value is greater than that the critical value resulting from the t-Student's distribution for the number of degrees of freedom $v_A$ and $v_B$ of estimators $S_{RA}$ and $S_{RB}$ and the agreed confidence level (usually 95%), then the bias resulting from the various test methods is statistically significant.

The $E_n$ test calculated according to the following formula recommended by ISO [11] (11) is very often used to determine the compatibility of results obtained in different laboratories (PT testing):

$$E_n = \frac{|\bar{x}_A - \bar{x}_B|}{\sqrt{U^2(\bar{x}_A) + U^2(\bar{x}_B)}} \quad (11)$$

where:

$\bar{x}_A$, $\bar{x}_B$ - results of tests of the same specimen obtained inlaboratories A and B,
$U(\bar{x}_A)$, $U(\bar{x}_B)$ - expanded uncertainties obtained inlaboratories A and B.

If the $E_n$ value is less than one, then the results obtained in the two laboratories are consistent.

Both the repeatability and the reproducibility developed at the method validation stage are a basis for estimating the expanded uncertainty of the test result $U(x)$. The measurement uncertainty is a parameter related to the measurement result, characterising the dispersion of values that can be reasonably assigned to the measured magnitude. Such a parameter can be a standard deviation or a coefficient of variation. The laboratory tests use a 95% confidence level and the interval around the test result is built for this probability [10].

## 5 Comparative Analysis of Volumetric Density Measurements

In order to validate the method proposed in the proprietary procedure, the results for specimens prepared from selected rock materials using the same method (hydrostatic), but of different shapes, were compared, i.e. the density results for regular specimens according to PN-EN 1936:2010) and the results in the test involving a direct measurement of the regular specimen sides(according to PN-66/B-04100). The determination was performed in the same laboratory, using the same equipment and personnel. The measurement accuracy measure used in the analysis was repeatability and intralaboratory reproducibility. In addition, the proficiency test was performed, involving the comparison of results from the proprietary procedure with the results from the other laboratory that measured the density according to the wire-basket method from PN-EN 1097-6:2013. The results were evaluated using the $E_n$ test.

Table 1 includes mean values with the standard deviation estimation for the volumetric density tests of the same materials tested with the use of the standardized method on cubic specimens and with the proprietary method on irregular specimens. These data were used to calculate the repeatability standard deviation for each rock material and the repeatability limit $r$, and also in the conformity assessment. The difference between the mean values of volumetric density determined using the hydrostatic method on regular specimens and on irregular specimens for the same rock materials is in the 2 - 45 kg/m$^3$ range at the repeatability limits from 29 - 176 kg/m$^3$. The absolute difference values between the results for individual rock materials did not exceed the values of corresponding repeatability limit $s_r$.

Table 2 presents the results of volumetric density measurements on rock specimens under the intralaboratory reproducibility conditions.

**Table 1.** Results under the repeatability conditions and the conformity assessments parameters.

| Rock type | Test method | Number of samples | $\rho_b$ [kg/m³] | $S_{rj}$ [kg/m³] | $S_r$ [kg/m3] | $r$ [kg/m3] | $\Delta x$ [kg/m³] | Assess-ment |
|---|---|---|---|---|---|---|---|---|
| Granite | proprietary | 50 | 2612 | 10 | 11 | 29 | 2 | yes |
| | PN-EN 1936:2010 | 10 | 2614 | 11 | | | | |
| Cergowa sandstone | proprietary | 50 | 2633 | 19 | 17 | 48 | 20 | yes |
| | PN-EN 1936:2010 | 36 | 2653 | 15 | | | | |
| Basalt | proprietary | 50 | 2890 | 64 | 63 | 176 | 32 | yes |
| | PN-EN 1936:2010 | 10 | 2922 | 62 | | | | |
| Krosno sandstone | proprietary | 8 | 2548 | 41 | 35 | 97 | 45 | yes |
| | PN-EN 1936:2010 | 7 | 2593 | 27 | | | | |

**Table 2.** Results under the intralaboratory reproducibility conditions and the conformity assessments parameters.

| Rock type | Test method | Number of samples | $\rho_b$ [kg/m³] | $S_{Rj}$ [kg/m³] | $S_R$ [kg/m3] | $R$ [kg/m3] | $\Delta x$ [kg/m³] | Assess-ment |
|---|---|---|---|---|---|---|---|---|
| Granite | proprietary | 10 | 2635 | 32 | 35 | 99 | 14 | yes |
| | PN-66/B-04100 | 10 | 2621 | 38 | | | | |
| Crystalline limestone | proprietary | 12 | 2628 | 32 | 40 | 111 | 30 | yes |
| | PN-66/B-04100 | 12 | 2599 | 46 | | | | |
| Thick-bedded sandstone | proprietary | 8 | 2548 | 41 | 39 | 108 | 25 | yes |
| | PN-66/B-04100 | 14 | 2523 | 36 | | | | |
| Thin-bedded sandstone | proprietary | 7 | 2593 | 27 | 38 | 106 | 2 | yes |
| | PN-66/B-04100 | 14 | 2591 | 46 | | | | |
| Bazalt | proprietary | 10 | 2922 | 62 | 62 | 174 | 18 | yes |
| | PN-66/B-04100 | 10 | 2904 | 63 | | | | |
| Limestone | proprietary | 10 | 2550 | 195 | 141 | 395 | 14 | yes |
| | PN-66/B-04100 | 15 | 2564 | 41 | | | | |

The $t$ statistics from the t-Student's distribution were also used to assess the significance of the difference between the mean volumetric density values determined with two independent methods (Table 2). The statistics parameters for individual rock materials are given in Table 3.

The difference between mean volumetric density values tested under the reproducibility conditions is in the 2-30 kg/m³ range, at the reproducibility limits from 99 to even 395 kg/m³. The absolute difference values between the results for individual rock materials did not exceed the corresponding values of reproducibility limits $R$. In addition, the analysis of the statistical significance of difference between the results did not show the impact of the test method on the results.

The $E_n$ test was used in the case of interlaboratory tests and its values are presented in Table 4.

**Table 3.** *t*-statistics.

| **Rock type** | $S_{\Delta\bar{x}}$ | $t$ | **Critical value from t-Student's distribution** | **Assessment** |
|---|---|---|---|---|
| Granite | 16 | 0.88 | For the number of degrees of freedom from 6 - 14, it is from 2.1448 to 2.4469 | No significant impact on the test results |
| Crystalline limestone | 16 | 1.83 | | |
| Thick-bedded sandstone | 17 | 1.41 | | |
| Thin-bedded sandstone | 16 | 0.13 | | |
| Basalt | 28 | 0.64 | | |
| Limestone | 63 | 0.22 | | |

**Table 4.** Results and conformity assessment in interlaboratory tests.

| | $\rho_b$ [kg/m$^3$] | $S_j$ [kg/m$^3$] | $U(x_j)$ | $E_n$ | **Conformity assessment** |
|---|---|---|---|---|---|
| Tests according to the proprietary method | 2332 | 45 | 49 | 0.36 | The $E_n$<1 results obtained in two laboratories are conforming |
| Test according to PN-EN 1097-6:2013 at an external lab | 2310 | - | 35 | | |

All analysed cases of the method precision assessment gave similar results, indicating that the suggested volumetric density testing method can be used for determination of these parameters, and the obtained results conform with the results obtained with the use of standardized methods.

## 6 Summary

The presented analysis aimed at showing the possibility of validation of a testing method for the rock material physical properties. The methodology and the possibilities of assessment of the results obtained using different testing methods, different equipment and even in different laboratories were presented on a simple example of volumetric density determination of rock materials.

The method validation did not show significant differences between mean volumetric density values obtained with the application of various testing methods. It was proven that the initially assumed factors (specimen shape, surface roughness, saturations methods) have no significant impact on the test results. However, as the rock material is very heterogeneous, a sufficient number of specimens of the same material has to be made.

The scientific research often aims at discovering new, hitherto unknown properties, using modern equipment, or testing at unstandardized external or environmental conditions, e.g. extremely high or low temperatures. The results of tests with a method not used to date can be unreliable and inconsistent. In order to support the obtained results and to ensure their correctness, all result-affecting factors must be identified and then the test method must be validated.

*This study was funded by statutory research funds of the AGH University of Science and Technology within framework of the research program No. 11.11.100.197.*

## References


1. T. Majcherczyk, *Zarys fizyki skał i gruntów budowlanych (Outline of the physics of rocks and building land)*, Library of Underground Mining School, Cracow, Series with Mining Light No. **5** (2000) (in Polish)
2. R. Ulusay, J.A. Hudson (eds), *The Blue Book. The Complete ISRM, Suggested Methods for Rock Characterization, Testing and Monitoring:1974-2006*, ISRM Turkish National Group, Ankara (2007)
3. E.M. Winkler, *Stone in Architecture, Properties, Durability Third, Completely Revised and Extended Edition Corrected 2nd printing*, Springer-Verlag Berlin Heidelberg (1997)
4. PN-EN ISO/IEC 17025:2018-02 *Ogólne wymagania dotyczące kompetencji laboratoriów badawczych i wzorcujących (General requirements for the competence of testing and calibration laboratories)* (in Polish)
5. PN-EN ISO 9000:2015 *Systemy zarządzania jakością - Podstawy i terminologia (Quality management systems - Fundamentals and vocabulary)* (in Polish)
6. PN-ISO 5725:2002 *Dokładność (poprawność i precyzja) metod pomiarowych i wyników pomiarów (część 1-6)* (*Accuracy (trueness and precision) of measurement methods and results (part 1-6))* (in Polish)
7. P. Burmistrz, K. Bytnar, K. Kogut, P. Rychcik, S. Stelmach, *Credibility of results of testing hard coal*, Mineral Resources Management **24**, 3/3, 33–47 (2008)
8. Z. Burnus, *Problematyka walidacji metod badań w przemyśle naftowym na przykładzie benzyn silnikowych (The problem of validation of test methods in the oil industry on the example of motor gasoline)*, Nafta-Gaz **2**, LXIX, 143-152 (2013) (in Polish)
9. A. Wyciślik, *Determinanty wyboru i walidacji metody analitycznej w badaniach materiałoznawczych (Determinants of selection and validation of the analytical method in materials science research)*, Inżynieria Materiałowa **3**, XXVI, 135–157 (2005) (in Polish)
10. EA-04/16 *Wytyczne dotyczące wyrażania niepewności w badaniach ilościowych (Guidelines for expressing uncertainty in quantitative research)* https://www.pca.gov.pl/publikacje/dokumenty/ea-/ (2003) (in Polish)
11. PN-ISO/IEC 17043:2011 *Ocena zgodności - Ogólne wymagania dotyczące badania biegłości (Conformity assessment - General requirements for proficiency testing)* (in Polish)
12. E. Volodarsky, Z. Warsza, L. Koshevaya, *System oceny statystycznej w badaniu biegłości laboratoriów badawczych (Statistical evaluation system in proficiency testing of research laboratories)* PAK **60**, 10, 816–821 (2014) (in Polish)
13. PN-EN 1936:2010 *Metody badań kamienia naturalnego - Oznaczanie gęstości i gęstości objętościowej oraz całkowitej i otwartej porowatości (Natural stone test methods. Determination of real density and apparent density, and of total and open porosity)* (in Polish)
14. PN-B-04100:1966 *Materiały kamienne - Oznaczanie gęstości objętościowej, gęstości, porowatości i szczelności (Stone materials - Determination of bulk density, density, porosity and tightness)* (in Polish)
15. PN-EN 1097-6:2013-11 *Badania mechanicznych i fizycznych właściwości kruszyw - Część 6: Oznaczanie gęstości ziarn i nasiąkliwości (Tests for mechanical and physical properties of aggregates - Part 6: Determination of particle density and water absorption)* (in Polish)

# The effect of micro- and nanosilica on the soil permeability coefficient under cyclic freezing and thawing conditions

*Zofia* Zięba*, *Kinga* Witek and *Jakub* Mońka

Wrocław University of Environmental and Life Sciences, Wrocław, Poland

**Abstract.** The paper presents the analysis of the permeability coefficient of frost-susceptible soil with microsilica (MS) and nanosilica (NS) addition. Tests were performed in a triaxial apparatus in three variants: on pure soil samples, on soil samples with a 5% MS and on soil samples with a 5% NS addition. Because of the frost-susceptible properties of analysed soil, the permeability coefficient was determined on unfrozen samples and on samples after 10 cycles of freezing and thawing. The preliminary test results demonstrated that both microsilica and nanosilica have beneficial properties related to subsoil sealing. These properties are considerably stronger for nanosilica. Also, in all cases, the permeability coefficient increased after 10 cycles of freezing and thawing, but the change trend remained the same. Nanosilica shows particularly good sealing properties in frost-susceptible soils. This confirms that it may be applied as a separate additive, which is not commonly used in engineering practice. Nano particles are usually used to extend the scope of micro additives' influence. In order to recognise the beneficial influence of analysed additives on soil permeability, the recognition of the changes in microstructure is necessary.

**Keywords:** microsilica, nanosilica, permeability coefficient, frost-susceptible soil, cyclic freezing and thawing

## 1 Introduction

The key factor affecting the suitability of soil for development is its geotechnical parameters. A special group of soils are frost-susceptible soils, whose capacity to increase their volume due to the freezing of the water contained in them is a frequent cause of damage or destruction to buildings, roads and hydrotechnical structures [1, 2].

In order to reduce the susceptibility of soils to freezing processes, various improvement methods are used [3, 4].

The frost heave of soils is closely related to the formation of ice lenses, which increase due to the freezing of migrating water [5, 6]. Water movement through the porous medium is a complex issue that depends on many factors, including its microstructure [7]. For this

* Corresponding author: zofia.zieba@upwr.edu.pl

reason, the methods of improving the properties of frost-susceptible soils can also include the decrease of their permeability coefficient.

In engineering practice, various stabilising admixtures are used to improve the soils properties [8, 9], including nano additives. Among them, nanosilica is the most commonly used [10, 11], although aluminium oxide [12], polymer compounds [13], organic silica [14], or nanocopper [15] have been also used.

The research on the effects of nanosilica on the formation of soils geotechnical parameters conducted so far has been mainly concerned with the study of basic physical and mechanical properties [10, 16, 17]. There is no research to determine the effects of a given nano additive on the permeability proprieties of frost-susceptible soils.

Research was also conducted on the beneficial changes in the parameters of fine-grained soils stabilised with microsilica [18]. It can therefore be assumed that with the same stabilising additive, but in a nanoscale, the change trend of soil parameters will be similar. However, they should occur to a much greater extent, which results from the properties of nanomaterials [19].

The goal of this work is to recognise the effect of micro- and nanosilica on the permeability process in frost-susceptible soil. The variability of the permeability coefficient, which is the basic parameter describing water movement in porous media, has been analysed in detail. In addition, the use of selected stabilising additives will allow the analysis of changes resulting from their transition from micro- to nanoscale.

## 2 Materials

The testing was conducted on the soil obtained from the area of Trzebnica Hills in the Lower Silesia province in Poland (Fig. 1), which shows susceptibility to frost heave. Its grain size distribution was determined by a standard hydrometer method according to PN-EN 1997-2:2007 [20]. Based on the resulting grain-size distribution curve, it was established that it is a sandy clayey silt (saclSi) [21].

**Fig. 1.** Physico-geographical regionalisation of Poland – Trzebnica Hills Mezoregion [22].

Two additives were used for stabilisation, which are often used in the construction industry, but these are not typical additives used in geotechnical engineering.

The first additive was microsilica (MS), also called silica fume. Its grain size distribution was determined by using a laser granulometer (Fig. 2a).

The second stabilising additive was nanosilica (NS) in the form of a colloid. To determine the size distribution of nanoparticles, tests were carried out using the Zetasizer Nano particle characterisation system (Fig. 2b).

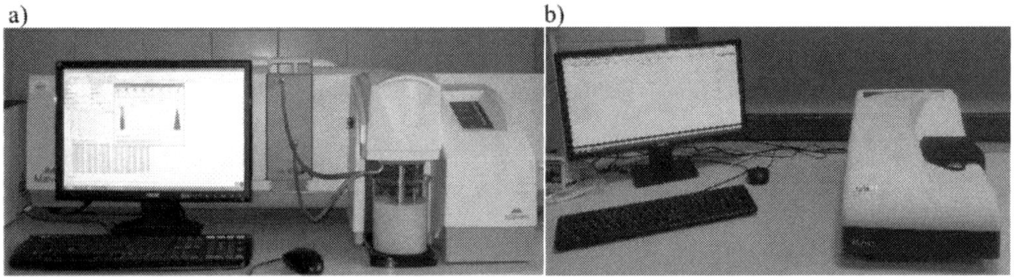


**Fig. 2.** The grain size distribution analysis of the stabilising additives: a) MS – laser granulometer; b) NS – Zetasizer Nano particle characterisation system.

The differences in grain size distribution of individual additives in relation to the analysed soil are shown in Figure 3.

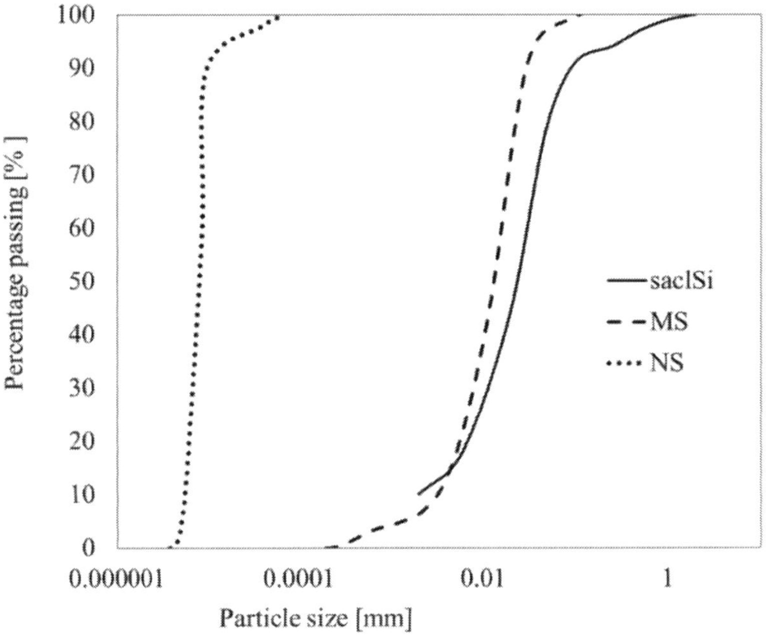


**Fig. 3.** Grain size distribution of the analysed materials.

## 3 Methods

Laboratory tests of the permeability coefficient were performed in three variants. The first included pure soil, the second was soil with a 5% admixture of microsilica, and the third was soil with a 5% admixture of nanosilica. For all variants, the tests were performed on 15 samples.

The contents of individual additives and methods for their mixing with soil were determined in the case of microsilica based on Kalkan's test results [18], and in the case of nanosilica based on Seyedi Gelsefidi and Mamaghanian's paper [16], as well as Bahmani and others' [10].

The main assumption of the tests was the analysis of samples at the maximum dry density $\rho_{dmax}$ and the corresponding optimum moisture content $m_{opt}$. These parameters were determined in the Proctor automatic compactor in accordance with ASTM D698 [23].

Unfrozen samples and samples after 10 cycles of freezing and thawing were tested for permeability. This allowed for the checking of the behaviour of individual soil mixtures in real conditions occurring in winter. This approach also made it possible to determine changes in the permeability coefficient depending on the stabilising additive used, as a result of the effects of cyclic freezing and thawing on the change in the structure of pore space.

Cyclic freezing and thawing were carried out according to PN-88/B-06250 [24] in the Weiss C 600 climate chamber. This device has a temperature range between -70 and +180°C, as well as two inspection holes to connect the external sensors. Thanks to this, the freezing and thawing temperatures could be monitored during the whole experiment. Additionally, the device is equipped with a panel, which enables the chamber control and automatic data acquisition.

The freezing and thawing process was carried out in a closed system, without any influx of water. This resembles the condition of low groundwater levels or artificially-induced lack of water in the frost zone. Most of all, research in such conditions allowed for the recognition of the fluctuation of the permeability coefficient, which results from the changes in the microstructure caused by the freezing of water in soil of a particular moisture content - in this case of optimum moisture content. Each cycle took place at -20°C/+20°C.

The testing of the permeability coefficient for both unfrozen samples and samples subjected to cyclic freezing and thawing were performed in the triaxial cell apparatus chamber in accordance with BS 1377-6:1990 [25]. Thanks to the apparatus used, the problem of the formation of privileged filtration pathways, which occur in typical permeameters, was eliminated. The tests were carried out on samples of 100 mm in diameter and 110 mm in height, with an effective cell pressure of 50 kPa and a hydraulic gradient of 20 [-].

The stages of preparation of a sample for testing are shown in Figure 4.

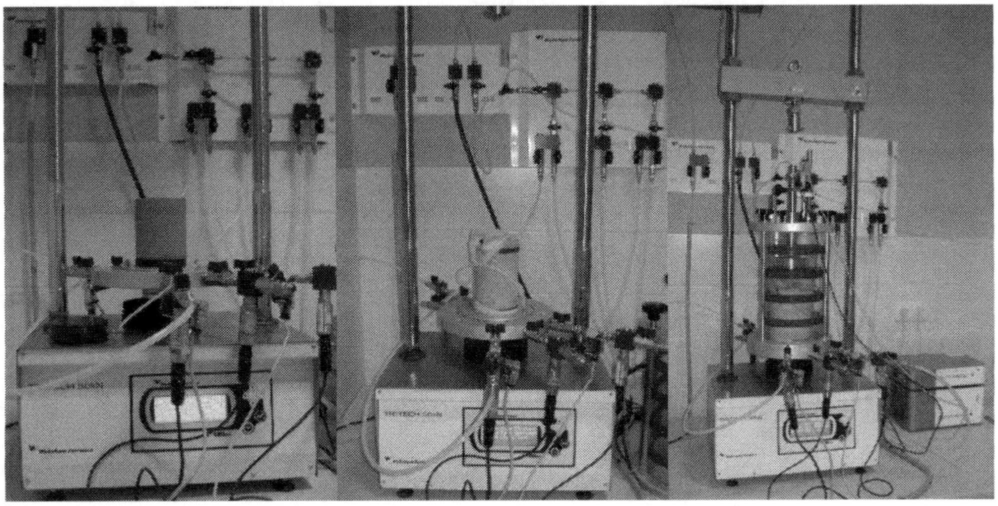

**Fig. 4.** The stages of sample preparation for permeability coefficient testing.

Before the tests, the samples underwent the saturation process to reach the value of the Skempton's coefficient (B-value) at a level of 0.95 [-]. To the permeability tests, the distilled water was used which was additionally deaerated using a vacuum pump. The flow of the water during the tests was upward.

# 4 Results and discussion

## 4.1 Compactibility testing

The compactibility testing of soil mixtures showed that the addition of microsilica causes a slight increase in the optimum moisture content and at the same time a decrease in the maximum dry density of soil structure. On the other hand, the soils stabilised with nanosilica show an increase in the maximum dry density of soil structure, while maintaining the optimum moisture content at a comparable level (Table 1).

**Table 1.** Summary of optimum moisture contents and maximum dry densities for individual soil mixtures

| Type of soil mixture | $\rho_{max}$ | $m_{opt}$ |
|---|---|---|
| [-] | [g·cm$^{-3}$] | [%] |
| saclSi | 1.882 | 11.80 |
| saclSi + 5% MS | 1.868 | 12.30 |
| saclSi + 5% NS | 1.912 | 11.60 |

## 4.2 Permeability coefficient testing

Fifteen samples from each variant of soil mixture were tested, both for unfrozen samples and samples subjected to cyclic freezing and thawing.

In order to recognise the arithmetic mean value as representative, a statistical analysis of the obtained results was carried out at a significance level of 0.05 [-]. First, the arithmetic mean value and standard deviation were calculated, and then the dispersion of the results in relation to the mean value based on the coefficient of variation was determined. It is assumed that if its value does not exceed 10%, the variation of a characteristic is statistically insignificant. Otherwise, extreme values must be rejected [26]. The analysis was carried out using the Statistica software. The values of the permeability coefficient averaged after the statistical analysis are shown graphically (Fig. 5).

The addition of 5% microsilica decreased the permeability coefficient of the analysed soil by 43%. This change trend coincides with the results of research by Kalkan and Akbulut [27] and Kalkan [18].

On the other hand, the addition of 5% nanosilica caused a decrease in permeability coefficient by almost 100%. No studies have been carried out so far to determine the effects of nanosilica addition itself on the permeability coefficient. Only mentions of the studies of permeability coefficient of soils stabilised with cement with an addition of nanosilica can be found in the literature [10].

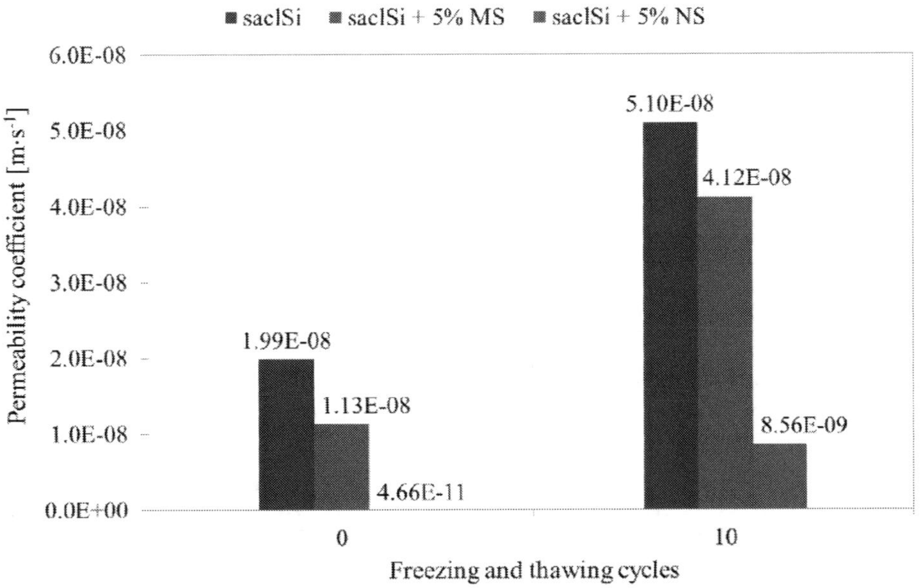


**Fig. 5.** Variability of the permeability coefficient depending on the stabilising additive before and after 10 cycles of freezing and thawing.

There was an increase in permeability coefficient observed as a result of cyclic freezing and thawing (10 cycles) in all cases. However, the stabilisation with nanosilica shows much lower values of the permeability coefficient (by 64%) than the stabilisation with microsilica. Also, in relation to the pure soil, the stabilisation with nanosilica showed a decrease in permeability coefficient by 83%. Thus, nanosilica has very favourable sealing properties for frost-susceptible soils.

The use of stabilising materials is strictly connected with changes in the structure of the pore space, which heavily influences both, the movement of the water in the soil, and the amount of bound water that is not involved in the flow. Additionally, changes in the microstructure also influence on the heat transfer and distribution of the temperature [28], which is crucial, especially in case of the stabilisation with nanomaterials in the process of freezing and thawing. For this reason, in the further stages of research it is necessary to recognise the changes in the pore space structure of non-frozen and frozen samples in each variant of soil mixtures.

## 5 Conclusions

The tests carried out allowed for the preliminary recognition of the effect of micro- and nanosilica on the process of water flow in frost-susceptible soils. Based on this, the following conclusions were drawn:

1. Both a 5% addition of microsilica and nanosilica reduced the permeability coefficient of frost-susceptible soil. However, nanosilica shows the capability of sealing frost-susceptible subsoil to a much greater extent.
2. After 10 cycles of freezing and thawing, all samples showed an increase in the permeability coefficient, however, samples stabilised with nanosilica were still characterised by the lowest values.
3. The tests carried out confirm possibilities of using nanosilica as an independent additive, which is not widely used.

4. In order to explain in detail the changes in the permeability coefficient, it is necessary to analyse the microstructure of both, the pure soil and the samples stabilised with individual additives.

## References

1. L. Rafalski, J. Wilczek, C. Kraszewski, *Ochrona przeciwmrozowa nawierzchni drogowych na przykładzie wybranych krajów (The case of antifreeze protection of road pavements in selected countries),* Drogownictwo **2**, 39-45 (2014) (in Polish)
2. A.A. Korshunov, S.P. Doroshenko, A.L. Nevzorov, Procedia Engineering, **143**, 682-688 (2016) doi : 10.1016/j.proeng.2016.06.100
3. I. Ickiewicz, *Zjawisko wysadziny zmarzlinowej – metody zapobiegania (The phenomenon of frost heave – methods of prevention)*, Izolacje **2**, 44-45 (2009) (in Polish)
4. K. Augustowski, *Możliwość zapobiegania powstawaniu wysadzin mrozowych (Possibilities of preventing soil movement with conjunction of frost action)*, Drogownictwo **7-8**, 264-268 (2012) (in Polish)
5. I. Ickiewicz, *Analiza zjawiska wysadziny zmarzlinowej dla celów inżynierskich (Analysis of frost heave phenomenon for engineering purposes)*, Tech Trans **5**, 63-70 (2009) (in Polish)
6. R.W. Style, S.S.L. Peppin, A.C.F. Cocks, J.S. Wettlaufer, Phys Rev E **84**, 4, 1-13 (2011) doi: 10.1103/PhysRevE.84.041402
7. Z. Zięba, J Hydrol Hydromech **65**, 1, 80-87 (2017) doi: 10.1515/johh-2016-0054
8. C. Onyelowe Ken, F.O. Okafor, *A comparative review of soil modification methods*, ARPN J Earth Sci **1**, 2, 36-41 (2012)
9. S.Z. Sharifah Zaliha, H. Kamarudin, A.M. Mustafa Al Bakri, M. Binhussain, M.S. Siti Salwa, Aust J Basic Appl Sci **7**, 5, 258-265 (2013) doi: 10.4028/www.scientific.net/KEM.660.298
10. S.H. Bahmani, B.B.K. Huat, A. Asadi, N. Farzadnia, *Stabilization of residual soil using SiO2 nanoparticles and cement*, Constr Build Mater **64**, 350–359 (2014)
11. A. Rezagholilou, H. Nikraz, *The Reasons for Introducing Nano-silica in Cementitious Layer in Pavement*, Electron J Geotech Eng **19**, 1761-1768 (2014)
12. H.L. Luo, D.H. Hsiao, D.F. Lin, C.K. Lin, Int J Transp Sci Technol **1**, 83-100 (2012) doi: 10.1260 2046 -0430.1 60 333
13. W.R. Azzam, Int J GEOMATE **7**, 927-937 (2014) doi: 10.21660/2014.13.29127
14. B.M. Lekha, S. Goutham, A.U. Ravi Shankar, *Laboratory investigation of soil stabilized with nano chemical*, Proc. of the IGC, 22-24, Roorkee (2013)
15. M.R. Taha, O.M.E. Taha, J Nanopart Res **14**, 10, 1-13 (2012) doi: 10.1007/s11051-012-1190-0
16. S.A. Seyedi Gelsefidi, J. Mamaghanian, Proc. of the 5th IYGEC, 134-137 (2013) doi: 10.3233/978-1-61499-297-4-134
17. R. Xiaochuan, H. Kai, Nanosci Nanotech Let **6**, 1010-1013 (2014) doi: 10.1166/nnl.2014.1857
18. E. Kalkan, Cold Reg Sci Technol **58**, 3, 130-135 (2009) doi: 10.1016/j.coldregions.2009.03.011

19. Z.H. Majeed, M.R. Taha, *A review of stabilization of soils by using nanomaterials*, Aust J Basic Appl Sci **7**, 2, 576-581 (2013)
20. PN-EN 1997-2:2007, *Geotechnical design – Part 2: Ground Investigation and testing*
21. PN-EN ISO 14688-2:2006/Ap2, *Geotechnical investigation and testing - Identification and classification of soil - Part 2: Principles for a classification*
22. Website: *https://commons.wikimedia.org/wiki/File:318.44_Wzg%C3%B3rza_Trzebnickie.png*
23. ASTM D698: 2003, *Standard Test Methods for Laboratory Compaction Characteristics of Soil Using Standard Effort*
24. PN-88/B-06250 *Beton zwykły (Ordinary concrete)* (in Polish)
25. BS 1377-6:1990, *Methods of test for soils for civil engineering purposes. Consolidation and permeability tests in hydraulic cells and with pore pressure measurement*
26. D.J. Sheskin, *Handbook of Parametric and Nonparametric, Statistical Procedures. 3rd ed.*, C&H/CRS, New York (2003)
27. E. Kalkan, S. Akbulut, *Eng* Geol **73**, 145-156 (2004) doi: 10.1016/j.enggeo.2004.01.001
28. J. A. Weaver, and R. Viskanta, *Melting of frozen, porous media contained in a horizontal or a vertical, cylindrical capsule,* Int J Heat Mass Transf, **29**, 1943-1951, (1986)
29. K. Domin, *Wpływ efektu nanouwięzienia na przejścia fazowe w nanoporowatych układach (Effects of Confinement on Phase Transitions in Nanoporous Systems)*, PhD thesis, Uniwersytet Adama Mickiewicza, (2016) (in Polish)

# S-wave velocity in samples of calcareous waste

*Elżbieta* Pilecka[1,*], *Jakub* Zięba[1], *and Rafał* Pilecki[2]

[1]Cracow w University of Technology, Faculty of Civil Engineering, Krakow, Poland
[2]Graduate from Cracow University of Technology, Faculty of Civil Engineering, Krakow, Poland

**Abstract.** The studies have been carried out in the calcareous waste area "White Seas" of the closed factory Solvay Sodium Plant in Krakow in southern Poland. The study was conducted to determine the S-wave velocity of samples of the anthropogenic subsoil. It was aimed at the possibility of using post-production waste for construction purposes. Due to the high humidity of the analysed samples, the tests were carried out on samples taken using Shelby type probes, from the 12m borehole. The saturation level of soil samples was estimated based on Skempton's law (B>0.95). The S-wave velocity was determined depending on the different degree of saturation as well as the different degree of consolidation of the soil sample. Only S-wave velocity was analysed because properties of samples caused difficulties in determining of the first arrivals of P-wave. The measurements were carried out in the triaxial apparatus equipped with Bender element (BE). Finally, the shear modulus G of the samples were calculated. Results showed the effect of S-wave velocity increase with increase of degree of saturation and with the increase of effective stress in tested samples. The S-wave velocity significant increased up to about 100 kPa, and then this increase was smaller.

**Keywords:** Solvay plant, calcareous waste, bender element test, P- and S- wave velocity

## 1 Introduction

In recent years, numerous geotechnical studies have been carried out in the area of the factory Solvay Sodium Plant in Krakow (which is currently closed) [1–7]. The factory produced sodium carbonate ($Na_2CO_3$) since 1906. As a result of the production, the studied waste was produced [6]. This waste developed in the form of a slurry and it was entirely deposited in sedimentation tanks. These tanks occupied a total area of over 30 hectares. The stored waste formed heap of nearly 20 meters high. In the 1990s, on the basis of hydrogeological and hydrotechnical studies, the environmental effect of the waste from the Solvay Plants was analysed [8], and then the reclamation of this area was decided. The designated plan [9] and further development plans in this part of Krakow forced the necessity of analysing the soil substrate and evaluating its suitability for construction purposes. Currently, in the area of "White Seas" in Krakow two construction projects are realized. These are the construction of

* Corresponding author: epilecka@pk.edu.pl

the Łagiewnicka Route by the enterprise "Trasa Łagiewnicka" [5] and the golf course implemented by the Golf & Spa Resort company. In both cases, post-production waste, which, as an anthropogenic soil occurs in the foundation level, was a problem related to the choice of the type of construction of future buildings.

The study was aimed at the possibility of using post-production waste for construction purposes. In addition, the Soil – Structure Interaction Division, Cracow University of Technology, also participated in research with the aim of determining the load capacity of piles [4] for future buildings included in the John Paul II Centre.

Due to the lack of data related to the deformation parameters of calcareous waste from the Solvay Sodium Plant in Krakow, the Soil – Structure Interaction Division, Cracow University of Technology, conducted research to determine the dynamic shear modulus G of the subsoil samples from the area of "White Seas" [6, 7]. Only dynamic shear modulus G was analysed because properties of samples caused difficulties in determining of the first arrivals of P-wave. The G modulus of the samples was calculated depending on the different degree of saturation as well as the different degree of consolidation of the soil sample. The tests were carried out on samples taken using Shelby type probes. In the study it was necessary to modify the probe to receive a sample with an undisturbed structure.

In the paper, a detailed modification of the sampler is presented. We also presented methodology of laboratory test and finally the results of the determination of the S-wave velocity and dynamic shear modulus G for the calcareous samples.

## 2 Undisturbed samples

In the study, for the P- and S- wave velocity determination by the laboratory method, the Shelby sampler was adapted to take samples with an undisturbed structure (Fig. 1).

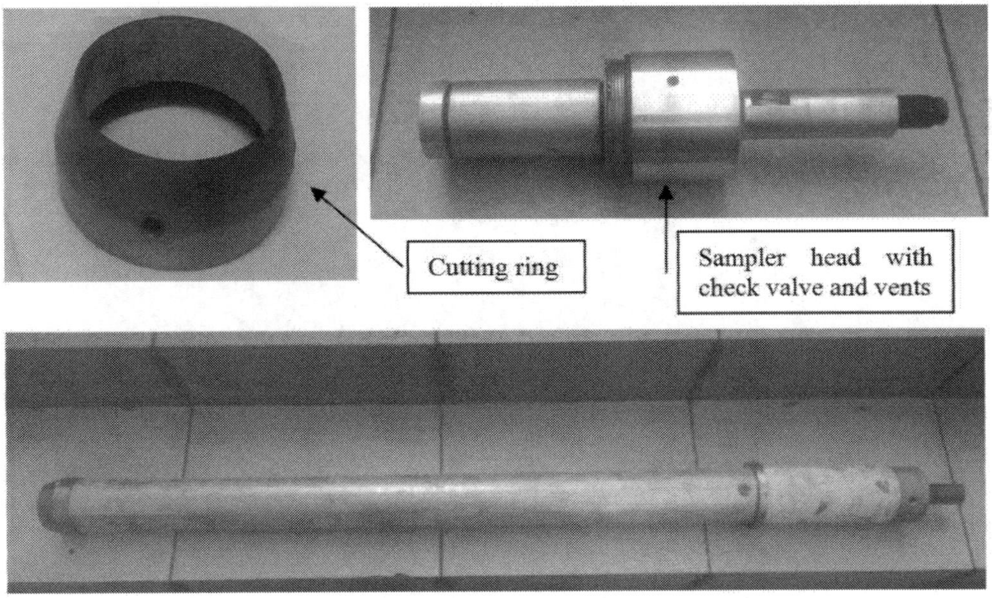


**Fig. 1.** Modified thin-wall open drive soil sampler.

Due to the high humidity of the analysed samples, i.e. around 200%, the soil exhibited large deformations even at low pressure. In addition, during the extrusion of the sample from the probe there was a significant consolidation of the soil sample and a significant loss of water from the tested soil. The possibility of receiving a sample with an undisturbed structure could therefore be achieved by matching the Shelby probe with a plastic insert, which could be cut in the laboratory as required.

Commercial solutions for probes with plastic inserts are mainly found for small diameters [10]. Finally, a probe with an exterior diameter of 80mm was chosen, which allowed for the installation of a plastic pipe with a diameter of 75mm. Therefore, the interior diameter of the probe was 71mm. The consequence of increasing the thickness of the sampler wall, consisting of a steel pipe and a PE pipe, was to adjust the cutting knife to avoid large deformations of the sample taken. The last modification of the sampler was to provide the probe with a cylinder, so that during the extraction of the probe from the borehole in the sampler a vacuum was created. Due to this modification, the probability of receiving the sample remaining undisturbed has been significantly increased. All modifications of the probe are shown in figure 1.

The probe is terminated with a screw, and because of this, it is possible to push the Pagani static probe (CPT) to a specific depth. Due to the anchoring of the probe, it is possible to take samples from depths of up to 20m. In the case of smaller sampling depths, it is possible to use a lightweight dynamic probe or impact hammer to drive the sampling probe into subsoil.

## 3 Materials and methodology

Samples for laboratory tests were taken from a borehole drilled to the layer at a depth of approximately 8 m up to around 12m with a high water content [4] (Fig. 2).

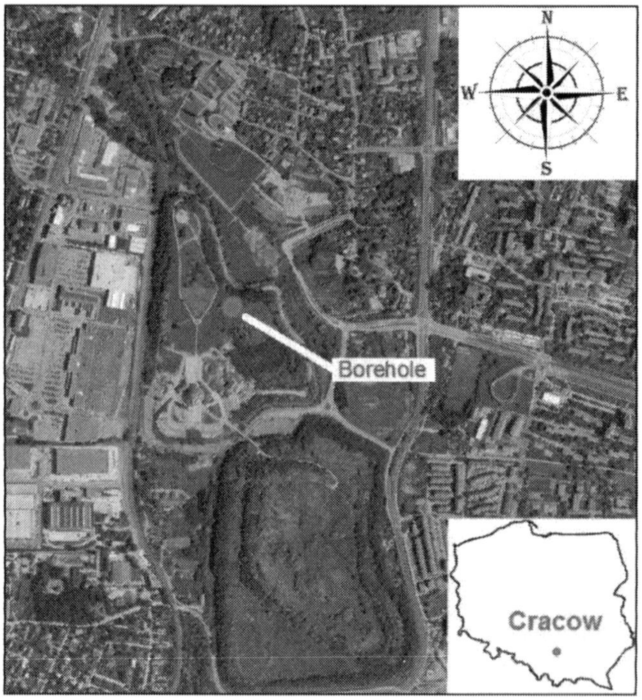


**Fig. 2.** The location of borehole on the research area of White seas.

Eight samples (S1 – S8) with dimensions of 50mm in diameter and an initial height of 75mm were cut from the probes discussed in chapter 2. The height before each test was determined individually [11]. The results of the particle size distribution and chemical composition of the tested samples did not differ from the samples taken from a depth of 1 to 5m, the results of which were previously presented by the authors [6]. The main element of the anthropogenic soil is calcium carbonate ($CaCO_3$), which occurs together with silicon oxides ($SiSO_2$) and other substances (i.e. $P_2O_5$, $CaSO_4$, $MgSO_4$, $BaSO_4$, NaCl, $CaCl_2$ [1]).

All tests were carried out in accordance with the European Standard [12], which also has the status of the Polish standard. The average physical parameters of soil from at a depth 8-12m are presented in table 1. It was also possible to calculate the following initial soil parameters, which are presented in table 2.

**Tab. 1.** The average physical parameters of the soil in the area of the "White Seas".

| Sample number | **Water content [%]** | **Bulk density [kg/m$^3$]** | **Specific density [kg/m$^3$]** |
|---|---|---|---|
| S1÷S8 | 200 | 1210 | 2680 |

**Tab.2.** The calculated avarage physical parameters of the soil in the area of the "White Seas".

| Sample number | **Dry mass density [kg/m3]** | **Porosity [-]** | **Total water content [%]** | **Degree of saturation [-]** |
|---|---|---|---|---|
| S1÷S8 | 400 | 0.85 | 210.5 | 0.95 |

The tests of determination of the P- and S-wave velocity were carried out using the VJ-Tech triaxial compression apparatus equipped with Bender element. The length of the piezo elements, i.e. the transmitter and receiver, is 6mm in total.

Shirley and Anderson [13] introduced bender ceramics in place of the shear plates for testing dry sand. Incorporation of bender elements in triaxial apparatus is arguably the most common practice, as demonstrated by Chan et al. [14].

The tests to determine the wave velocity were carried out for the following eight sample states:

- for undisturbed structure sample (sample S1),
- for samples fully saturated (sample S2),
- for samples fully saturated and then isotropically consolidated with the values of the effective stress being 50, 100, 150, 200, 250, 300kPa (samples S3 - S8).

All tests were performed using a sinusoidal wave, generated with time intervals of 15 seconds. This time was determined on the basis of the total time of signal suppression and the time necessary for registering the reply. The registered signal was filtered into a continuous function to better evaluate wave transition time. For this purpose, a moving average method was used with an eight-point window. For each sample, twenty designations were made on the basis of which the result was averaged. Filtering were performed in order to obtain the exact time of P- and S-wave arrivals.

By using the "back pressure" method [15] it was possible to fully saturate soil samples using a triaxial compression test apparatus. "Back pressure" was applied to the top of the sample. It is possible to determine the saturation level of a sample in the form of calculated Skempton parameter B:

$$B = \frac{\Delta\delta_1}{\Delta\delta_3} \quad (1)$$

where:
$\Delta\delta_1$ - change in pressure value in the triaxial cell [kPa],
$\Delta\delta_3$ - change in pressure in the soil sample [kPa].

Then a consolidation step was carried out by varying the pressure difference in the cell and in the sample. Six levels of consolidation were used, i.e. 50, 100, 150, 200, 250 and 300kPa. At each state of the sample, the determination of the P- and S-wave velocity was carried out. Finally, only S-wave velocity was analysed because properties of samples caused difficulties in determining of the P-wave first arrivals (Fig. 3).

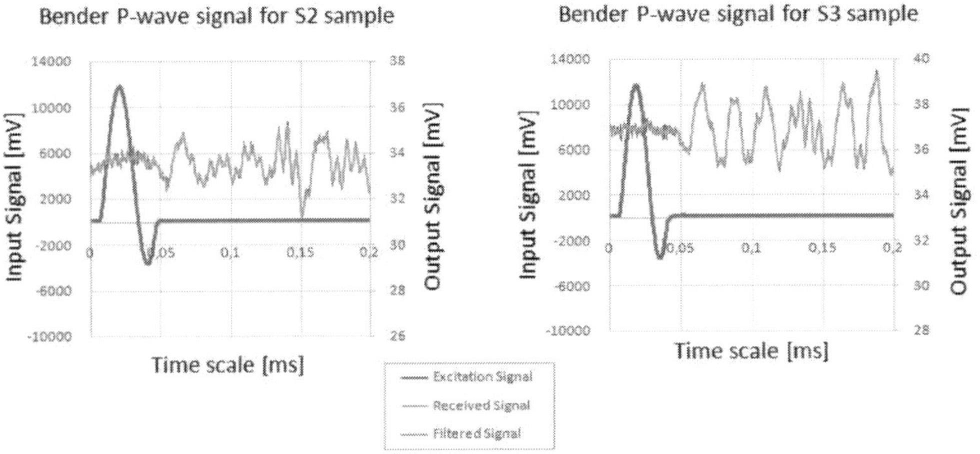


**Fig. 3.** The P-wave excitation and received signal for S2 and S3 sample.

In the study, an additional element the Local Strain Transducers LVDT (linear variable differential transformer) was applied which provided small strain measurements of axial and radial strain Fig. 4). Due to the use of sensors that were mounted directly on the ground sample, it was possible to control the change in sample height at each stage of the triaxial test [16, 17]. The sensitivity of the LVDT displacement transducers was 0.001mm and the measuring range of LVDT was 50mm.

The advantage of the small deformation sensors used was the precise determination of the height changes and then time of the wave propagation from the transmitter to the receiver.

## 4 Results and analysis

The waveforms of the input and output signal for S-wave are shown in figure 5. The S-wave velocity changes from 49.9 m/s for sample in undisturbed structure up to 86.9 m/s in fully saturated conditions. In conditions of effective stress increase, S-wave velocity increases from 94.1 m/s for 50 kPa to 168.2 for 300 kPa of fully saturated samples.

The results of the S-wave velocities for eight sample states are presented in table 3.

**Fig. 4.** A calcareous waste sample in the triaxial cell equipped with micro-scale transducers.

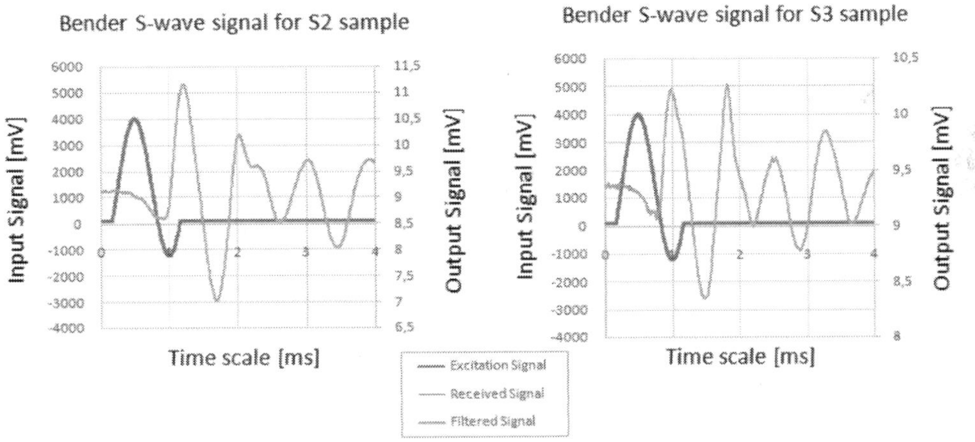


**Fig. 5.** The S-wave excitation and received signal for S2 and S3 sample. (Red line coincides with green line).

**Tab. 3.** The average values of the parameters of the tested samples for the S-wave test.

| Sample states | Undisturbed structure (S1) | Fully saturated (S2) | Effective stress in consolidation [kPa] (fully saturated) | | | | | |
|---|---|---|---|---|---|---|---|---|
| | | | 50 (S3) | 100 (S4) | 150 (S5) | 200 (S6) | 250 (S7) | 300 (S8) |
| S-wave velocity [m/s] | 49.9 | 86.9 | 94.1 | 132.9 | 142.0 | 154.8 | 151.9 | 168.2 |

As a result of analysing the wave velocity, it may be noticed that with the increase of effective stress in tested samples the S-wave velocity increases according to figure 6. The S-wave velocity significant increases up to about 100 kPa, and then this increase is smaller.

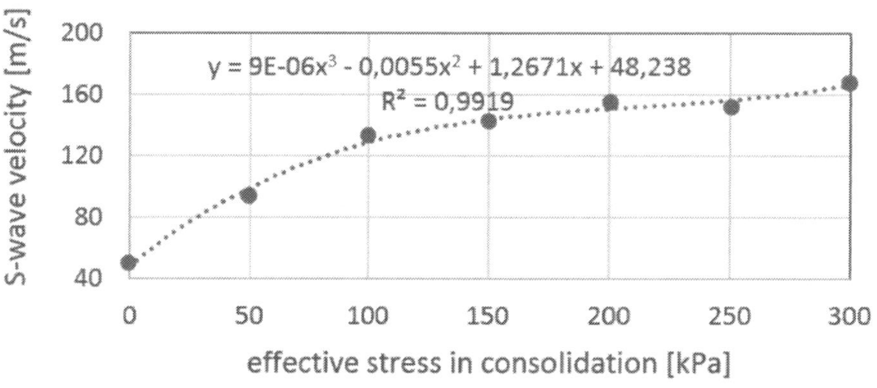


**Fig. 6.** S-wave velocity to effective stress in calcareous waste samples.

In addition, the measurements of deformation of the sample with LVDT sensors allowed for the determination of changes in the volume of the sample. In effect, it was possible to precisely determine the changes in the porosity of the tested sample. It is necessary to note the very large outflow of water at the consolidation stage. Volumetric distortions registered during the test for 50 kPa amounted to about 18.6%, while for the 300kPa test, volumetric strains reached a water pressure of above 32.8% at the time of full dissipation. Finally, the calculated dynamic G modulus was presented in table 4.

**Tab. 4.** The average values of the dynamic shear modulus G of the soil calcareous waste samples.

| Sample states | Undisturbed structure (S1) | Fully saturated (S2) | Effective stress in consolidation [kPa] (fully saturated) | | | | | |
|---|---|---|---|---|---|---|---|---|
| | | | 50 (S3) | 100 (S4) | 150 (S5) | 200 (S6) | 250 (S7) | 300 (S8) |
| Dynamic modulus G [MPa] | 2.92 | 9.07 | 10.63 | 21.19 | 24.21 | 27.69 | 28.76 | 33.94 |

## 5 Conclusions

The study presents the results of the determination of S-wave velocity and dynamic G-modulus conducted by the Soil – Structure Interaction Division, Cracow University of Technology in the area of " White Seas " in Krakow. The samples taken represented a subsoil of post-production waste of the Solvay Sodium Plant. The S-wave velocity of the samples was measured depending on the different degree of saturation as well as the consolidation of the soil sample. Using the triaxial apparatus, due to which the wave velocity was determined in the saturated and consolidated medium, it was possible to calculate the dynamic modulus.

In general, we may observe the effect of S-wave velocity increase with increase of degree of saturation of the tested samples. It may be noticed that with the increase of effective stress in tested samples the S-wave velocity significant increases up to about 100 kPa, and then this increase is smaller. The same effect was observed with modulus G changes.

## References

1. I. Krzak, *Zagospodarowanie terenów poprzemysłowych Krakowskich Zakładów Sodowych „Solvay" (Post - industry reclamation of Cracow Soda Industry „Solvay")*, Problemy Ekologii Krajobrazu, **17**, 283–287 (2005) (in Polish)
2. B. Czado, *Analiza nośności pali fundamentowych na podstawie polowych badań gruntów sondą statyczną (Analysis of bearing capacity of foundation piles based on field cone penetration tests results)*, Rozprawa doktorska, Biblioteka Politechniki Krakowskiej (2014) (in Polish)
3. M. Gliniak, M. Pawul, W. Sobczyk, *Wpływ transportu i składowisk poprzemysłowych byłych Krakowskich Zakładów Sodowych „Solvay" na stan i jakość wody rzeki Wilga w Krakowie (Impact of the transport and postindustrial landfills of Cracow Soda Works "Solvay" on the status and quality of water in Wilga River in Krakow)*, Logistyka **4**, 4295–4302 (2014) (in Polish)
4. B. Wrana, *Pile load capacity – calculation methods*, Studia Geotechnica et Mechanica **37** (4), 83–93 (2015) doi: 10.1515/sgem-2015-0048
5. http://www.trasalagiewnicka.krakow.pl/
6. E. Pilecka, J. Zięba, E3S Web Conf. **24**, 02006 (2017) doi: 10.1051/e3sconf/20172402006
7. J. Zięba, M. Bazarnik, E3S Web Conf. **36**, 03006 (2018) doi: 10.1051/e3sconf/20183603006
8. *A. Ślęzak, Wpływ składowiska odpadów Krakowskich Zakładów Sodowych na wody (The impact of the Krakowskie Zakłady Sodowe waste landfill on water),* Przewodnik III Konferencji Sozologicznej PTG. Wyd. AGH, Kraków (1993) (In Polish)
9. *Biuro Rozwoju Krakowa, Miejscowy plan zagospodarowania przestrzennego terenów Krakowskich Zakładów Sodowych „Solvay" w likwidacji, w rejonie ulic: Zakopiańskiej, Myślenickiej i Podmokłej (The local development plan for the areas of the Krakow Soda Plant "Solvay" in liquidation, in the area of: Zakopiańska, Myślenicka and Podmokła Streets),* Kraków (1994) (In Polish)
10. http://www.szkurlat.pl/
11. B. Wrana, *Laboratory testing of soil mechanics,* Cracow University of Technology Publishing House (2015)
12. EN ISO 17892-1,2,3:2014 Geotechnical investigation and testing - Laboratory testing of soil - Part1, 2, 3
13. Shirley DJ, Anderson AL, *Acoustical and engineering properties of sediments*, Report no. ARL-TR-75–58, applied research laboratories. University of Texas, Austin (1975)
14. Chan KH, Boonyatee T, Mitachi T *Effect of bender element installation in clay samples.* Ge´otechnique **60**(4), 287–291 (2010) doi: 10.1680/geot.7.00135
15. Black, D.K. and K.L. Lee., *Saturating laboratory samples by back pressure*, J. Soil Mech. & Found. Division, ASCE. **99**, 75–93 (1973)
16. T. Cuccovillo, M.R. Coop, *The measurement of local axial strains in triaxial tests using LVDTs,* Geotechnique **47** (1), 167–171 (1997) doi: 10.1680/geot.1997.47.1.167
17. G. Da Re, M.C. Santagata, J.T. Germaine, *LVDT Based System for the Measurement of the Prefailure Behavior of Geomaterials,* Geotechnical Testing Journal **24** (3), 288–298 (2001) doi: 10.1520/GTJ11347J

E3S Web of Conferences **66**, 02006 (2018) https://doi.org/10.1051/e3sconf/20186602006
*AG 2018 – 4th International Conference on Applied Geophysics*

# Long-term study of the impact of temperature changes, material aging and service load on the strains of a reinforced concrete structure

*Adam* Kanciruk*

Strata Mechanics Research Institute of Polish Academy of Sciences, Cracow, Poland

**Abstract:** The article discusses five year long measurements of strains of a concrete floor of a hangar. That hangar, originally meant for servicing light military aircraft, was rebuilt with a view to making it fit for servicing larger and heavier passenger planes. As part of that redevelopment, a new floor – reinforced concrete slab, capable of withstanding the weight of the planes – was constructed. In the areas of the floor where the greatest loads occur, ie. in the areas of the concrete slab on which the wheels of the three undercarriage legs rest, three strain rosettes were installed so that the slab strains could be measured. The rosettes were connected to two meters - dataloggers. The latter were programmed in such a way as to register the measured strains and additionally temperatures many times during every 24 hours. The monitoring process, conducted in such an automatic way, demonstrated the occurrence of strains resulting from service load, aging of the reinforced concrete, as well as changes in its temperature.

**Keywords:** construction, metrology, strain rosette

## 1 Introduction

The research location is a hangar originally built for servicing light military aircraft whose take-off mass does not exceed 9,100 kg. That hangar is a hall whose dimensions are 84×42.5 m, with one of the longer sides being a sliding door. Inside the hangar, two planes of the size corresponding to an Airbus A320 or Boeing 737 can easily be fit, which is why the place was rebuilt so that it could become apt for servicing passenger aircraft, mostly the types mentioned above. Figure 1 presents the inside of the hangar before the redevelopment. The mass of a plane such as the Airbus A320 approximates 42 000 kg (without passengers, cargo and fuel) [1], which is why the load from the wheels of its undercarriage legs damaged the floor that was unfit for such loads (Fig. 1). Additionally, the laboratory research of the samples of that floor demonstrated considerable irregularities of the construction. Therefore, as part of the hangar redevelopment, a decision was made to partly replace the floor.

*Corresponding author: kanciruk@img-pan.krakow.pl

**Fig. 1.** The inside of the hangar. Damages to the floor are visible.

## 2 Construction of the new hangar floor

Figure 2 presents, in a simplified way, the construction of the eastern part of the hangar floor after redevelopment of the place. The western part is its mirror image. The construction of the roof was left largely unchanged, which – due to its height – makes it impossible to park an entire plane inside the hangar. It was assumed that the control surfaces and the back of the fuselage will be protruding outside the gate during servicing.

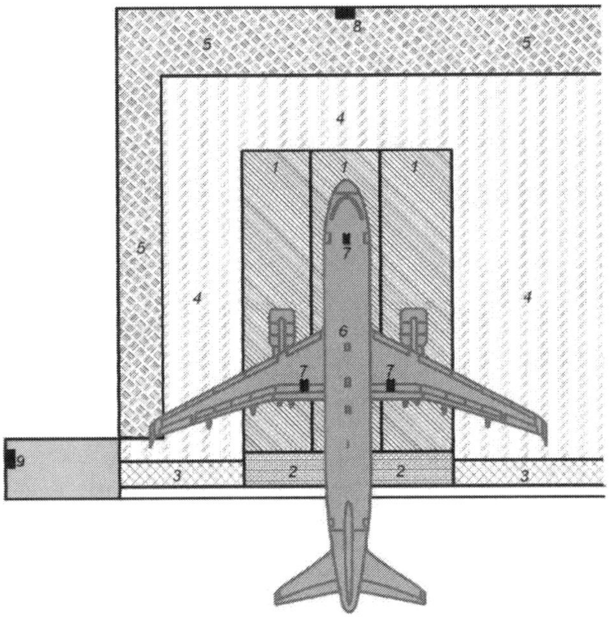


**Fig. 2.** Scheme of the hangar floor after the redevelopment, the eastern part: 1- constituent slabs of the main slab, 2,3 – slabs in the door area, 4 – buffer slab, 5 – fragment of the previous slab, 6 – silhouette of an Airbus A320, 7 – undercarriage wheels, 8,9 – measuring instruments – recorders.

In the construction process of all the slabs, i.e. the main one, the buffer one and the slabs in the door area, the same concrete was used. Each of the slabs is 30 cm thick, and they all rest on the same base, having a gliding layer underneath them. The slabs differ only with respect to the concrete reinforcement. The main slab, whose dimensions are 26.25×18 m, was constructed as three separate 6m wide slabs, surrounded by construction joints. Additionally, each of these components of the main slabs is cut up with contraction joints.

## 3 Rationale for installing research equipment in the hangar

Taking into consideration the damages to the previous hangar floor, it was decided that measuring equipment would be installed in the hangar with a view to determining the strains of the new floor during its exploitation. These strains might be caused by contraction and drying up [2] of the concrete used to build the floor, as well as by changes in the temperature and the load from the planes serviced in the hangar. That load poses a particular threat to these parts of the floor on which the wheels of the three undercarriage legs rest (Fig. 2). The prescribed location spot of the front leg (in the nose) during parking is precisely determined. At the same time, the wheels of the main legs (under the wings) may rest in different places, depending on the type of aircraft being serviced, although the latter happens mostly to be the Airbus A320. Also, it is not necessary that the axis of the fuselage of a plane parked in a hangar should overlap with the axis of the main slab (Fig. 2).

It is obvious that, due to the works conducted in the hangar, measuring the floor strains in a manual way cannot be allowed. It is also impossible to apply instruments that would hamper servicing of the planes in any way. Therefore, it was decided that an automatic measurement system would be installed in the hangar. The system's objective is to perform measurements of the strains to the floor with a frequency sufficient to determine the impact of various factors upon these deformations. Of the different types of measurement transducers, vibrating wire extensometers were selected – simple instruments characterized by high reliability, solid long-term metrological parameters, as well as their adjustment to immersion in a semi-liquid concrete mix before it solidifies.

## 4 Installing measuring rosettes in the hangar's main (eastern) slab

The measuring system (Fig. 3) consists of three equilateral strain rosettes and two meters – strain recorders. The rosettes were installed in the spots where the biggest load from the undercarriage wheels of the serviced Airbus A320 planes is expected. For clarity's sake, the rosettes presented in Figure 3 are magnified; as a matter of fact, their real dimensions correspond to the dimensions of the small triangles pictured inside them. Each of the rosettes is built of three identical vibrating wire transducers that had been set in the floor's body, some 5 cm above its lower surface adjacent to the gliding layer (Fig. 4). In order to ensure proper identification of particular extensometers, they were marked with three colored „jackets" (yellow, green and blue) pulled over their wires (Fig. 3, 4). To measure the temperature of the concrete in the place where the rosettes are set, three thermistors were used. Thermistors of the same type were applied to measure the temperature of the air inside the hangar.

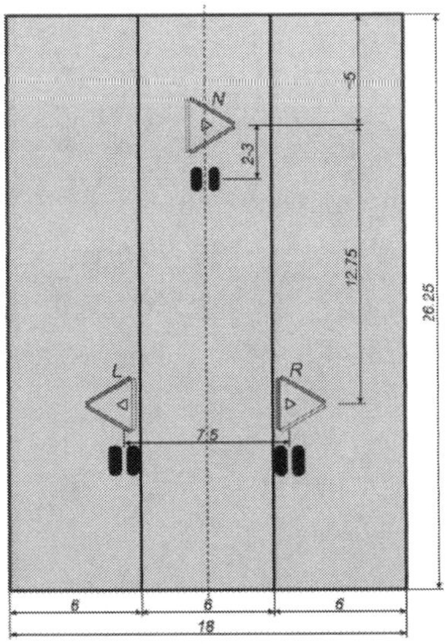


**Fig. 3.** Distribution of the strain rosettes in the main slab.

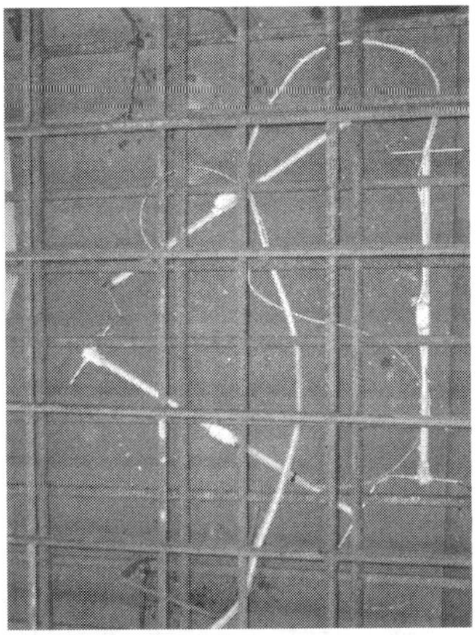

**Fig. 4.** The rosette installed in the area of the load from the wheels of the left main undercarriage leg.

Unfortunately, the refurbishment of the hangar roof that followed the construction of the floor, along with the installation of the new gate, necessitated a minor correction of the prescribed location of the planes in the hangar, which, however, was of great importance in the light of the conducted measurements. Under the correction, the planes are parked in the hangar in such a way that the undercarriage wheels do not stop above the installed rosettes, but some 2 meters above them, looking from the plane's nose (Fig. 3, 5). Thus, the deformations calculated on the basis of the readings of the extensometers are smaller than in the places of the wheels' resistance, and they are of the opposite sign (compression instead of stretching).

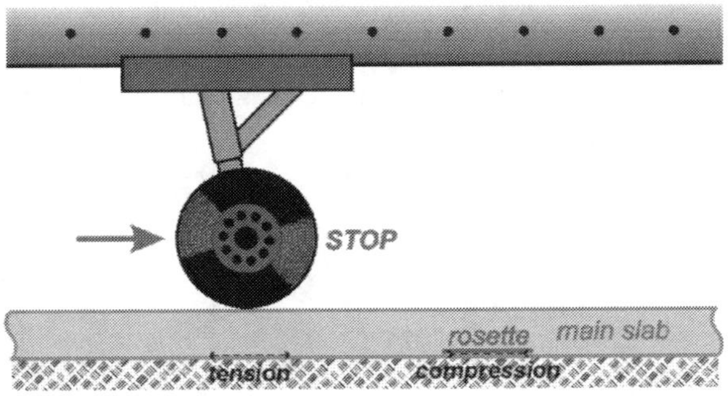


**Fig. 5.** The position of the plane's undercarriage wheels in relation to the rosettes.

## 5 Measurements and recording the results. Interpreting the obtained data

All the extensometers and thermistors of the rosettes were connected to two meters - recorders (8,9 – Fig. 2). The period of measurement repetition for these instruments can be readjusted. In the first period of its activity, lasting ca. 3 years, the system performed 50 measurement a day, a number which was later reduced to 25 measurements a day. Figure 6 presents the values of strains calculated on the basis of the readings of particular extensometers of the left rosette, as registered in November 2011.

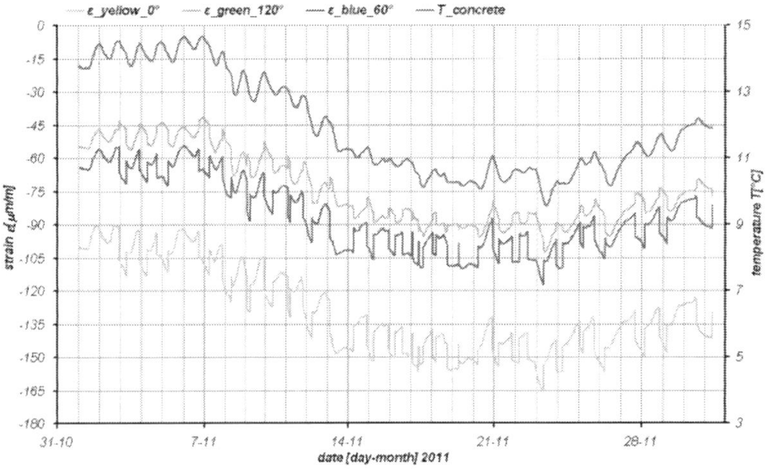


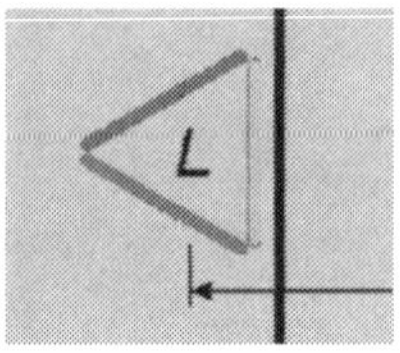


**Fig. 6.** Left strain rosette. The values of strains calculated on the basis of readings of particular extensometers ("yellow", "green" and "blue" one), under the temperature of the bottom layer of the floor (the temperature of concrete). November 2011.

All the extensometers reacted to changes in the temperature of concrete and exploitation loads in the way they were supposed to. Every time the plane was driven into the hangar and out of it, sudden changes in the strains of the main slab of the floor occurred, especially in the area where that slab was impacted by the main undercarriage legs' wheel load. These changes might be determined by means of a software developed specifically for that purpose. That software's task is to single out the deformations caused by the aforementioned load and, on that basis, establish the values of the major strains and their directions. Figure 7 presents the established values of the major strains for each case of parking the plane inside the hangar in November 2011. The plane driven into the hangar some time after midnight on November 11 caused compression strains of 20.25 μm/m and stretching strains of 6.35 μm/m. The values of the strains obtained by means of the rosette placed in the area of the load from the wheels of the right undercarriage leg, in the same period of time, are almost two times smaller. The moments of driving the plane into the hangar and out of it coincide.

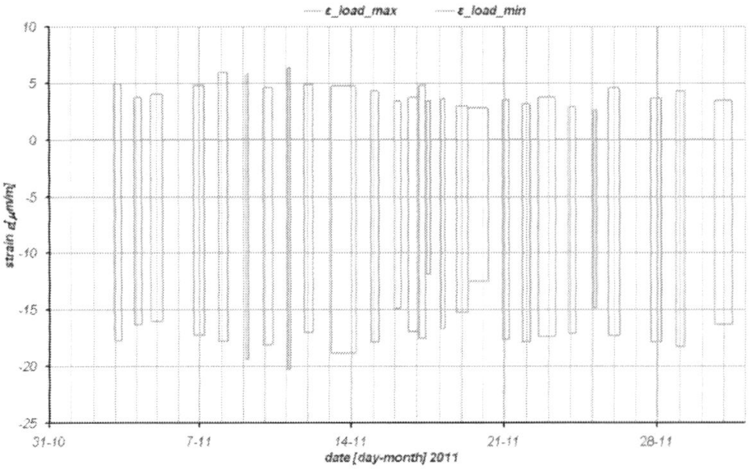


**Fig. 7.** Left rosette. The values of the major strains: maximum (*epsAmax*), minimum (*epsAmin*), caused by the load exerted on the main slab of the hangar floor by a plane. November 2011.

Figure 7 presents just the values of the major strains of the floor caused by the wheel load from the main undercarriage legs. In order to depict graphically both the values and directions of the major strains of the hangar floor in the spots where the left and right rosettes were installed, the method explained in Figures 8-10 was adopted:

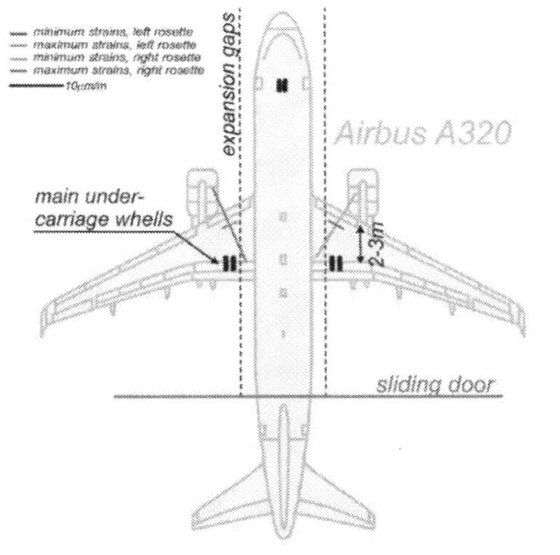


**Fig. 8.** Graphic interpretation of the measurement results: values and major strains for a single registered case of parking a plane inside the hangar.

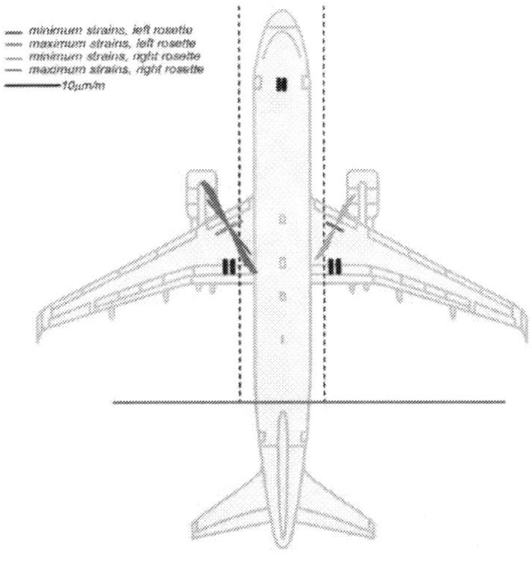


**Fig. 9.** Graphic interpretation of the measurement results: values and major strains for all the registered cases of parking a plane inside the hangar in November 2011.

Figure 8 presents an outline of a plane, with dotted lines marking longitudinal expansion gaps of the slabs making up the main slab. In front of the wheels of the main left undercarriage (in the direction of the plane's nose), two segments intersecting at a right angle were drawn. The intersection spot marks the place where a rosette is installed. The longer segment, drawn with a purple line, is oriented in accordance with the direction of the minimum, compression strains. The shorter, "red" segment, is oriented in accordance with the direction of the maximum, stretching strain. The lengths of the segments are proportional to the values of the deformations. Similarly, the segments represented by green and blue lines depict the major strains determined by the rosette installed in front of the wheels of the right main undercarriage. It is clearly visible that the outermost components of the main slab are dented towards the middle slab. It may certainly be attributed to the fact that each particular slab, as presented in Figure 3, is 6m wide, and the spacing between the main undercarriage legs of an Airbus A320 is ca. 7.5 m. Figure 9 presents all the registered cases of parking the plane inside the hangar in November 2011.

After removing the plane outline from Figure 9, adding the time axis to that Figure and properly locating all the cases of parking the plane in time, we obtain a clear temporal graph illustrating the values and directions of major strains calculated on the basis of the readings of both rosettes' extensometers (Fig. 10). The points where the pairs of segments intersect correspond, time-wise, to the midpoints of the periods when the plane was parked in the hangar. It can clearly be seen that the strains measured by the left rosette are larger than those measured by the right rosette.

Just like in Figure 10, the graphs were also drawn for each of the observation months that followed, from June 2011 to April 2016. Strains of the hangar floor are still being measured.

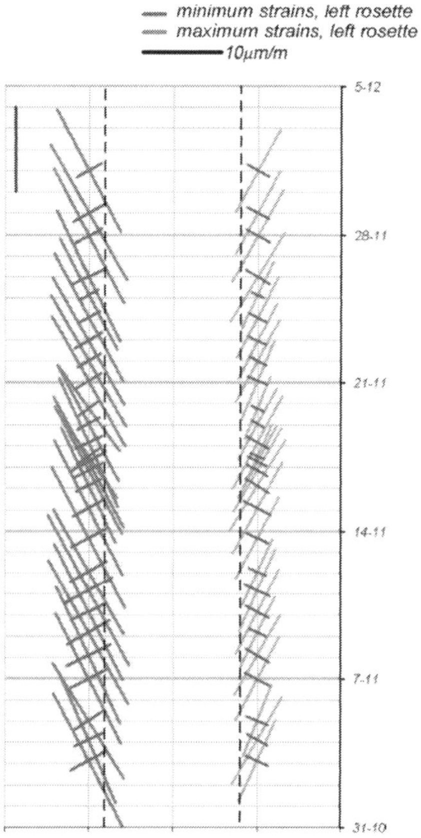


**Fig. 10.** Major strains of the components of the main slab in the rosettes' installation spots, for each case of parking the plane in the hangar in November 2011.

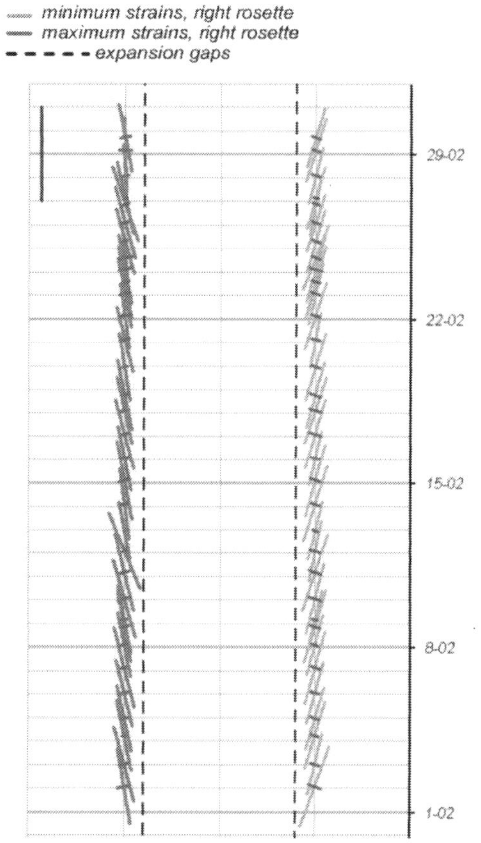


**Fig. 11.** Major strains of the slabs making up the main slab in the installation spots of the rosettes, for each case of parking a plane in the hangar in February 2016.

Figure 10 presents the strains recorded in November 2011, nine months after the main slab of the hangar floor was built. With time, due to the stabilization of the slab and its foundation, the values of the strains became considerably smaller. Figure 11 presents the strains of the slab in February 2016.

The deformations measured by both rosettes are not only smaller than in November 2011 (on February 2, 2016, the right rosette showed a compression strain of 9.09 μm/m and a stretching strain of 1.37 μm/m), but also their values are very similar.

The tendency of temporal reduction in the values of strains caused by planes parked in the hangar is demonstrated by Figure 12 (left rosette) and Figure 13 (right rosette). Each pair of points – one above the dotted line and the other below it – represent the values of the strains: stretching and compression ones, respectively, caused by a single case of parking a plane in a hangar. The total number of such cases is 1,442.

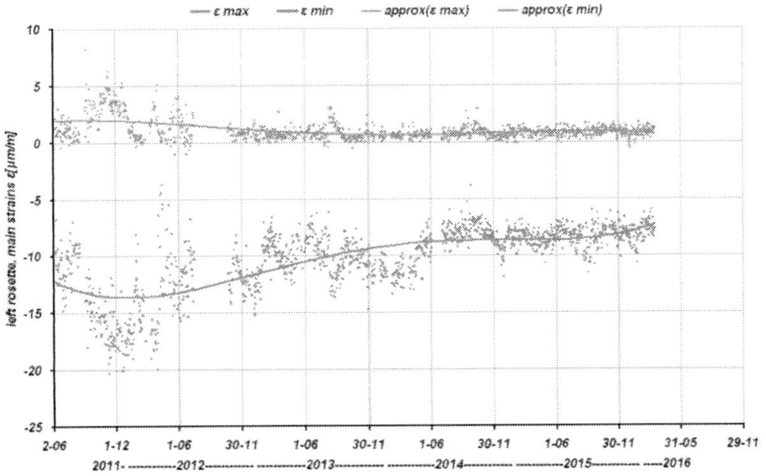


**Fig. 12.** Left rosette, the tendency of reduction in the values of maximum strains *(ε_ max)* and minimum strains *(ε_ min)* for each case of parking a plane inside the hangar.

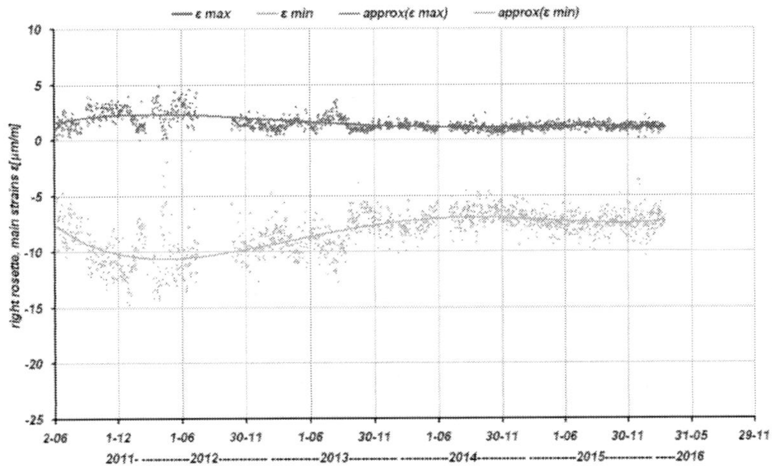


**Fig. 13.** Right rosette, the tendency of reduction in the values of maximum strains *(ε_ max)* and minimum strains *(ε_ min)* for each case of parking a plane inside the hangar.

Figures 12 and 13 clearly show that in the initial, year-long period of using the hangar, the values of the strains, module-wise, displayed a tendency to grow. That tendency later changed into a tendency to decrease, resulting in a stabilization of the strain values. Apart from the aforementioned reason for the stabilization of strains, there is also another potential reason, i.e. building a ca. 30 cm thick concrete parking slab of an area of 11 ha in the vicinity of the hangar. That investment was carried out in the summer months of 2012. During that time, the hangar was not used, which is visible in Figures 12 and 13. for sure, the load from the constructed parking slab had an impact on migration of waters, which resulted in an increase in humidity and bearing capacity of the ground beneath the hangar. This is confirmed by increased water content in the servicing wells of the hangar, as well as by a small, several millimeter uplift of the floor along the gate.

As it was shown in Figure 5, in the current, prescribed location of the plane during the time of its parking, the wheels of the main undercarriage legs exert a load on the hangar floor in spots other than those in which the strain rosettes are installed. Still, of all the 1,442 registered cases of parking the plane in the hangar, there was only one short instance when the aircraft was not parked in a prescribed manner. Figures 14 and 15 present the major strains of the hangar floor in the spots where the rosettes were installed.

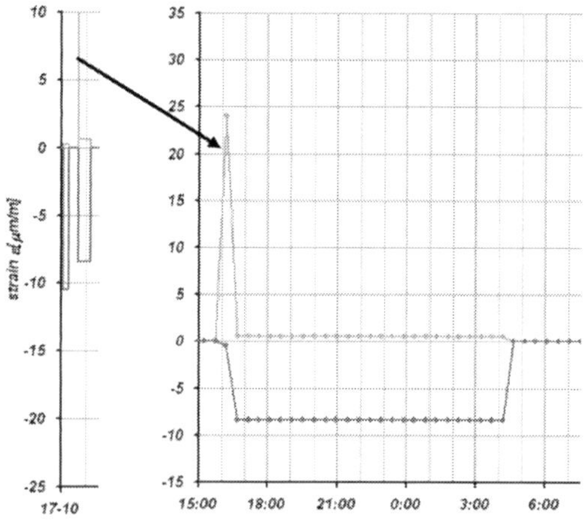


**Fig. 14.** Values of the major strains registered on October 17 and October 18, 2013. Left rosette.

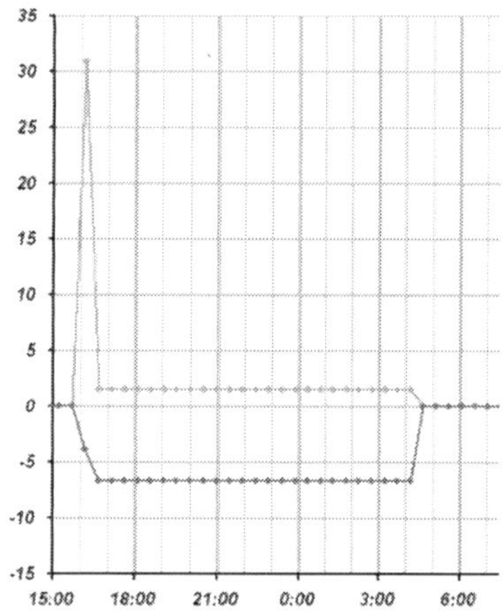


**Fig. 15.** Values of the major strains registered on October 17 and October 18, 2013. Right rosette.

On October 17, 2013, at 16:00, the values of the major strains for the left rosette were: 24 µm/m (stretching) and 0.5 µm/m (compression), and for the right rosette – 30.9 µm/m (stretching) and 3.9 µm/m (compression). The strains registered half an hour later were: 8.4 µm/m (left rosette) and 6.7 µm/m (right rosette). Therefore, module-wise, they are 2.9 and 4.6 times smaller, respectively, than the values registered at 16:00. Most likely, in the initial stage of parking, the plane was driven considerably further into the hangar (it might have been the case that the wheels of the main undercarriage leg stopped in the rosettes' installation spot), and then pushed back to the prescribed position. It may be expected that, if such a situation had occurred on November 11, 2011, the registered stretching strains would have reached the values: $20.25 \times 2.9 \approx 59$ µm/m for the left rosette, and $10.8 \times 4.6 \approx 50$ µm/m for the right rosette.

The rosette located in the area of the load exerted by the wheels of the nose leg occupies the central position in relation to the longitudinal expansion gaps of the main slab, and thus it demonstrates considerably smaller usage strains as compared to the other rosettes, especially that the nose undercarriage leg bears not more than 15% of the plane's total weight. When analyzing the total strains, it can be noticed that the usage strains connected with the wheel load are a few µm/m, at best, so they cannot result in scratches. Thus, they pose no threat to the slab.

As each rosette was equipped with a thermistor, it is possible to estimate the thermal strains of the slabs in the spots where the rosettes were installed. These strains fall within the range of 10÷14 µm/m, so they approximate the values of the concrete strains as specified in the tables [3]. The hangar is heated in wintertime, and in summer, the temperature in its inside may reach up to 30°C, which results in a temperature gradient on the floor slabs. That gradient causes the slabs to dent, especially on the edges, which was demonstrated by measurements [4]. The aging of the material of which the slabs were built, i.e. reinforced concrete, results in its slow contraction. After an almost six year long observation period (cf. Fig. 16), that contraction reached the values within the range of 120 µm/m (rosette *P*, minimum deformation) – 230 µm/m (rosette *N*, maximum deformation).

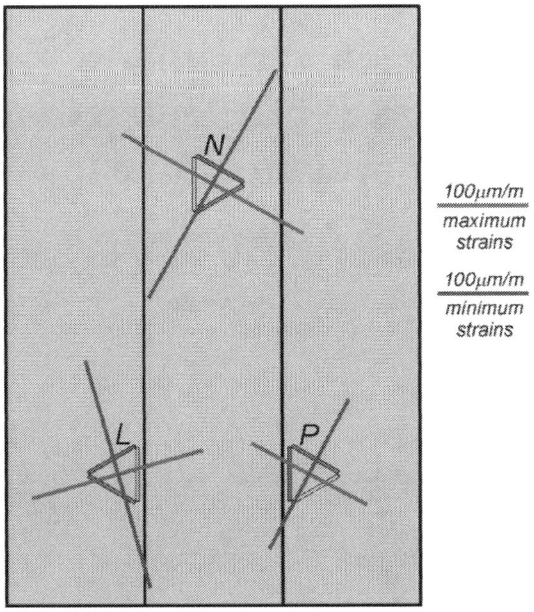


**Fig. 16.** Contraction of the slab's building material (reinforced concrete) in the installation spot of strain rosettes.

## 6 Summary

The vibrating wire-type measuring equipment used in the measurements demonstrated a high level of reliability and good metrological parameters (including high sensitivity), in spite of its prototypical nature. Thanks to over 50,000 automatically performed strain measurements (for each extensometer), main slab strains were registered for all the instances of parking aircraft inside the hangar that have occurred so far, along with the slab strains caused by changes in the temperature of that slab and the aging of reinforced concrete. Monitoring of the usage strains demonstrated that the strains of the floor slabs in the spots impacted by the load from the wheels of the main undercarriage legs are highly likely to result in discontinuity of concrete (emergence of scratches) in the spots that cannot be accessed for direct observation.

## References

1. *http://pl.wikipedia.org/wiki/Airbus_A320* (available Jun.01.2018).
2. *Budownictwo ogólne. Tom 1. Materiały i wyroby budowlane (General construction. Vol. 1. Construction materials and products)*, Praca zbiorowa pod kierunkiem prof. dr hab. inż. Bogusława Stefańczyka. Arkady Warszawa (2007) (in Polish)
3. *Tablice fizyczno-astronomiczne (Physical and astronomical tables)*, Wydawnictwo Adamantan, Warszawa (1995) (in Polish)
4. A. Kanciruk, *Metody tensometryczne w badaniach przemieszczeń, deformacji i zjawisk dynamicznych w gruncie i obiektach budowlanych (Tensometric methods in studies of displacements, deformations and dynamic phenomena in soil and building objects)*, Archiwum Górnictwa, Wydawnictwo Instytutu Mechaniki Górotworu Polskiej Akademii Nauk, Kraków (2012) (in Polish)

# Interpretation of rock mass thermal conductivity at the design stage of heat pump installation and its impact on system efficiency (COP)

*Michał* Kaczmarczyk[1] *Magda* Kaczmarczyk[1] *Konrad* Thürmer[2] *Magdalena* Klich[3]

[1]AGH University of Science and Technology, Krakow, Poland
[2]BTU CS Brandenburg University of Technology Cottbus – Senftenberg, Cottbus, Germany
[3]IWSÖ Institut für Wasserwirtschaft, Siedlungswasserbau und Ökologie, Coudray, Germany

**Abstract.** The recognition of geological and thermal conditions of the rock mass in the case of designing a vertical borehole heat exchanger as the ground source for heat pump installations is a key issue affecting the efficiency of the heating/cooling system operation. This is especially important for large-sized buildings with a high demand for thermal power, which affects into the size of the ground source installation. The aim of the article is to indicate the difference in the obtained results concerning thermal calculations at the design stage of the brine/water heat pump installation with the vertical heat exchanger, in relation to the theoretical values of the rock mass thermal conductivity and the real (measured) values obtained during the thermal response test (TRT). For this purpose, calculations of thermal efficiency from one meter of the current rock mass were made, with particular emphasis on the change in the value of the thermal conductivity coefficient in the tested drilling profiles. Correspondingly, heat pump coefficients of performance (COP) were calculated, which allowed to analyze the influence of the over/undersizing phenomenon of the ground source on the technical parameters of the heat pump's operation and the economic effect of the investment.

**Keywords:** heat pumps, thermal conductivity, thermal response test

## 1 Introduction

The heat pump market in Europe and in Poland develops dynamically, which is confirmed by statistical data published both by the European Heat Pump Association [1], as well as the Polish Association for Heat Pump Technology and Development [2]. It is especially important according to the importance of using the low temperature geothermal resources to the heat production [3–4]. When analyzing this data, attention should be paid at two main issues. The first one is the stability of the ground-source heat pumps market and its share in the total sales of heat pumps. The second issue is the percentage sales change each year, which indicates that while in the European market in the analyzed interval 2010–2015, there

[1] Corresponding author: mkz@agh.edu.pl

was a sale decrease in 2012 compared to 2011 by 7%, in Poland in the period of 2012–2017 the market was constantly growing at the level of 5–22% annually (Fig. 1, 2). In view of the above, it should be assumed that the heat pump sector will continue to develop, and hence requires a precise approach to the design of installations. This applies in particular to brine/water heat pumps with a vertical borehole heat exchanger (BHE), not only due to the fact that the implementation of the heat source is a significant contribution to the costs of the entire installation, but also due to the need of properly estimate the coefficient of thermal conductivity of the rock mass, which is crucial parameter for the proper design of the BHE.

To determine the correct configuration of the vertical borehole heat exchangers for the heat pump system, it is reasonable to conduct a thermal response test (TRT). The implementation of TRT is necessary especially in the case of locations with poorly recognized geological and hydrogeological conditions, and in the case of heat pumps with higher cooling power (i.e. over tens of kW) [5–10]. Unfortunately, in practice, the coefficient of thermal conductivity is often selected by the designer based on the theoretical values for individual layers in the lithological profile. This may lead to errors in the calculation the total length of BHE, and consequently to oversizing (much more common) or undersizing of the heat source. TRT allows for the empirical determination of the coefficient of thermal conductivity, thus obtaining precise information on the geological/hydrogeological conditions prevailing at the site of the planned installation [11–14].

The paper presents calculations for an exemplary vertical borehole heat exchanger for which a thermal response test was carried out and compared with the theoretical models for the selection of BHE, based on literature data [15]. Based on calculations, the consequences resulting from over/undersizing of the heat source were indicated, both from an economic point of view and the effects that such practice can cause at the heat pump's operating stage and the efficiency of its operation.

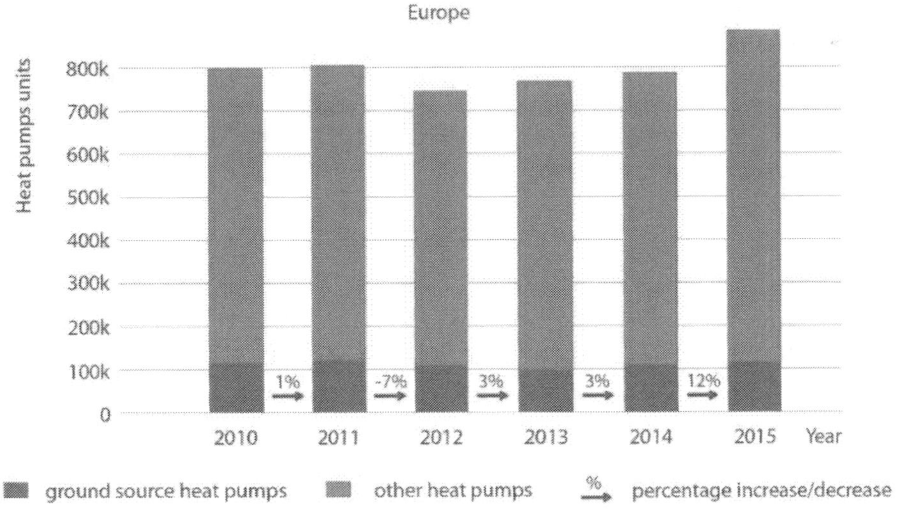


**Fig. 1.** Share of ground source heat pumps in total heat pump sales in Europe (based on [1]).

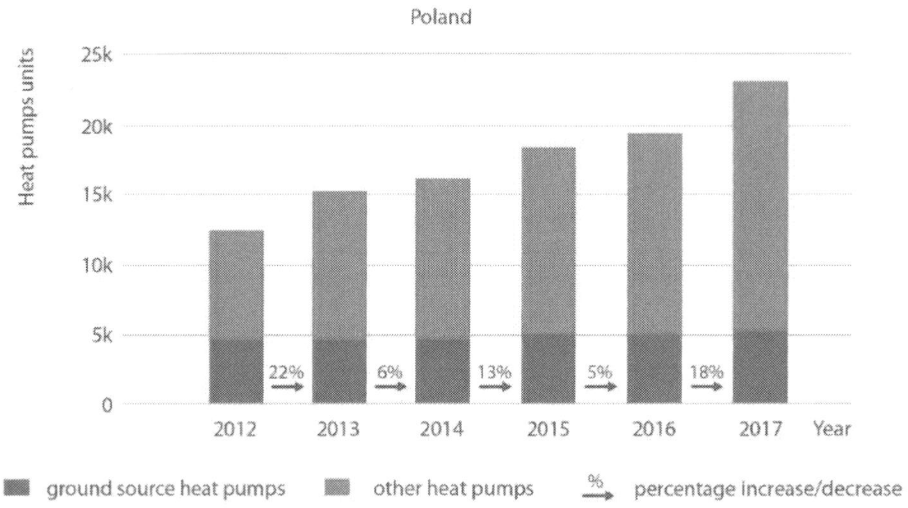


**Fig. 2.** Share of ground source heat pumps in total heat pump sales in Poland (based on [2]).

## 2 Analyzed vertical borehole heat exchanger

As mentioned in the introduction, the performance of the thermal response test is economically justified especially for large heat pumps installations. For this reason, a BHE was analyzed with a total depth of 210 m. The technical parameters of the BHE are listed in Table 1 and the lithological profile is shown in Figure 3. The theoretical values of the coefficient of thermal conductivity were assigned to each layers in the lithological profile [15], taking into account the minimum and maximum values. It allowed to calculate the average (based on weighted average) value of the coefficient of thermal conductivity in the whole profile of the tested borehole, which was assuming the minimum values of 0.96 W/mK, and for the maximum values 2.15 W/mK. It should be pointed out that the discrepancy depending on the adopted values is significant.

**Table 1.** Parameters of the analysed borehole heat exchanger.

| Parameter | Value |
|---|---|
| depth of BHE | 210 m |
| type of BHE | U-tube |
| tube | fi 40 x 3.7 mm |
| material of BHE | PE HD, gravel filling |
| borehole diameter | 70 mm |
| test duration | 49 h |
| injected power (Q) | 5.70 kW |
| ground temperature | 12.99°C |

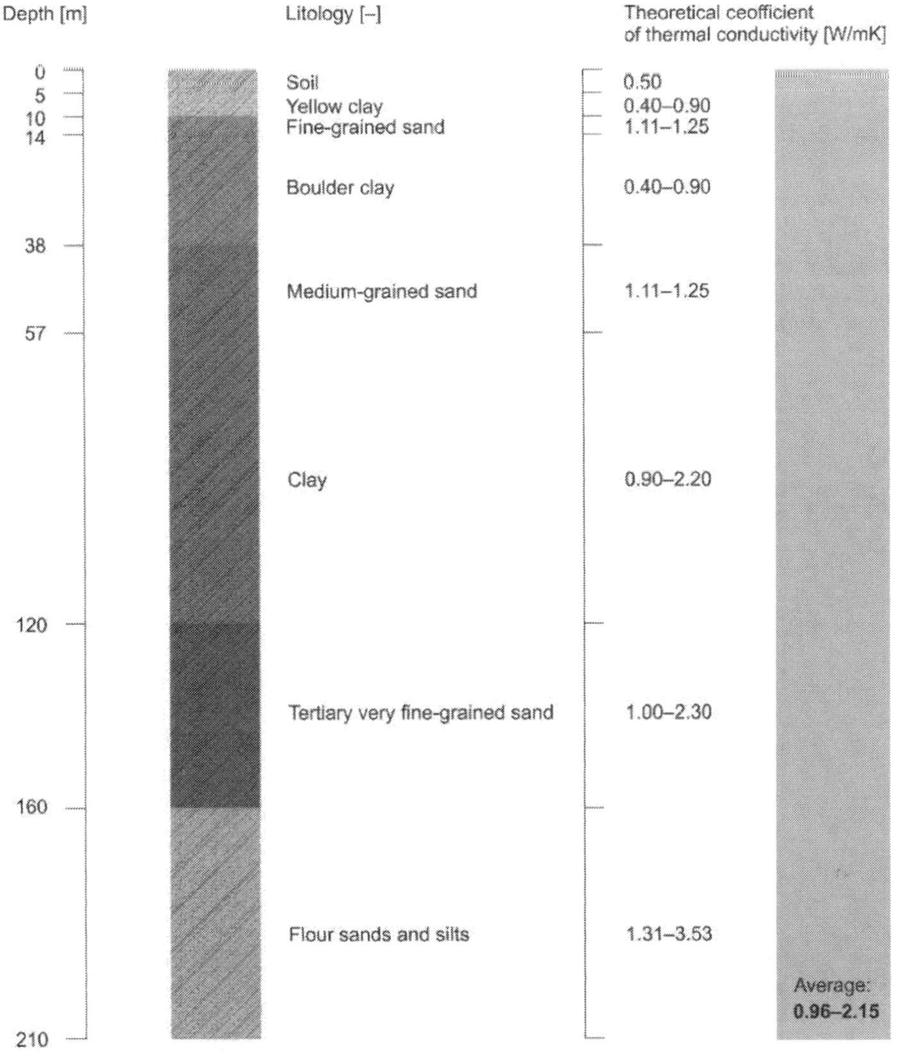


**Fig. 3.** Lithological profile with theoretical coefficient of thermal conductivity values (theoretical coefficient of thermal conductivity based on [15]).

## 3 Methodology and results of calculation

As mentioned above, the aim of TRT is to test the vertical borehole heat exchanger in order to determine in an empirical way, the amount of heat that can be obtained from the ground. To make calculations of the effective coefficient of thermal conductivity, during the thermal response test the following parameters were measured: fluid temperature (water/glycol) on input and output as a function of time (which allowed to determine the average value $T_f$) and unsettled ground temperature $T_{ground}$ [16]. The change of temperature $T_f$ in relation to undisturbed ground temperature $T_{ground}$ allowed to determine the supplied thermal energy Q/H. The coefficient of the effective thermal conductivity $\lambda_{eff}$ and the total thermal resistance of the heat exchanger $R_b$ according to the linear model were described by the formula (1) [17]:

$$T_f(t) = (Q/H \cdot \pi \cdot 4 \cdot \lambda_{eff})\ (ln(4 \cdot \alpha \cdot t/r^2) - \gamma) + (Q/H) \cdot R_b + T_{ground} \tag{1}$$

$T_f(t)$ – the average temperature of the fluid in the BHE [K]
Q/H – thermal power supplied [W]
$\lambda_{eff}$ – effective coefficient of thermal conductivity [W/mK]
$\alpha$ – thermal diffusivity of the ground [$m^2/s$]
t – duration of the test [s]
$\gamma$ – Euler constant, $\gamma = 0{,}5772$
r – radius of the borehole [m]
$R_b$ – total thermal resistance of the BHE [$m^2K/W$]
$T_{ground}$ – unsettled ground temperature [K]

The coefficient of thermal conductivity depends mainly on the thermal parameters of the rock mass surrounding the BHE, and thermal resistance of the BHE depends on the quality and technology of its implementation. Equation 2 is a simplified mathematical record of the logarithmic function of equation 1, where k parameter is the directional coefficient of the curve created by the semi-logarithmic presentation of the temperature changes of the medium during the heating phase of the test.

$$T_f(t) = k \cdot ln(t) + m \tag{2}$$

From equations (1) and (2) results that:

$$k = Q/(4 \cdot \pi \cdot H \cdot \lambda_{eff}) \tag{3}$$

and

$$m = R_b \cdot (Q/H) + T_{ground} \tag{4}$$

The coefficient of effective thermal conductivity was calculated from the formula (5):

$$\lambda_{eff} = Q/(4 \cdot \pi \cdot H \cdot k) \tag{5}$$

From the logarithmic curve fit to TRT measurements (Tf(t)), k and m parameters were obtained. The effective coefficient of thermal conductivity $\lambda_{eff}$ calculated using the above methodology was 2.19 W/mK. The heat exchanger efficiency per meter length was obtained considering the dependence of the $\lambda$ on the performance specified in SIA 384/6 (Fig. 4), for the curve 2,0 MJ/$m^2$ on the assumption that working time of the BHE is 1850 h/year.

Calculations concerning the selection of the total length of the vertical borehole heat exchanger were carried out with the assumption of 3 variants of the heat pump cooling capacity: 9, 18 and 40 kW. Calculating total length of BHE is the result of the quotient of heat pump cooling capacity and performance based on SIA 386/4. Calculating heat pump capacity based on theoretical values of thermal conductivity is the result of the multiplication of total length of BHE and measured coefficient of thermal conductivity. The obtained results are presented in Table 2 and indicate that in the analyzed lithological profile, regardless of whether the designer adopts the theoretical values of the coefficient of thermal conductivity at the minimum level (0.96 W/mK) or maximum level (2.15 W/mK), will be lower than the value calculated on the basis of measurements from the thermal response test – 2.19 W/mK. It has to be mention that it is not recommended to choose the maximum values [15] which gave a result close to the calculated one. This results in oversizing the heat source in each of the analyzed cases. In the case of a 9 kW heat pump, instead of the total length of the exchanger of 257 m, the length of 474 m (for the minimum theoretical value of the coefficient of thermal conductivity) and 272 m (for the maximum value) were calculated. This is an error at the level of 6–84%. Consequently the power that the heat source will have can be sufficient for a heat pump with a cooling capacity of 9.5–16.5 kW.

For heat pumps with 18 and 40 kW cooling capacity, the total length values of the vertical borehole heat exchanger obtained in the calculations change proportionally in relation to the increase of the cooling power. Designed powers of the heat source based on the theoretical parameters are respectively: instead of 18 kW – 19–33 kW, instead of 40 kW – 42.5–73.5 kW (in both cases an error at level 6–84%).

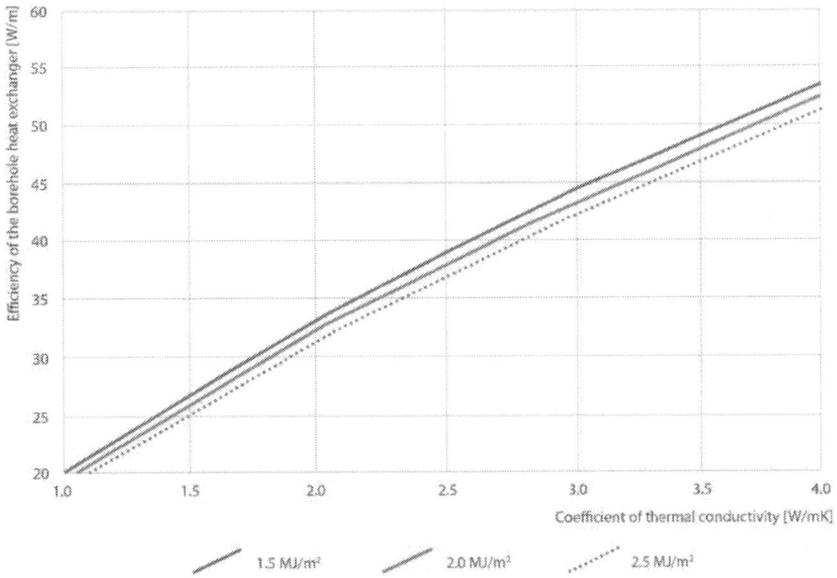


**Fig. 4.** Dependence of the borehole heat exchanger performance per 1 length meter and coefficient of thermal conductivity (based on [18]).

**Table 2.** Results of the TRT calculation in comparison to theoretical values.

| | **Lowest theoretical coefficient of thermal conductivity, 0.96 [W/mK]** | **Highest theoretical coefficient of thermal conductivity, 2.15 [W/mK]** | **Measured coefficient of thermal conductivity (TRT), 2.19 [W/mK]** | ***Calculating heat pump capacity based on theoretical values of thermal conductivity*** |
|---|---|---|---|---|
| Performance based on SIA 386/4 [W/m] | 19.00 | 33.00 | 35.00 | – |
| Calculating total length of BHE for **9 kW** cooling power [m] | 474 | 272 | 257 | *9.5–16.5 kW* |
| Calculating total length of BHE for **18 kW** cooling power [m] | 947 | 545 | 514 | *19.0–33.0 kW* |
| Calculating total length of BHE for **40 kW** cooling power [m] | 2105 | 1212 | 1143 | *42.5–73.5 kW* |

## 4 Summary and conclusions

Presented results of the calculations indicate that in the tested lithological profile of a vertical borehole heat exchanger, the value of the coefficient of thermal conductivity calculated on the basis of the thermal response test performed is significantly different from the theoretical values given in the literature. The real value is 2.19 W/mK, while the literature values range from 0.96–2.15 W/mK. In this particular case, the coefficient of thermal conductivity is underestimated at 2% to 228%. The consequence resulting from the selection of the heat source based on theoretical data will be its oversizing. The natural effect of oversizing the heat source will be higher investment costs.

The cost of realization the 1 meter of the vertical borehole heat exchanger depends on the geological structure in the location where the investment is being made. However, it can be assumed that the average cost of the BHE in Poland is 100 PLN per 1 meter (ca 23.5 EUR). As a consequence in the analyzed case the investor would pay more in a variant with a 9 kW pump by 1,500–21,700 PLN (ca 64–923 EUR), for a 18 kW pump, between 4,000–40,600 PLN (ca 170–1,728 EUR), and for a 40 kW pump by 6,900–96,200 PLN (ca 294–4,094 EUR).

It should be pointed out, that oversizing of the BHE causes, that too little heat load makes the compressor work in short periods of time. Lowering the assumed temperature in building activates the heat pump which due to the high installed capacity quickly heats the building. Therefore, it is switched off in a short time. Such cycles forcing the continuous start and stop, the compressor is strained. This process is accelerated by the fact that the start-up is accompanied by increased mechanical stress. Moreover, any such start-up brings with it an abrupt increase in energy consumption, which results from the necessity to overcome inertia of moving parts. It results in a greater electricity intake, and the heat pump's coefficient of performance (COP) decreases, which results from the change in the proportion of electric energy taken by the compressor to heat energy taken from the ground.

Despite the fact that in the analyzed case study there was no situation in which the heat source is undersizing, it should be added that such a situation would also causes decrease of the COP. Undersizing of the BHE results in longer compressor working time, which shortens the service life and the investor is exposed to additional costs. It is recommended to assume that the compressor will operate during the year between 1800 and 2500 hours, so that it does not wear out after 20 years of use. The heat pump works longer during the day to provide the required amount of heat which increases operational costs. During cold winters this effect is intensified, because the insufficient power of the device is supported by an electric heater. COP is lower than expected. In addition to extending the compressor's work cycle, the consequences is also that the ground around the heat exchanger cools down so much that it can't provide the designed cooling power. Such a situation can lead to emergency shutdowns of the device as a result of the evaporator low temperature. During mild winters this defect may not be revealed immediately.

The basic conclusion that should be emphasized from the analyzed example is that on the basis of literature data, for a brine/water heat pump is unable to precisely determine the total length of the vertical borehole heat exchanger. As a result of such a procedure, the investor incurs additional costs both at the investment and operational stages, and the efficiency of the heat pump's decrease. Therefore, especially in the case of large investments, the TRT method should be used as a means of obtaining information about the coefficient of thermal conductivity in an empirical way. In addition, it is worth considering adding information from the TRT to the results of i.e. electrical resistivity method, which may be an interesting implementation of geophysical methods to the field of heat pumps.

*The paper has been prepared under the AGH-UST statutory research grant No. 11.11.140.031.*

## References

1. EHPA, http://stats.ehpa.org/hp_sales/story_sales/, access: 29.08.2018
2. PORT PC, http://portpc.pl/spektakularny-wzrost-rynku-pomp-ciepla-polsce-2017-roku/, access: 29.08.2018
3. B. Tomaszewska, L. Pająk, J. Bundschuh, W. Bujakowski, Desalination **435**, 35–44 (2018) doi: 10.1016/j.desal.2017.12.032
4. L. Pająk, B. Tomaszewska, *Porównanie efektów energetycznych, ekonomicznych i ekologicznych wykorzystania pompy ciepła typu woda/woda i solanka/woda do ogrzewania domu jednorodzinnego (Comparison of energy, economic and environmental effects of the use of water/water and brine/water heat pumps for heating single-family house),* Ciepłownictwo Ogrzewnictwo Wentylacja **47**, 4, 152–157 (2016) (in Polish)
5. B. Nordell, *Thermal Response Test (TRT) State-of-the art 2011* (IEA ECES ANNEX 21, 2011)
6. M. Wajman, *Metody pomiarowe badania warunków pracy pionowych sond gruntowych – testy TRT i DTRT (Measuring methods in test of working conditions of vertical ground heat exchangers-Thermal Response Tests (TRT and DTRT)),* Technika Chłodnicza i Klimatyzacyjna **12**, 438–444 (2011) (in Polish)
7. M.A. Kaczmarczyk, M. Kaczmarczyk, *Analysis of the thermal conductivity of a borehole heat exchanger measured during TRT compared with theoretical calculations and its impact on the installation parameters* (29046), Proceedings World Geothermal Congress, Australia – New Zealand, (2015)
8. G. Pełka, W. Luboń, J. Kotyza, *Design and development of portable, low-cost thermal response test device,* 2'nd International Conference Renewable Energy Sources: engineering, technology, innovations – scientific conference, 511–518, (2015)
9. G. Ryżyński, P. Czarniak, P. Sobótka, *PGI-NRI experiences in geothermal parameters measurements* (GEOPLASMA-CE: Knowledge exchange workshop on workflows and standards for calibration of thermal Response Test devices, (2017)
10. T. Śliwa, A. Sapińska-Śliwa, A. Gonet, Z. Jezuit, A. Bieda, T. Kowalski, J. Ozimek, A. Złotkowski, *Specifying the number of borehole heat exchangers based on thermal response test and geoenergetic analysis,* AGH Drilling, Oil, Gas **34**, 1, 273–290 (2017)
11. B. Sanner, *Kann man eine Erdwärmesonde mit Hilfe von spezifischen Entzugsleistungen auslegen? (Is it possible to design BHE using the thermal efficiency of the rock mass?),* Geothermische energie. Mitteilungsblatt der Geothermischen Vereinigung e.V. **26/27**, 3/4, 1–4 (1999) (in German)
12. B. Sanner, G. Hellstrom, J. Spitler, S. Gehlin, *Thermal Response Test – Current Status and World-Wide Application*, Proceedings World Geothermal Congress, Turkey, 1436, (2005)
13. B. Sanner, E. Mands, M.K. Sauer, E. Grundmann, *Thermal response test, a routine method to determine thermal ground properties for GSHP design*, 9th International IEA Heat Pump Conference, Switzerland, 20–22 (2008)
14. M.A. Kaczmarczyk, *Testowanie otworowych wymienników ciepła dla celów optymalizacji parametrów i układu wymienników dolnego źródła dla gruntowej pompy ciepła (Testing of borehole heat exchangers to optimize parameters and configuration of borehole heat exchangers for ground source heat pump),* Dokonania naukowe doktorantów: nauki inżynieryjne, 106–115 (2013) (in Polish)
15. PORT PC, *Wytyczne projektowania, wykonania i odbioru instalacji zasilanych pompami ciepła. Część 1 (Guidelines for designing, realization and receiving installations powered by heat pumps. Part 1)*, Kraków (2013) (in Polish)

16. J. Acuña, *Improvements of U-pipe Borehole Heat Exchanger*, Licentiate Thesis in Energy Technology, KTH Royal Institute of Technology, Sweden (2010)
17. C. Eklöf, S. Gehlin, *TED-a mobile equipment for thermal response test: testing and evaluation,* Master's Thesis, Lulea University of Technology, Sweden (1996)
18. SIA 384/6:2010, Erdwarmesonden, Schweizer Norm 546 384/6, Zurich (2009)

# Assessment of parting rock weak zones under the joint and downward mining of coal seams

*Volodymyr* Bondarenko[1], *Iryna* Kovalevska[1,*], *Hennadii* Symanovych[1], *Mykhailo* Barabash[2] and *Vasyl* Snihur[3]

[1]National Mining University, Underground Mining Department, Dnipro, Ukraine
[2]LLC "DTEK Energy", Department of Coal Production, Kyiv, Ukraine
[3]MM "Pershotravenske", PJSC "DTEK Pavlohradvuhillia", Pavlohrad, Ukraine

**Abstract.** The aim of the forecasting effort is to identify troublesome zones of stability loss by a parting lengthwise of the extraction panel under the joint and downward mining of coal seams. Analyses have been carried out of active stress component curves for a 3-D model computational experiment compared with the strength characteristic of each lithotype of a parting. An algorithm has been developed for the stability assessment of a parting lengthwise along the extraction panel. The relationship patterns have been estimated between the sizes of the parting rocks discontinuity zones and the main geomechanical parameters. A scientifically grounded basis has been created for the detection of the parting rock weak zones lengthwise along the extraction panel for the calculation of the mounting and security systems of the development works. A complex of underground instrumental observations was made, which was used to set up a correspondence of patterns to indicate the variation in rock pressure manifestation intensity and the tendencies for changes in the parting structure. All of this confirms the adequacy of the techniques for parting state forecasting, which is recommended for use in the engineering documentation for the joint and downward mining of coal seams.

**Keywords:** coal seams, joint mining, computational experiment, stability

## 1 Introduction

For several years in the Western Donbas mines research and development work has been carried out in order to study the structural transformation of parting rock caused by the joint and downward mining of two or three coal seams in series. These transformations were researched from the viewpoint of rock pressure manifestations control, and at the same time as a way to implement the significant concentration and intensification of coal-face works with the benefit of obtaining the corresponding technical-and-economic parameters. So far, the main scientific and practical results achieved are as follows [1 – 4]:

1. It was established that under the joint mining of coal seams (not classified as contiguous grade) abnormal areas of rock pressure are manifested [1, 2] at the coal-face

* Corresponding author: kovalevska_i@yahoo.com

operation in a wide range of parting thicknesses up to 20 – 50 m.

2. The parting rock structure systematization under the condition of joint and downward mining of coal seams in Western Donbas made it possible to allocate the five most common geological structures [3]. They have been provided as the basis for (using the finite element method) the calculation of abnormal areas of rock pressure manifestation, the subsequent assessment of the parting stability and corresponding decision-making as to the development, fastening, and maintenance of both mine workings and coal-face works.

3. A comprehensive methodology for carrying out multivariate computational experiments has been developed. It consisted of a three staged logical detection and account of the structural transformations of the parting rocks under the process of the joint and downward mining of coal seams. This provided an increasingly proficient degree of geomechanical process modelling in order to obtain more exact results for the parting rocks stress-strain state calculation by means of representation in the models of the most significant structural disturbances of each lithotype in the coal-bearing massif for the whole period of the mining operation: the preparation of extraction panels on the overlying seam, its mining with simultaneous development works on the underlying seam and subsequent coal-face works on it.

4. As a result of the stress-strain state calculations of the parting rocks, it has been established that the vertical stress concentrations penetrate the overlying seam bottom to the middle of the parting thickness and deeper, and constitute a danger of weakening the soft wet argillite and siltstone at a depth of 200 – 300 m, a discontinuity occurs at a depth of 400 m in their naturally wet state. Below that depth, the vertical stress concentrations extend to the underlying seam and create rock pressure, which increases by 20 – 50% on the support construction of existing works [1]. This makes it possible to develop in advance the measures for the strengthening of development works to support construction in dangerous areas [4].

Accordingly, the aim of the article is the assessment of the parting rock weak zones under the joint and downward mining of coal seams. The studies have been carried out on the basis of a computational experiment by means of the finite elements method, the mining observations of rock pressure manifestations and their comparative analysis.

## 2 Calculation of the weak zones of parting rocks

The calculation principle of the weak zone is to compare each of the active stress components $\sigma_y$, $\sigma_x$ and $\sigma_z$ with a condition of absolute unloading (criteria $\sigma_x = 0$, $\sigma_z = 0$ and $\sigma_y = 0$), when the process of parting rock segregation with tensile forces begins. But in areas where compressive stresses take precedence, the intensity of stresses $\sigma$ is equal to the calculated ultimate compressive strength value $R_i^{R,B}$ of one lithotype or another [3].

The calculation algorithm of discontinuity areas of the parting lithotypes presumes the isoline contours $\sigma_{y,x,z} = 0$ for the determination under each of curves $\sigma_y$, $\sigma_x$ and $\sigma_z$ stress components; also the isolines $\sigma \geq R_i^{R,B}$ are plotted for each lithotype by the curve $\sigma$. For this purpose in each particular area of the model, at a specific value of $R_i^{R,B}$ an integrated application is used, this makes it possible to plot the contours of pointed isolines taking into account a certain defined accuracy. Then, the contours of all isolines ($\sigma_{y,x,z} = 0$ and $\sigma = R_i^{R,B}$) are united in a section of the model and a single overall external contour of the weak zone of the rock is plotted in general.

Some fragments of bottom rock discontinuity areas in the mining of an overlying seam $C_{10}{}^t$ are demonstrated by a depth increase $y_B$ of their distribution with an increase in depth $H$ of coal seam mining (Fig. 1). The relationship between depth $y_B$ of bottom rocks weakening during the mining of seam $C_{10}{}^t$ with strength characteristics $R_i^{R,B}$ may be observed for all the structures of a parting, but the tendency of $y_B$ to decrease due to increases in $R_i^{R,B}$ is based on the influence of thickness $m^B$ and depth $h^B$ of the stratification in the sandstone bottom, which adds to the pattern of significant ambiguity, which in turn does not allow for approximation by any function.

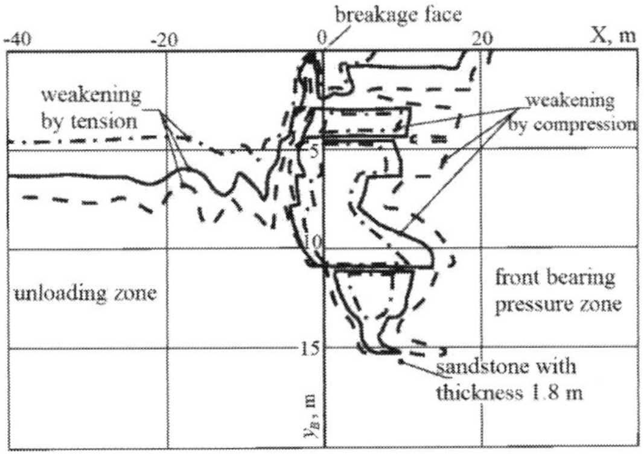


**Fig. 1.** The curves of the fragments of mining depth have influence $H$ on the discontinuity area sizes of the seam $C_{10}^{t}$ bottom: - · - $H = 300$ m; —— $H = 400$ m; - - - $H = 500$ m.

The similar tendencies of height $y_R$ of the roof rock weak zone relationship with geomechanics parameters $H$ (Fig. 2) and $R_i^{R,B}$ are observed in the roof of the seam $C_8^{l}$; these patterns are also violated by the ambiguous influence of parameters $m_i^R$ and $h_i^R$ on sandstone stratification. Special attention is given here to the tendencies towards the influence of sandstone stratification parameters. Therefore, the sandstone in the main roof with average strength characteristics (in Western Donbas conditions) of lithotypes of a parting is capable of localising the upper part of the weak zone even at its average thickness, and detaching it from the main one, this characteristic is positively affected by the parting's stability. With the occurrence of thick sandstone (3.0 – 5.0 m), not only weak zone decomposition occurs into separate "foci", there is a decrease in the overall height of the disturbances extension by 2.2 – 2.3 times.

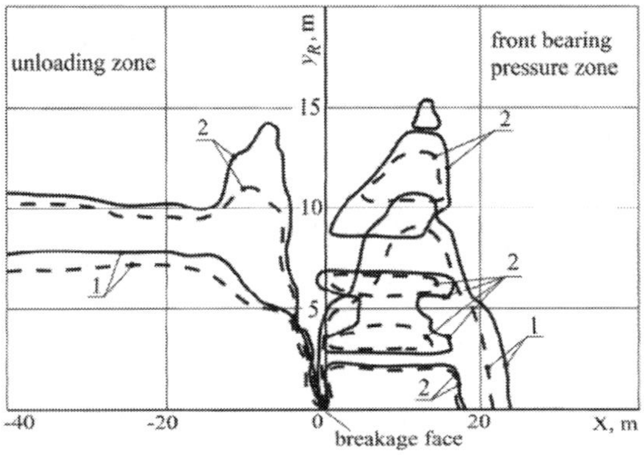


**Fig. 2.** The curves of fragments of mining depth $H$ and the parting structure type has an influence on the discontinuity area sizes of the roof rocks in the underlying seam: 1 – coarse-grained and medium-bedded structure; 2 – medium- and thin-bedded structure; - - - $H = 300$ m; —— $H = 500$ m.

Thereafter, the volumes of weakened rocks are decreased even more intensively (by 3.5 times). In the unloading zone behind the longwall, the reduction in the height of the main roof rock segregation goes as far as 3 times with sandstone with a thickness of 5.0 m.

Summing up the results of the studies, it is necessary to acknowledge (in terms of the roof rocks of underlying seam) in a formulated form before thesis (in terms of the overlying seam bottom) about the complexity and ambiguity of the geomechanical parameters that have an influence over weak zone sizes and the assessment of the degree of the parting rocks stability. Nevertheless, due to the wide variety of the parting rocks and stress-strain state calculations, the base for the development of specific and, at the same time, resumptive predictive assessments of the parting stability under the joint and downward mining of two or three coal seams, has been basically developed.

With regard to the most troublesome zones of parting rock stability probable loss, it is estimated that at approximate indifference of weak zones sizes in different segments of a parting along the length of the works, its stability should be assessed, at least, in three zones: in the front bearing pressure zone in advance of the longwall at a distance of $x \leq 7-9$ m; in unloading zone after the longwall withdrawal at a distance from the entrance of $z \leq 15-20$ m in direction of the mined-out space, and in the lateral bearing pressure zone of the virgin coal seam.

The next stage of the study was concerned with the criterion of substantiation, which compares the overall thickness $y_B + y_R$ of the weak zones in the overlying seam bottom $y_B$ and in the underlying seam roof $y_R$ with the thickness $H_m$ of a parting, which loses its stability at:

$$y_B + y_R \geq H_m . \tag{1}$$

In view of geometric and physical inhomogeneity of the parting structure, the weak zones often have an "intermittent" form with the formation of a range of local subareas. To resolve the situation with regard to the stability assessment of undisturbed space intervals between adjacent local weak zones, supplemental criteria have been developed, which take into consideration the distances between surfaces of weakness and rheological factor impact. Therefore, the strength characteristics of the roof and bottom rocks, as well as coal $\sigma_{compr}^{R,B,C}$ are accepted according to [5 – 8] and through the data of the Geological Survey of mines; their analysis made it possible to substantiate the following averaged ranges of compressive resistance variation (in the specimen) by lithotypes: coal – 30 – 40 MPa; argillite – 10 – 20 MPa; siltstone – 15 – 25 MPa; sandstone – 30 – 60 MPa. As a result, in view of problem structures (in terms of stability) of a parting [3] three variants may be distinguished for the computational values distribution of the strength characteristics of lithotypes:

– variant A – partially saturated rocks along the lower boundary of compressive resistance: coal $R^{coal} = 20$ MPa, argillite $R_i^{R,B} = 5$ MPa, siltstone $R_i^{R,B} = 10$ MPa, sandstone $R_i^{R,B} = 20$ MPa;

– variant B – saturated rocks along the upper boundary of compressive resistance: coal $R^{coal} = 30$ MPa, argillite $R_i^{R,B} = 10$ MPa, siltstone $R_i^{R,B} = 15$ MPa, sandstone $R_i^{R,B} = 45$ MPa;

– variant C – unsaturated rocks along the upper boundary of compressive resistance: coal $R^{coal} = 40$ MPa, argillite $R_i^{R,B} = 20$ MPa, siltstone $R_i^{R,B} = 25$ MPa, sandstone $R_i^{R,B} = 60$ MPa.

For the stability criterion of direct use (1) and in view of the extreme ambiguity and unreliability of the functional relationship of coordinates $y_B$ and $y_R$ with the main influential factors, families of graphs are plotted for the parameters $y_B$ and $y_R$ calculations (Fig. 3 and 4). These characteristic curves in conjunction with criterion (1) allow us to detect potential problems over lengthwise areas of the mineable panel.

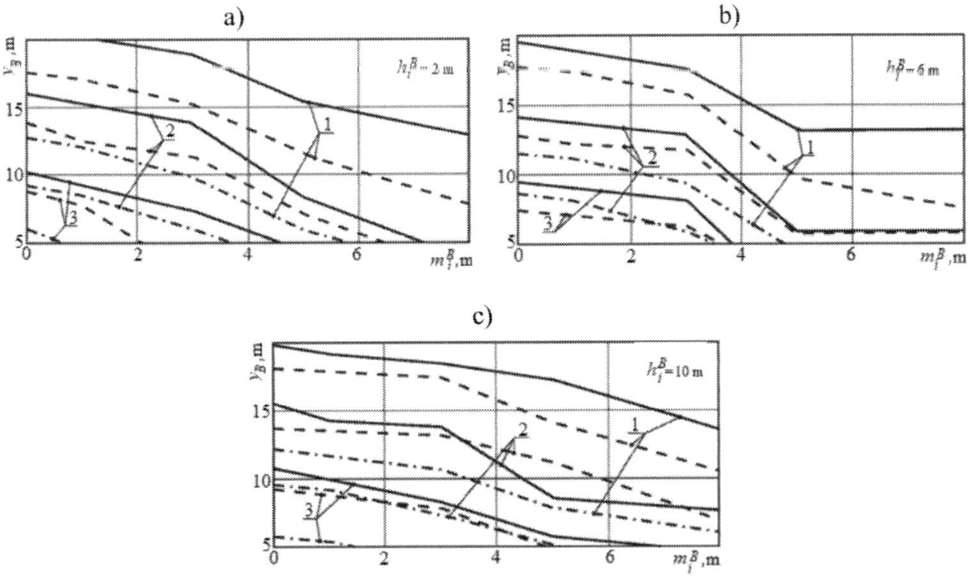


**Fig. 3.** The curves of the relationship between the depth yB of the bottom rocks weakening of the overlying seam with thickness $m_i^B$ and the depth $h_i^B$ of sandstone occurrence: a) $h_i^B = 2$ m ; b) $h_i^B = 6$ m ; c) $h_i^B = 10$ m; 1 – variant A; 2 – variant B; 3 – variant C; – · – $H = 200$ m; – – – $H = 400$ m; —— $H = 600$ m.

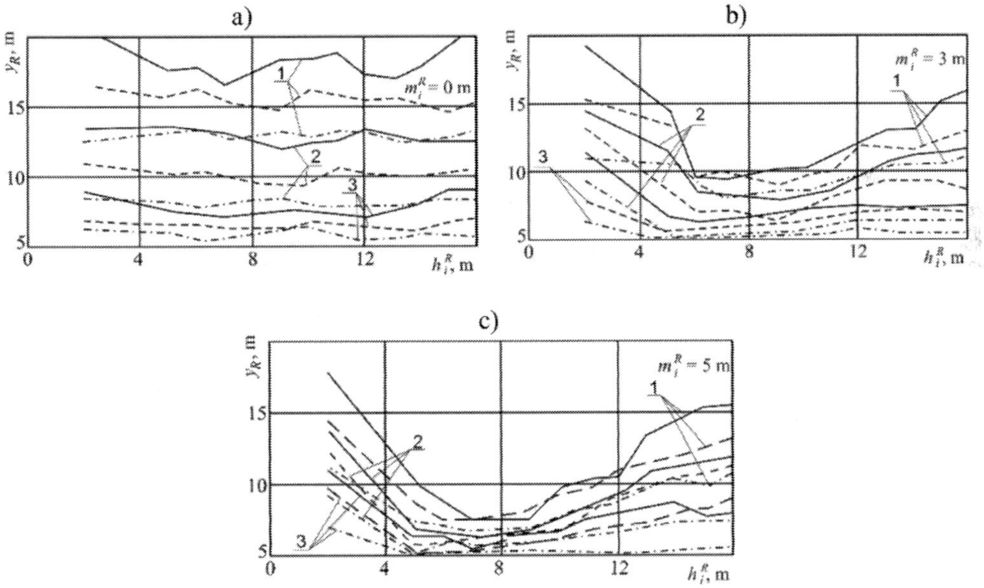


**Fig. 4.** The curves of the relationship between the height $y_R$ of roof rocks weakening of the underlying seam with height $h_i^R$ and thickness $m_i^R$ of sandstone occurrence: a) $m_i^R = 0$ m; b) $m_i^R = 3$ m; c) $m_i^R = 5$ m; 1 – variant A; 2 – variant B; 3 – variant C; – · – $H = 200$ m; – – – $H = 400$ m; —— $H = 600$ m.

The patterns described above make it possible to outline the areas of mining-and-geological conditions to definitively determine the stable state of a parting:

– due to the factor of the overlying seam roof rock weakening – the depth of sandstone occurrence is about 6 m where its thickness is not less than 3.0 m;

– due to the factor of the underlying seam roof rock weakening – the height of sandstone occurrence is in the interval of 6 – 9 m where its thickness is not less than 3.0 m.

Thus, these ratios of rock weakening distribution vary according to the thickness of the parting and are determined not only by the mining depth and strength characteristics of lithotypes, but also by the sandstone thickness and height (depth) of its occurrence relative to mineable seams; the parting is stable (even in conditions of saturated rocks) with a sandstone thickness of more than 3 m and its location in the roof (bottom) of the seams at a distance of 6 – 9 m. This provides a way to establish the range of mining-and-geological conditions, where downward and joint mining may be performed without considering the risk of stability loss by a parting.

## 3 Forecasting the effect of troublesome zones on parting weakening

Using graphs in Fig. 3 and 4 in combination with criterion (1), it is possible to make an assessment of parting stability. It is, however, connected with a number of calculations, extrapolations, and specifications of the sandstone sheet (sheets) location relative to the overlying and underlying seams. This complicates, in some ways, the algorithm for the stability assessment of a parting, especially with respect to the way in which this assessment is performed lengthwise along the mineable panel. Therefore, for the operational assessment of parting stability, a numerical analyses of curves has been performed (Fig. 3 and 4) in conjunction with criterion (1). This results in the determination of interface areas of the stable and unstable state of the parting rocks, which are shown in Fig. 5.

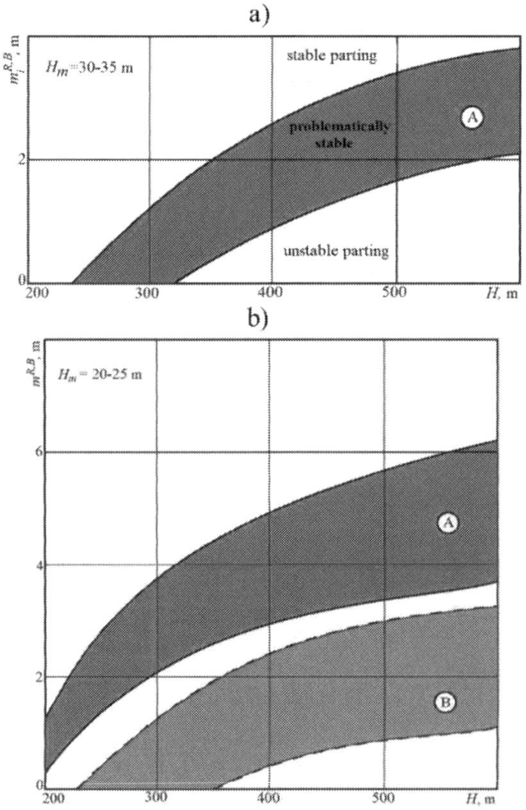


**Fig. 5.** Predictive areas of rock state of a parting with thickness: a) $H_m$ = 30 – 35 m; b) $H_m$ = 20 – 25 m; A – saturated rocks along the lower boundary of hardness; B – the rocks of average compressive resistance (as applicable to Western Donbas conditions).

These interface areas make it possible to predetermine, at a preliminary stage, three variants in the mining-engineering situation: a parting is stable, a parting is unstable and problematically stable (it is possible that its stability loss may occur due to certain more specific geomechanical parameter ratios).

## 4 The algorithm of the parting stability assessment

The assessment of the parting stability lengthwise along the extraction panel is made in the following sequence:

1. The preliminary separation of a mineable panel length into areas of a stable parting and areas with a possible loss of stability are performed. This is done at the scientific assessment level, and the emphasis is on the areas with thin- and medium-bedded structure, saturated rocks (according to prediction), increased mining depth and decreased thickness of a parting.

2. Based on the mining-and-geological forecasting data and with the use of normative technique regulations [9] the strength characteristics of the parting rocks belong to one of three categories: A, B or C.

3. The structure of a parting in a specific area from a range of selected ones is studied, with the determination of sandstone thickness in the $m_i^{R,B}$ variation range, the depth $h_i^B$ of its occurrence in the overlying seam bottom and height $h_i^R$ of its distribution in the underlying seam roof. If there are two sheets of sandstone in the structure of a parting, the parameters $h_i^B$ and $h_i^B$ are determined for the upper sheet, but parameters $m_i^R$ and $h_i^R$ are specified for the lower sheet of sandstone.

4. With the use of Figure 5 a general assessment of the parting stability is made:

– in the category of "Stable parting", the studied area is excluded from further consideration, and mining in the underlying seam is performed without consideration of the influence of coal-face operation in the overlying seam;

– in the category of "Unstable parting", the mining and engineering measures are developed in the area to consider the influence of the coal-face operation in the overlying seam, such as, to improve the mounting and security systems in the underlying seam;

– in the category of "Problematically stable", it is necessary to study in greater detail the depth $y_B$ and height $y_R$ of disturbance extensions with the use of curves in Figures 3 and 4. Thereafter, the values $y_B$ and $y_R$ are calculated according to the curves in Figures 3 and 4 (with the use, if necessary, of an extrapolation between two adjacent single lines) and inserted into criterion (1):

– if the criterion is met, the studied area lengthwise along the mineable panel falls into the category of "Unstable parting" with the corresponding mining and engineering measures developed to improve the safety and reliability of mining operations in the underlying seam;

– in cases where criterion (1) is not met, this area falls into the category "Stable parting", and there is no need for any change of parameters for technological operations.

5. The stated algorithm of the parting stability assessment is performed throughout the length of the mineable panel with the addition of the necessary structural-and-technological measures to the engineering documentation of mining operations.

## 5 Analysis of the underground investigation results

To confirm the adequacy and reliability of the parting stability forecasting, a methodology for carrying out comprehensive underground instrumental observations has been developed. Two types of experimental characteristic curves have been obtained, which characterize the rock pressure manifestations in mining, for example, of the seam $C_9$: directly in the breakage face by a relative ratio $\Delta$ of the extension variation of mechanized support for hydraulic props and in 914 development works by maximal $(u_v)_{max}$ and average $(u_v)_{aver}$ which encounter the velocity of roof and bottom rocks in the area of intersection with 914 longwall faces (Zakhidno-Donbaska Mine, Western Donbas) [7].

An analysis of the characteristic curves (Fig. 6) confirmed a correspondence of patterns for indicator variation in rock pressure manifestation intensity and tendencies of changes in the parting structure lengthwise along the 914 mineable panels of the seam $C_9$ longwall face. In such a way, the underground investigation results made it possible to recommend the technique of parting stability forecasting for use in engineering documentation for the joint and downward mining of coal seams. For example, recommendations were developed concerning the choice of parameters for mounting and security schemes of the works in case of its reuse. These recommendations are based on the performed analyses of mining-and-geological conditions for the maintenance of the works, as well as on the computational experiment results of the geomechanical system stress-strain state calculations (Fig. 7).

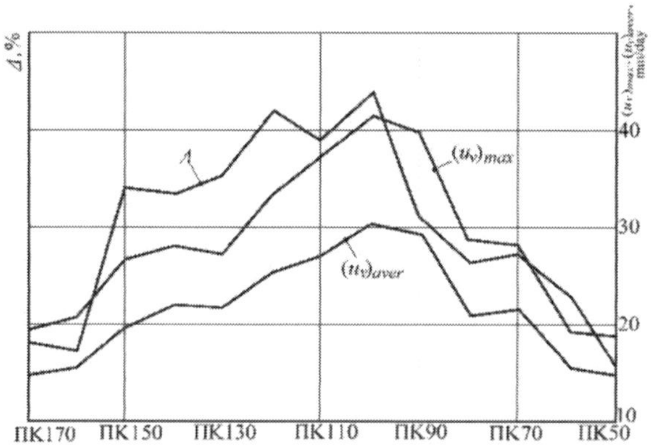


**Fig. 6.** Characteristic curves of $\Delta$, $(u_v)_{max}$, $(u_v)_{aver}$ indicators variation on the working unit with thin-bedded structure of parting and saturated rocks in mining of the $C_9$ seam.

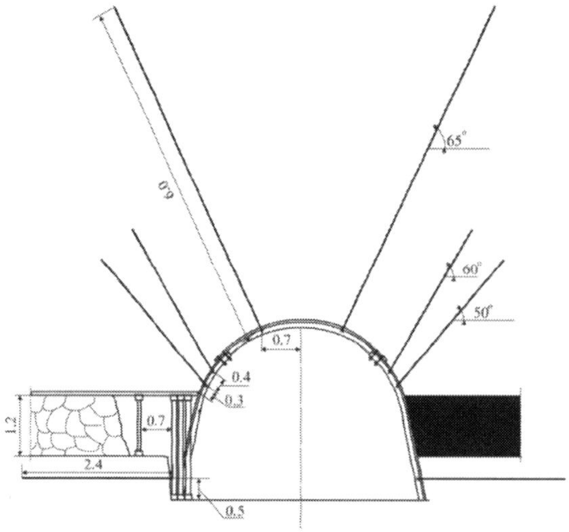


**Fig. 7.** The model of the recommended scheme for the maintenance of the works.

## 6 Conclusions

1. It has been established, that in the front pressure zone of the longwall end areas, 7 – 9 m to the maintained works and behind the longwall at a distance up to 27 – 31 m from the mine works, the superposition of rock pressure manifestations occurs, this is caused by face works and the maintained works by itself. This provides grounds for the allocation of the longwall end area with a width of up to 20 m (from the drift towards the mined-out space) as one of the most troublesome areas from the viewpoint of the parting stability loss. Therefore, the unloading and bottom rocks segregation area deepens by 1.5 – 4.5 times more than throughout the rest of the mined-out space. The account of this phenomenon allows for the making of a provision in the extraction panel mining passport for measures directed toward the adjacent massif strengthening in order to provide stability to the junction between the longwall and the works, when mining the troublesome zones.

2. The criteria and algorithms for the comprehensive assessment of parting rock stability have been substantiated.

A comprehensive assessment is provided by the account of segregation under the impact of tensile stresses $\sigma_{y,x,z} \geq 0$ in all three directions (Y, X, Z) of each parting lithotype occurring in space; the compression failure areas $\sigma \geq R_i^{R,B}$ also have a spatial representation and take into consideration the impact of factors, which weaken the rock, such as fracturing, water saturation and rheology. This allows for the determination of the patterns of discontinuity area development and their location in space of a parting taking into account its variable structure and strength characteristics, fracturing intensity and water saturation and liability to the creep deformation of lithotypes, which form a parting.

3. The main influential geomechanical parameters have been systemized and the patterns have been determined for their relationship with the depth of weakening in the overlying seam bottom and with the height of structural disturbances in the underlying seam roof. Patterns have been obtained in the form of graph families of the relationship between the sizes of parting rock discontinuity areas and the main influential factors: depth of mining $H$, parting thickness $H_m$ and compressive resistance $R_i^{R,B}$ of its lithotypes, thickness $m_i^{R,B}$ and vertical coordinate $h_i^{R,B}$ of sandstone distribution within the bounds of a parting.

4. A forecasting technique for the troublesome zones in a parting lengthwise along the mineable panel has been developed, which is distinguished by operational calculations and availability for use. This technique presents the possibility to detect in advance the problem areas for performing warning measures of safe and highly-productive mining in underlying seams.

*The authors express their gratitude to senior management of "Pavlohradvuhillia DTEK" PLC and Mine Administration for the assistance in research development.*

## References

1. M.V. Barabash, *State analysis of overworked and underworked parting between contiguous seams and during their simultaneous top-down mining*. Min. Miner. Depos. **10**, 2, 34-39 (2016) doi.org/10.15407/mining10.02.034
2. V.I. Bondarenko, I.A. Kovalevskaya, G.A. Simanovich, M.V. Barabash, V.G. Snigur, *Vliyanie gorizontal'nykh napryazheniy na strukturnye preobrazovaniya porod mezhduplast'ya sovmestno otrabatyvaemykh ugol'nykh plastov* $c_{10}^{v}$ *i* $c_9$. Ugol' Ukrainy **4**, 35-39 (2017)
3. I. Kovalevska, M. Barabash, V. Snihur, *Development of a research methodology and analysis of the stress state of a parting under the joint and downward mining of coal seams*. Min. Miner. Depos. **12**, 1, 76-84 (2018) doi.org/10.15407/mining12.01.076

4. I.A. Kovalevskaya, G.A. Simanovich, M.V. Barabash, *Ekspertnaya otsenka vybora krepezhnoy i okhrannoy sistem povtorno ispol'zuemykh vyrabotok pri otrabotke plasta* $c_8^n$, Ugol' Ukrainy **12**, 17-20 (2015)
5. B.M. Usachenko, V.P. Cherednichenko, I.E. Golovchansky, *Geomekhanika okhrany vyrabotok v slabometamorfizirovannykh porodakh* (Kyiv, Ukraine, 1990)
6. B.M. Usachenko, *Svoystva porod i ustoychivost' gornykh vyrabotok*. Kyiv: Naukova dumka (1979)
7. V.I. Bondarenko, I.A. Kovalevskaya, G.A. Simanovich, A.V. Martovitskiy, A.F. Kopylov, *Metody rascheta peremeshcheniy i uprochneniya prikonturnykh porod gornykh vyrabotok shakht Zapadnogo Donbassa* (Dnipropetrovsk, Ukraine, 2010)
8. I. Kovalevs'ka, G. Symanovych, V. Fomychov, *Research of stress-strain state of cracked coal-containing massif near-the-working area using finite elements technique*. Min. Miner. Depos., 159-163 (2013) http://www.scopus.com/inward/record.url?eid=2-s2.0-84892146430&partnerID=MN8TOARS
9. SOU 10.1.001.85790.011:2007, *Pidhotovchi vyrobky na polohykh plastakh. Vybir kriplennia, sposobiv i zasobiv okhorony* (Donetsk, Ukraine, 2008)

# Identification of zones of increased permeability using numerical modelling of temperature distribution – an ISMOP case study

*Maciej* Dwornik*, *Anna* Franczyk and *Andrzej* Leśniak

AGH University of Science and Technology, Faculty of Geology, Geophysics and Environmental Protection, Kraków, Poland

**Abstract.** The physical condition of earthen levees changes over time. Levee weakening can be caused by the leaching of fine grains of soil or by animal activity. Weakened areas can potentially cause destabilization or even collapse. Assessment of the physical condition of levees using GPR and geodetic measurements does not always correctly detect areas of increased permeability.
Thermal anomalies, caused by existing zones of high permeability, are observed in levees during the transition of flood waves. They are generated by rapid infiltration of water with a different temperature than the temperature inside the levee. Therefore, inner thermal measurements may be used to detect places with higher permeability. This research was realized with 2D numerical modelling. Main aim was test possibility of using temperature sensors in weak zone detection. Test with different theoretical permeabilities proved that it is possible only in specific condition (i.e. thermal contrast). The geometry and geomechanical properties of the studied levee were taken from the ISMOP project (polish acronym: Computer system for monitoring river embankments).

**Keywords:** levee, temperature measurements, weakened area, numerical modelling

## 1 Introduction

Earthen levees are one of the most popular methods of protecting human life and properties against floods; however, they must be in good condition in order to offer proper protection. There are many methods of checking the condition of levees. These methods are mainly based on cyclic measurements in parts of levees (i.e. GPR [1], ERT [2], geodetic measurements). The long-time gaps between measurements mean that new weak zones are often not detected before floods.

There are only a few approaches to continuously monitoring large parts of levees [3-5]. There were mainly used water pore pressure sensors to recognize faster flows inside levee. The main problem is the huge length of these constructions, which is problematic when it is

* Corresponding author: dwornik@geol.agh.edu.pl

considered that predictive systems should cheaply and easily determine any weak zones. One option is to use a dense grid of temperature sensors inside levees. This idea is based on two phenomena: thermal property differences between dry and wet soil, and larger zones of temperature changes if water whose temperature is different than the soil temperature infiltrates weak zones. So, if there is a temperature contrast between soil and water, it should be possible to determine weak zones by comparing neighbouring sensors.

In this paper, the idea of temperature sensors will be tested. Main aim was recognize possibility and restrictions of temperature sensors in detection weak zone. Thermal sensors were used previously, for example in monitoring roadway embankments [6]. Two-dimensional numerical modelling is used to determine the temperature distribution inside a levee during a flood. The modelling is performed with parameters used in polish acronym: Computer System for Monitoring River Embankments (ISMOP) projects (flood wave shape, soil properties, weather conditions, levee geometry) [7].

The location of the weak zone is similar to the location of the buried pipe in the levee. A leaking zone appeared along this pipe during the conducted experiments. Different theoretical permeabilities of the weak zone and the temperature condition were also used in the numerical models.

## 2 ISMOP Project

The ISMOP (a Polish acronym) Project was carried out by AGH University of Science and Technology and two companies (Neosentio and Budokop), and co-funded by the National Centre of Research and Development. A full-size levee was built for tests and experiments (fig. 1). The levee has a stadium shape with a water reservoir inside. Different parts of the levee are constructed of a variety of materials and the slopes are similar to the levees found on Polish rivers. Several sensor networks were located inside the levee: six sections with 5–6 temperature and pore pressure sensors, 74 sections of 14 temperature sensors, two loops of optical fibre with temperature sensors with a 1m offset along the fibre, 6 strain sensors and 33 piezometers (fig. 2) [8]. Additionally, GPR, ERT and LIDAR measurements were conducted periodically. Flood waves were simulated by changing the water level in the reservoir.

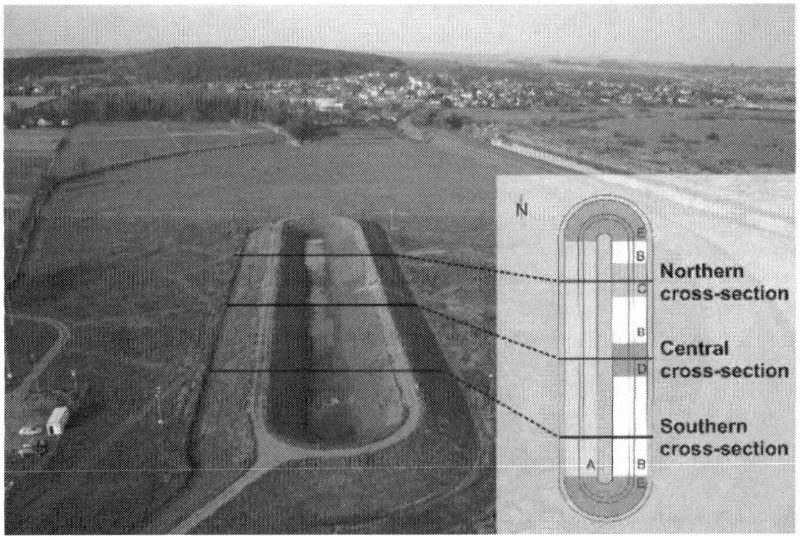


**Fig. 1.** Earthen construction built for the ISMOP Project (200 x 50 x 4.5m). A-E: type of material used [9].

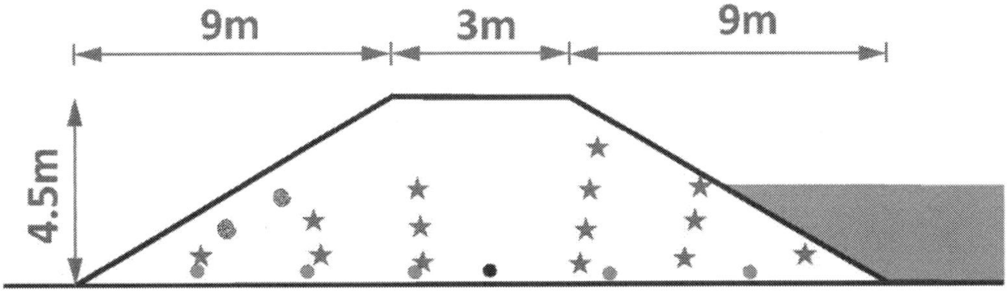


**Fig. 2.** Location of sensors inside a section in the NW part of the levee. Red stars – Neosentio network (temperature only); dots – Budokop network (blue – temperature & water pore pressure; black – strain & temperature); gray dot – optical fibre. The bottoms of the vertical piezometers were located near the blue dots.

## 3 Numerical modelling

Sixteen numerical models were created in Itasca Flac 2D 7.0 software [10-12]. Each model was built using 0.1 m square cells (fig. 3). Values of geomechanical and hydrogeological properties are presented in table 1. Flood wave modelling was performed assuming the same variable water level for all the 16 numerical models (fig. 4). Material properties, wave shape and temperature came from a real experiment in the ISMOP project. The experiment took place in April 2017. The initial soil temperature decreased linearly from 8°C at the bottom edge to 5°C at the top edge of the model.

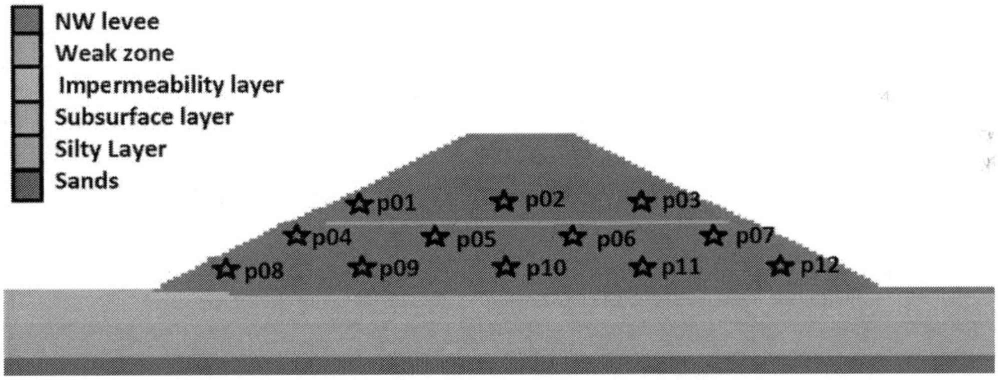


**Fig. 3.** Physical model used in the calculations. Points p01-p012 were used in fig. 8-9.

**Table 1.** Geomechanical parameters used in the numerical modelling [13,14].

| | NW levee, (Weak zone) | NE levee | Subsurface, (Impermeab ility layer) | Silty | Sands |
|---|---|---|---|---|---|
| Density [kg/m3] | 1900 | 1960 | 2100 | 1890 | 1850 |
| Cohesion [KPa] | 12.50 | 15.43 | 10.30 | 13.70 | 10.00 |
| Friction [°] | 30.04 | 35.20 | 32.90 | 22.50 | 36.20 |
| Bulk Module [MPa] | 8.53 | 7.25 | 7.25 | 16.20 | 36.30 |
| Shear Module [MPa] | 3.27 | 3.35 | 3.43 | 6.63 | 21.80 |
| Porosity [%] | 37 | 32 | 27 | 40 | 35 |
| Filtration factor [m/s] | $1.83*10^{-5}$ (vary) | $5.24*10^{-5}$ | $1.52*10^{-5}$ ($1.52*10^{-12}$) | $1.35*10^{-5}$ | $5.60*10^{-6}$ |

Calculations were performed for the various initial and boundary temperature conditions (fig.5):

- real temperature recorded during the experiment
- air temperature plus 5°C;
- water temperature = air temperature plus 5°C;
- water temperature plus 5°C.

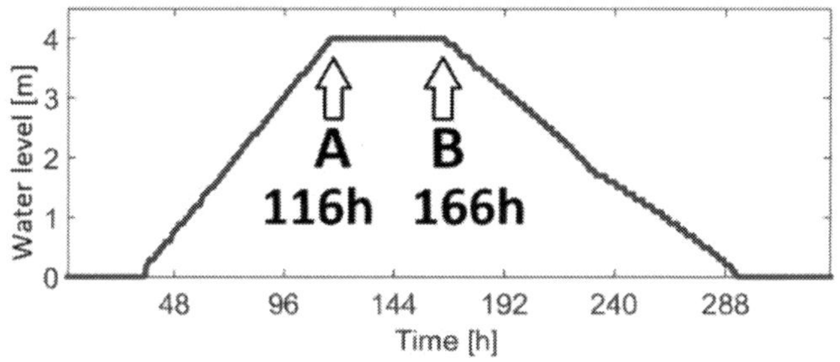


**Fig. 4.** Water level inside the reservoir used in the numerical modelling. Time points A,B were used for plots in fig. 6-7.

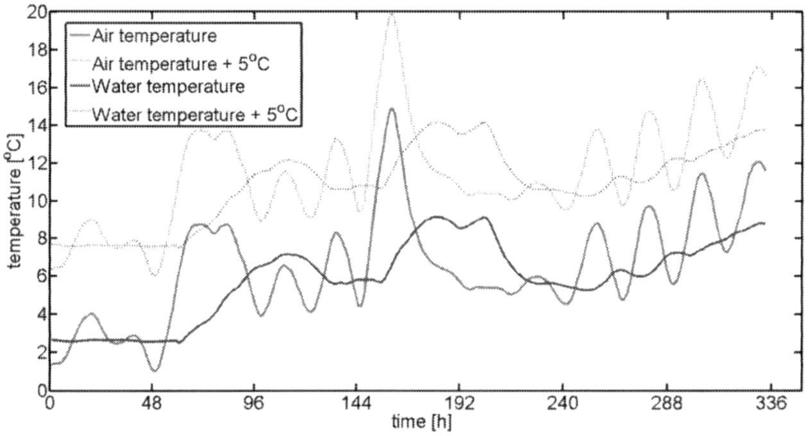


**Fig. 5.** Water and air temperature used in the numerical modelling (solid line – measured temperature, dots – measured temperature plus 5°C).

The weak zone was represented as a horizontal body (0.1m height and 11.5m width). The geomechanical properties of this zone were the same as for the levee. Only permeability and porosity were modified in the numerical model. These modifications are presented in table 2.

Numerical modelling was performed in several steps.

1) Calculation of the equilibrium of the geological medium.
2) Calculation of the equilibrium of the geological medium under levee stress.
3) Iterative calculation of:
   a. fluid transport;
   b. thermal transport;
   c. mechanical equilibrium in a variety of water level and air temperature conditions.
4) Factor of Safety (FoS) calculation.

Iterative calculation were done with 1h step. In every step temperature and water level values were fixed on the top edges of the levee model according to fig. 4-5[10].

**Table 2.** Parameters of numerical modellings with calculated FoS value.

| ID | Ratio of permeability of weak zone (WZ) and levee (L) | Increasing porosity of weak zone due to levee | Temperature changes due to first modelling | Factor of Safety (FoS) |
|---|---|---|---|---|
| 1 | WZ = 1 x L | ---- | ---- | 2.308 |
| 2 | WZ = 10 x L | +0.05 | ---- | 2.314 |
| 3 | WZ = 100 x L | +0.05 | ---- | 2.305 |
| 4 | WZ = 1000 x L | +0.05 | ---- | 2.267 |
| 5 | WZ = 1 x L | ---- | Air + 5°C | 2.308 |
| 6 | WZ = 10 x L | +0.05 | Air + 5°C | 2.314 |
| 7 | WZ = 100 x L | +0.05 | Air + 5°C | 2.305 |
| 8 | WZ = 1000 x L | +0.05 | Air + 5°C | 2.267 |
| 9 | WZ = 1 x L | ---- | Water = Air + 5°C | 2.308 |
| 10 | WZ = 10 x L | +0.05 | Water = Air + 5°C | 2.311 |
| 11 | WZ = 100 x L | +0.05 | Water = Air + 5°C | 2.305 |
| 12 | WZ = 1000 x L | +0.05 | Water = Air + 5°C | 2.267 |
| 13 | WZ = 1 x L | +0.05 | Water + 5°C | 2.308 |
| 14 | WZ = 10 x L | +0.05 | Water + 5°C | 2.311 |
| 15 | WZ = 100 x L | +0.05 | Water + 5°C | 2.305 |
| 16 | WZ = 1000 x L | +0.05 | Water + 5°C | 2.267 |

## 4 Results

The numerical experiment was focused on determining the possibility of detecting temperature changes caused by liquid infiltration in the weakened zone. Fig. 6 and fig. 7 present differential maps of temperature values obtained from the numerical modelling.

In the left column, the differences between temperature values modelled for the two ambient temperature boundary conditions are presented. The differences between the temperature values modelled for the two different boundary conditions of the infiltrated water temperature are presented in the right column. The boundary values of the ambient temperature and the temperature of the infiltrating water assumed during numerical modelling are plotted in red and blue in Fig. 5. In both cases the difference between the considered boundary conditions of the ambient and infiltrated liquid temperature was 5 degrees.

Numerical computations were carried out for the various permeability values of the weakened zone. The temperature differences obtained for the base permeability value (table 2) are presented in plots a and e (fig. 6 and 7). In the subsequent plots, the temperature differences obtained for the permeability value of the weakened zone were increased ten times (plots b and f), one hundred times (plots c and g) and one thousand times (plots d and h).

Temperature changes at several monitoring points are presented in fig. 8 and fig. 9. Temperature changes were visible only for curves with high permeability (at least 100x more than the levee) and are mainly located just under the weak zone. When the soil and water temperatures were similar, there were no significant differences between the curves (fig. 8). The maximum absolute difference between the results of models 1–4 was more than 1.5°C at only one sensor.

When the difference between the soil and water temperatures was bigger (water temperature plus 5°C, fig. 9), the curves diverge after 100h (water level above the weak zone). The maximum absolute difference was as high as 4°C at the sensors just under the weak zone. The differences between the curves were mainly visible close to the weak zone in the centre of the levee. Small differences in the bottom line of sensors were mainly caused by the existing water residue above the impermeable layer.

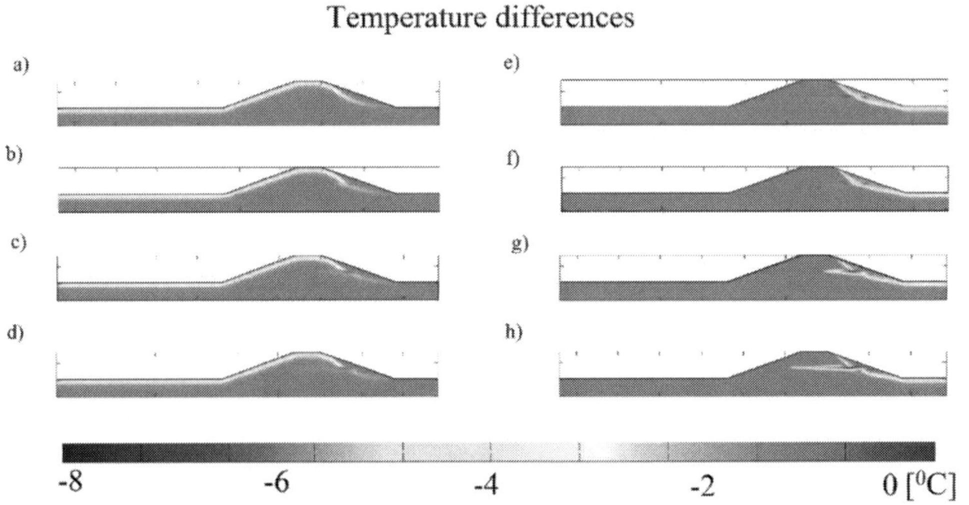


**Fig. 6.** The differences between the modelled temperature fields for different boundary conditions of the ambient and water temperature obtained for flood wave accumulation (point A in fig. 4). The temperature differences modelled for various ambient and water temperatures are presented in the left and right columns, respectively. Table 2 lists the permeability values used in calculations carried out for the weakened area (a and e), and for the weakened area with permeability 10 (b and f), 100 (c and g) and 1000 (e and h) times higher.

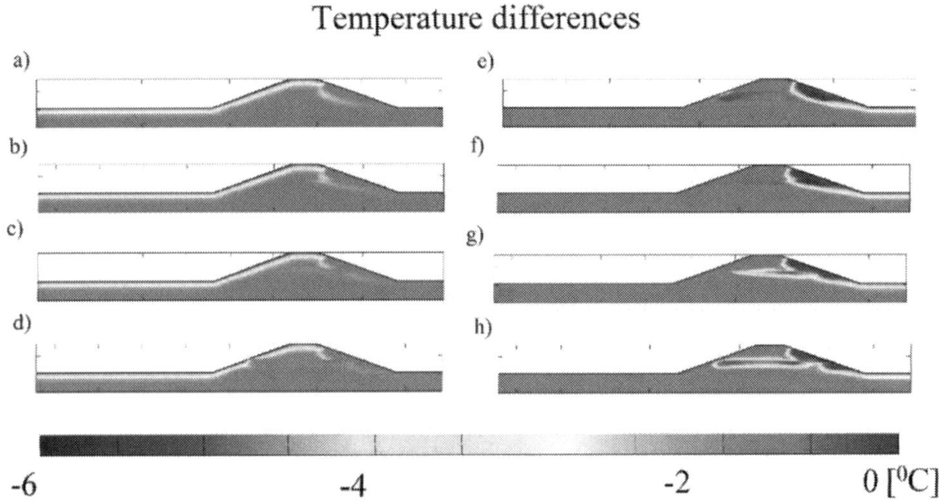


**Fig. 7.** The differences between modelled temperature fields for different boundary conditions of ambient and water temperature obtained for flood wave accumulation (point B in fig. 4). Temperature differences modelled for various ambient and water temperatures are presented in the left and right columns, respectively. Table 2 lists the permeability values used in calculations carried out for the weakened area (a and e), and for the weakened area with permeability 10 (b and f), 100 (c and g) and 1000 (e and h) times higher.

The range of air and water temperatures do not significantly influence the FoS value. Table 2 shows that the FoS value decreases slightly with the increasing permeability of the weak zone, but it was still above 2.2 (stable state).

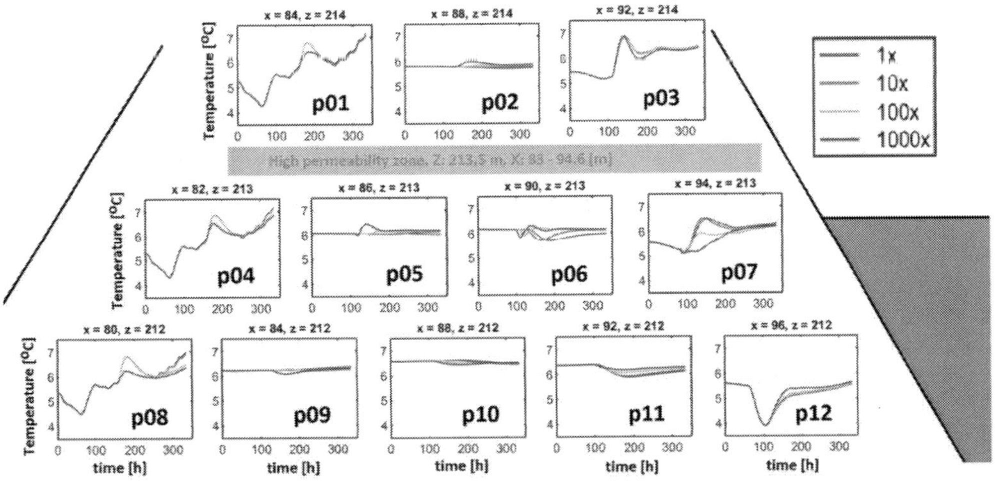


**Fig. 8.** Temperature changes for model 1-4. Points location were presented on fig. 3. Line 1x – 1000x mean how many times permeability of weak zone are bigger than levee in numerical model.

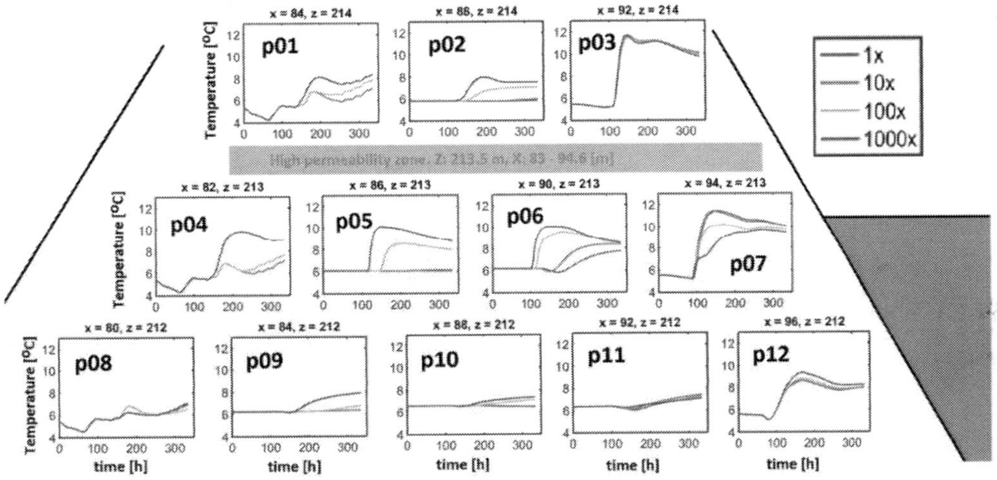


**Fig. 9.** Temperature changes for model 1-4. Points location were presented on fig. 3. Line 1x – 1000x mean how many times permeability of weak zone are bigger than levee in numerical model.

## 5 Conclusions

Numerical modelling shows that under some conditions it is possible to use temperature sensors to detect high permeability zones in levees. Temperature sensors can be used when a big (at least 5°C) temperature difference between the soil and the water is observed. It was also observed that air temperature exerts a minor influence on sensors close to surface area. Moreover, temperature sensors should be located near potential weak zones. Huge (greater than 4°C) temperature differences were visible for the temperature sensors located 0.5m under the weak zone in the centre of the levee. In this area, the effect of air temperature was the smallest and had no influence on the sensor measurements registered during the filtration-related processes.

*This work was financed by AGH University of Science and Technology, Faculty of Geology, Geophysics and Environmental Protection, as a part of statutory project no 11.11.140.613. Data used in this work were collected as part of the ISMOP project, sponsored by NCBiR under Grant no. PBS1/B9/18/2013.*

## References

1. M. Di Prinzio, M. Bittelli, A. Catellarin, P. Rossi Pisa, J. Appl. Geophys. **71**, 2-3, 53–61 (2010), doi: 10.1016/j.jappgeo.2010.04.002
2. I.K. Cho, J.Y. Yeom, Geophysics **72**, 2, G31–G38 (2007)
3. M.A. Van, C. Zwanenburg, A.R. Koelewijn, H. van Lottum, Proc. 17th Int. Conf. on Soil Mech. and Geot. Eng., Alexandria, Egypt, 2048–2051 (2009) doi: 10.3233/978-1-60750-031-5-2048
4. V.V. Krzhizhanovskaya, G.S. Shirshov, N.B. Melnikova, R.G. Belleman, F.I. Rusadi, B.J. Broekhuijsen, B.P. Gouldby, J. Lhomme, B. Baliś, M. Bubak, A.L. Pyayt, I.I. Mokhov, A.V. Ozhigin, B. Lang, R.J. Meijer, Procedia Comput. Sci. **4**, 106–115 (2011) doi: 10.1016/j.procs.2011.04.011
5. F. Cejka, V. Benes, F. Glac, Z. Boukalova, International Journal of Environmental Impacts **1** (3) 267–278 (2018) doi: 10.2495/EI-V1-N3-267-278
6. M.M. Darrow, Cold Regions Science and Technology **65**, 3, 474–487 (2011) doi: 10.1016/j.coldregions.2010.11.00
7. ISMOP – Computer System for Monitoring River Embankments (in Polish: Informatyczny System Monitorowania Obwałowań Przeciwpowodziowych), www.ismop.edu.pl
8. K. Sekuła, A. Borecka, D. Kessler, P. Majerski, Computer Science **18**, 4, 357–384 (2017) doi: 10.7494/csci.2017.18.4.2220
9. B. Bukowska-Belniak, M. Dwornik, A. Leśniak, Computer Science **18**, 4, 385–397 (2017) doi: 10.7494/csci.2017.18.4.2221
10. M. Dwornik, K. Krawiec, A. Franczyk, A. Leśniak, E3S Web Conf. **7**, 03021 (2016) doi: 19.1051/e3sconf/20160703021
11. Itasca Consulting Group Inc., *FLAC Fast Lagrangian Analysis of Continua and FLAC/Slope – User's Manual* (2011)
12. K. Krawiec, B. Bukowska-Belniak, A. Leśniak, D. Kessler, E3S Web Conf. **7**, 03018 (2016) doi: 10.1051/e3sconf/2016
13. J.W. Mościcki, G. Bania, M. Ćwiklik, A. Borecka, Studia Geotech. Mech. **3**, 1, 63–70 (2014) doi: 10.2478/sgem-2014-0008
14. J. Stanisz, A. Borecka, Z. Pilecki, R. Kaczmarczyk, E3S Web Conf. **24**, 03002, AG 2017 (2017), doi: 10.1051/e3sconf/20172403002

# The importance of pre-tensioning cable bolts used to reinforce of arch yielding support

*Zbigniew* Rak[1,*], *Jerzy* Cieślik[1], *Jerzy* Stasica[1], *Pavel* Dvořák[2]

[1]AGH University of Science and Technology, Cracow, Poland
[2]Minova Bohemia s.r.o., Ostrava, Czech Republic

**Abstract.** Widespread in Polish hard coal mining, cable bolts are used to reinforce the arch yielding support of roadways. This article attempts to prove that change in approach to long bolting technology can bring significant benefits in terms of improving the stability of roadways, including limiting the development of the fracturing zone in the roof, and ultimately also increasing the frame spacing of arch yielding support. The article contains the results of numerical calculations using the finite element method showing the effect of pre-tensioning on the work of cable bolts installed in a situation when a zone of a fractured rock mass appeared in a roof excavation. As a result of the calculations, the dependence of the final vertical displacements of the roadways roof on different pre-tensioning variants of the cable bolts was presented. The second part of the article presents the practical aspects of the use of cable bolts with pre-tensioning in reinforcing the rock mass and arch yielding support of roadways. In addition, the article identifies the developments in bolting technology which could lead to an increase in work safety and reduce the costs of arch yielding support.

**Keywords:** reinforcement of support, arch yielding support, cable bolts, pre-tensioning

## 1 Reinforcing the arch yielding support with the use of cable bolts in coal mining in Poland and the Czech Republic

Since the 1990s, roof boltings have been used in Polish and Czech hard coal mining (in Upper Silesian and Lublin Coal Basins) to reinforce the sets of arch yielding supports. Rock bolting the support arches is performed mainly to increase the load capacity of sets, which results in improved stability of supports of excavations. This is proven not only by observations and results of excavation measurements, but also by site surveys performed during extraction, which demonstrated that appropriately used bolts greatly increase the load capacity of arch support [1]. The main objective of using cable bolts in practice is, however, the possibility of remove sidewall arches of roadway arch supports at longwall-gateroad junction. Cable bolts eliminate the necessity of the use of props, which is very labour-consuming and not always practicable due to intersection dimensions.

* Corresponding author: zrak@agh.edu.pl

Supports of excavations are mostly reinforced by rock bolting the arches of support to the rock mass using cable bolts. Furthermore, sometimes the rock mass is reinforced in the fields between the arch support using rock bolts. This system is analogous to the one used, for example, in deep ore mines (Fig. 1) [2, 3]. In Poland and the Czech Republic, cable bolts are connected to the arch support using V-stringer, usually made from V29 or V32 steel profiles. The binding joists are usually made from long straight connectors, i.e. 4.5 to 6.0 m or short sections, i.e. from 0.3 m to 0.5 m longer than the span of the arch support (Fig. 2) [4]. The steel stringers are connected to the support using screws, further stabilising the structure of the support. As a rule, cable bolts are installed using two or three resin cartridges (usually Lokset with a length of 0.6 m each), thus the length of the bonding section is approx. 1.5 m. The Polish and Czech mining normally uses cable bolts with a length of 6.0 m to 11.0 m. The length of these bolts depends on the range of the anticipated failure zone above the excavation. This is especially important for excavations which outline the longwall. In this case cable bolts are used not only to support set arches, but also facilitate the disconnecting of several (usually 4 to 6) sidewall arches of the arch support at an intersection with the wall.

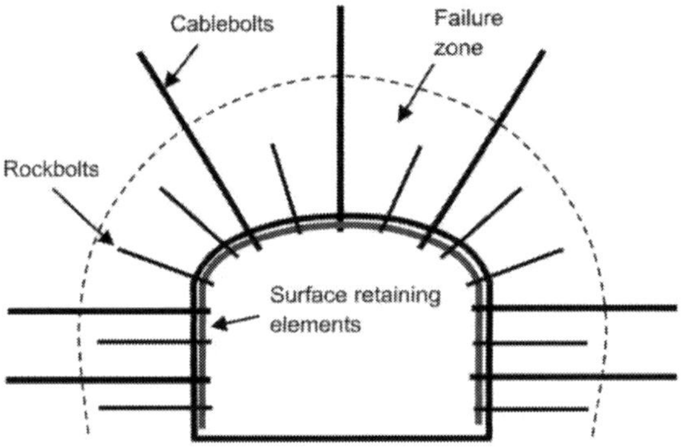


**Fig. 1.** Methodology of rock bolting in deep metal mines [2].

After disconnecting the sidewall arches at longwall-gateroad junction, cable bolts are the only reinforcement of an incomplete arch support.

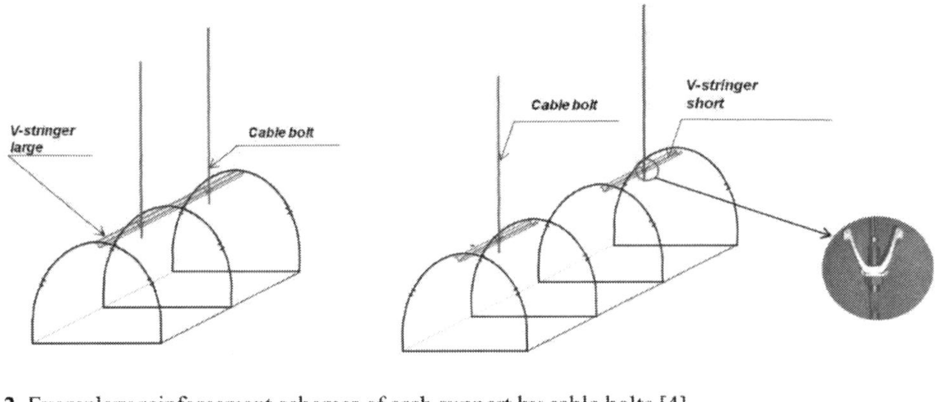


**Fig. 2.** Exemplary reinforcement schemes of arch support by cable bolts [4].

In the Polish and Czech mining industry, IR cable bolts are used the most. Such bolts are composed of several untwisted spring steel wires clamped with several bands along the length of the strand. The bolt head is a threaded rope socket, whose internal, conically profiled surface is used to fasten the individual wires using a single steel wedge (Fig. 3).

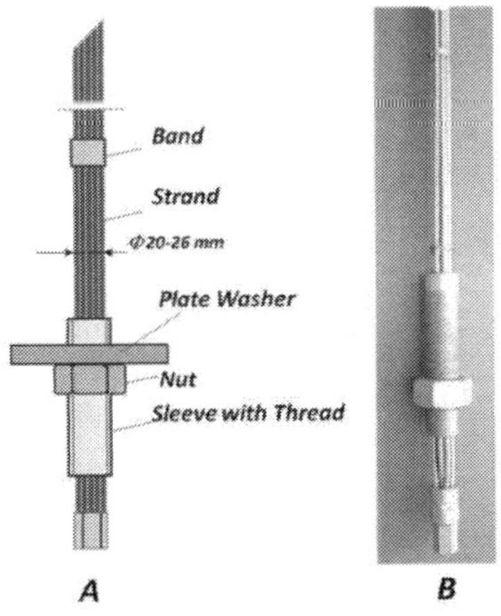


**Fig. 3.** IR cable Bolt; A) scheme , B) photo: www. interram.pl.

The load capacity of these bolts depends on the type and ranges between 280 and 470 kN. For several years now, there has been a cable bolt solution in place facilitating its installation along the entire length using the injection technique, employing liquid two-component resin or mineral binding agents. These, however, are used relatively infrequently. The typical method of securing roof excavations in the Bogdanka Mine is to reinforce the arch support using IR bolts and reinforce the rock mass with rock bolts, as shown in Fig. 4. Due to the problems with the mechanisation of rebar bolts installation in the foreheads of driven excavations, in mines are used only cable bolts. The installation of bolts takes place several dozen to several hundred metres away from the face of the driven excavation, thus in an already fractured rock mass. Therefore, cable bolts play an incredibly important role in the quality of excavation maintenance.

## 2 The conditions of interaction between IR cable bolts, the support and the rock mass

Filling the void between the arches of the support and the rock mass and pre-tensioning the bolts guarantee the proper interaction between the bolts, support and rock mass.

### 2.1 Filling the void between the arches of the support and the rock mass

One of the proven methods of filling the void between the support and the rock mass is the use of a low compressible lining usually based on the bullflex system [5, 6, 7]. A correctly constructed excavation lining with bolted roof arches is presented in Fig 5. Bags extended along the length of roof arches are filled with mineral and cement binder with a relatively short binding time and uniaxial compressive strength exceeding 20 MPa. Pumps for mineral binder grout, located near the face, are used for this purpose. In the case of a driven excavation, the lining, in order for it to serve its purpose, should be constructed on an ongoing basis. Such a solution makes it possible to apply the support and obtain a uniform load on roof arches very quickly. Fig. 5 presents the pattern of interaction between the rock bolt, the rock mass and support in the case of the lining being constructed correctly (item A).

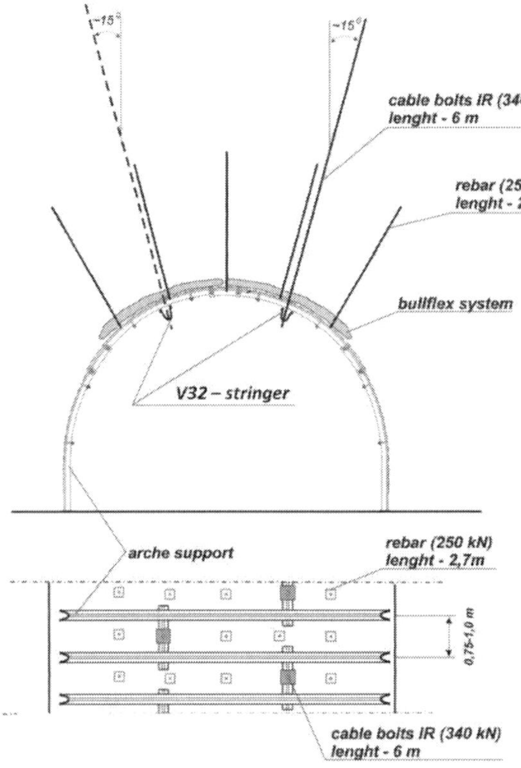


**Fig. 4.** Exemplary bolting scheme of maingate in LW Bogdanka mine.

Manual lining or lack thereof results in the bolt, directly after installation, not interacting with the rock mass. It is impossible to pretension the bolt, as this could result in raising the support. The interaction between the bolts and the support starts only after rock mass cracks and additional load being placed on the set arches. Therefore, rock mass reinforcement and the reduction of its cracking through bolting is not achieved.

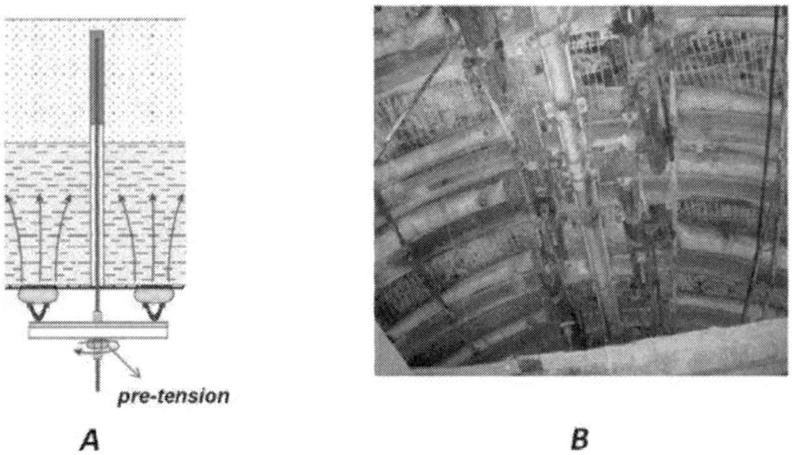


**Fig. 5.** Bolted roof yielding arches in the roadway with bullflex system; A) scheme , B) a general view.

## 2.2 Pre-tensioning of cable bolts

Cable bots reinforcing the support and the rock mass are usually installed several dozen to several hundred metres away from the forehead of the driven excavation. After their installation in the rock mass, they are pretensioned by applying the correct torque to the nut holding the support, by V-stringer. This generates compression forces in the roof (Fig. 5). The discussion regarding the application or non-application of pre-tensioning in roof bolting has been ongoing for more than 20 years. The opinions in this respect are strongly divided as regards bolts installed in the roof when there are no cracks. Theoreticians and practitioners alike are more and more often indicating the validity of pre-tensioning in the case of cracked rock masses [8]. Such a situation happens in the case of a roadway support which allows substantial roof deformation before the installation of cable bolts, which is usually performed at a significant distance to the driven excavation. In this case the pre-tensioning of bolts results in the interlocking of rock mass cracks and strata, thus increasing its load capacity. In order to illustrate the point, the results of numerical modelling of the excavation with support reinforced by cable bolts, given the varied values of pre-tensioning, are presented below.

### *2.2.1 The principles of numerical model design*

The FEM (Finite Element Method) analysis, utilising the Abaqus calculation system, was conducted on a two-dimensional 55 x 45 m model of a rock mass, for a rectangular 2 x 4 m roadway. In a homogeneous rock mass model, a 1.5 m fractured and cracked layer constituted the roof of the excavation. Within this layer, rock mass divisibility was introduced in the form of two sets of discontinuity. The vertical discontinuity distance was 0.2 m, while the horizontal discontinuity distance was 1 m, and created irregular, flat blocks (Fig. 6). The contact between the blocks was described using the classic Coulomb's law with friction coefficient ($\mu$) equal to 0.6. A linear elastic physical model of the rock mass was adopted for the calculation. It should be noted that the adopted linear elastic model of the rock mass is greatly simplified in relation to the actual conditions; however, for the objective of calculations, it is sufficient and makes it possible to characterise the phenomena observed in mines in qualitative terms. Displacement and stress boundary conditions were applied in the calculation, i.e. zero displacement of vertical and bottom horizontal edge of the model – ($u_x$=0), ($u_y$=0) respectively, normal stress was applied to the top edge with a vertical direction corresponding to the component of the original stress for a depth of 800 m. The support of roadway constituted the primary support reinforced by cable bolts with a length of 5.0 m. The cable bolts are represented in the model by rod elements. The primary, support of roadway is not directly represented in the model. It was assumed that it has a passive impact on the rock mass in the form of pressure on the roadway roof, allowing its significant deformation. Such calculation conditions correspond to the situation in which a zone of fractured, cracked and stratified rock mass is created above the roadway. This is usually observed in real mining conditions above the yieldable support as early as several dozen metres of the face of a driven excavation. Cable bolts were pretensioned some time after installing the support, treated as the basic support of the roadway. In the calculation model, it was assumed that cable bolts are installed in the excavation after reaching 80% of final convergence which is maintained just in a prop support. Such an approach made it possible to analyse the interaction between cable bolts and the rock mass in the case of pre-tensioning and lack thereof. In the calculations it was assumed that the pretension of bolts (N) was 64 kN, 100 kN and 130 kN, respectively, which corresponded to the preliminary axial stress on bolts of 203 MPa, 318 MPa and 414 MPa. Assuming the load capacity of cable bolts at 200 kN, which corresponds to normal stresses in the bolt of 636 MPa, the pretension was approximately at 32%, 50% and 65% of the load capacity of cable bolts, respectively.

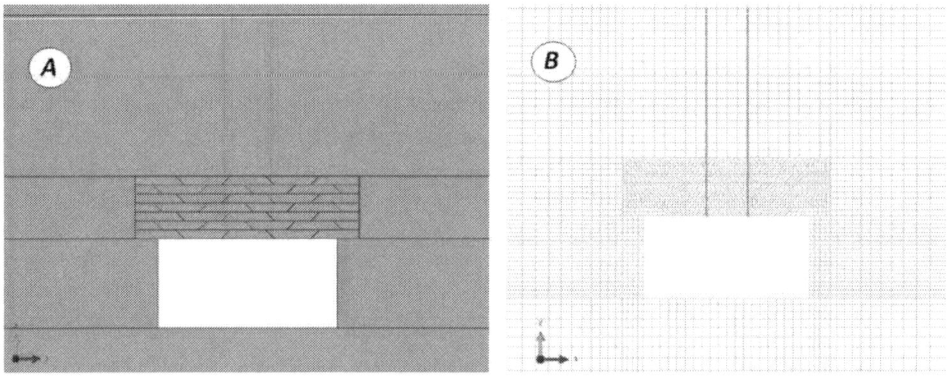


**Fig. 6.** FEM geometry and mesh of the roadway calculation model.

### *2.2.2 The results of numerical calculations*

Delaying the installation of cable bolts until 80% of final displacement on the outline is reached resulted in the effect of cracked and fractured roadway roof being observed during the installation of the bolts with clear stratification and gaps with a crack with up to 2 mm (Fig. 7). When cable bolts are not pretensioned, the width of cracks in the roadway roof for its final deformation was nearly 3.9 mm locally. In this case cable bolts are supporting only the rock mass which has been provisionally slackened by the flexibility of the yielding support. The load on not-pretensioned cable bolts in relation to their load capacity is fairly low. The value of equivalent stress according to the Huber-Mises-Hencky criterion ($\sigma_{HMH}$) amounts to nearly 170 MPa, which correspondents to its effort at 25%. If not pretensioned, the bolts work within a small range of their capacity. If the bolts are pretensioned, the rock mass behaves differently. The pre-tensioning of cable bolts in the excavation roof resulted in bracing the blocks of the rock mass into a package of sorts, which is less susceptible to deformation than in the case of no pre-tensioning.

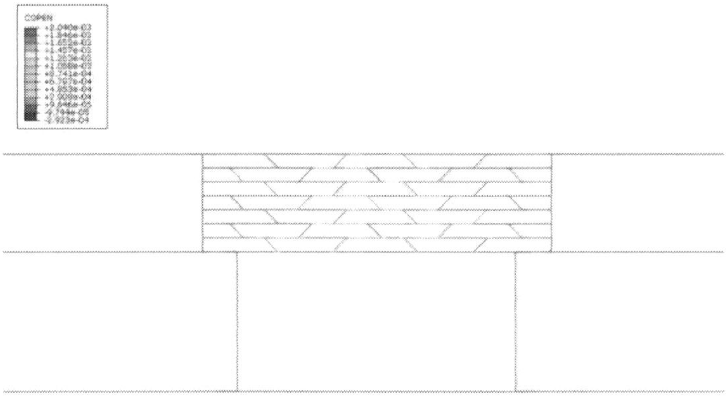

**Fig. 7.** The crack width (COPEN) map in the context of individual rock mass blocks during the installation of cable bolts.

In the case of pre-tensioning to the maximum level of 130 kN, the maximum crack width in the rock mass for the final deformation of roadway outline equals 2.3 mm, and is significantly lower than when no pre-tensioning is applied (Fig. 8). The effort of cable bolts, determined based on equivalent stress ($\sigma_{HMH}$) is, in this case, as much as 481 MPa (Fig. 8), which corresponds to an effort of 75%.

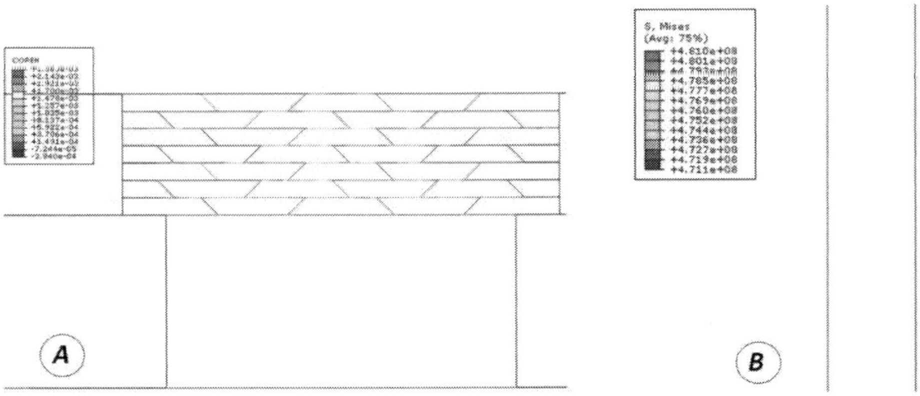


**Fig. 8.** The crack width (COPEN) map in the context of individual rock mass blocks with pre-tensioning (130 kN) of cable bolts (A) and HMH equivalent stress in the bolts (B).

As a result of the pre-tensioning of bolts, the cracked rock mass is characterised by greater load capacity than in the case of no pre-tensioning. The greater the tension, the more visible this effect is, and it results in smaller vertical displacements of the excavation roof (Fig.9). Pre-tensioning results in the reduction of cracks and rock mass bracing, therefore increasing its load capacity. A markedly smaller roadway roof displacement is obtained at the cost of greater effort on the part of cable bolts and full utilisation of their load capacity. In summary, pre-tensioning facilitates effective interaction between the cable bolts, cracked rock mass and support.

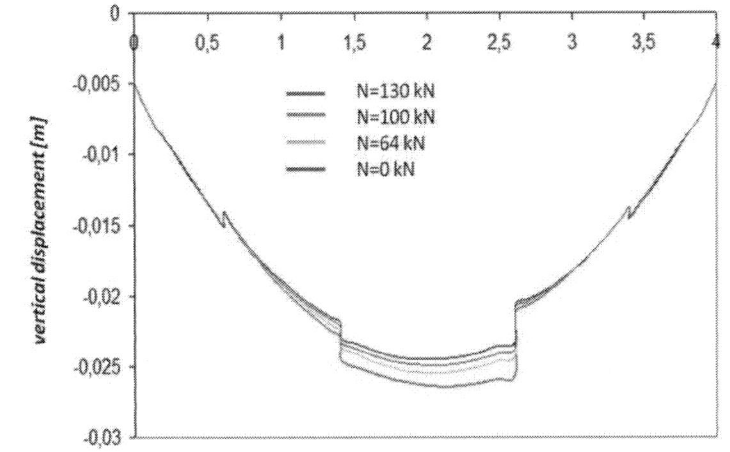


**Fig. 9.** A chart of final vertical displacements of the roadway roof for various variants of cable bolt pre-tensioning (N).

## 2.3 Pre-tensioning of IR cable bolts in practice

The observations and measurements of the extent of cracks conducted in excavations secured with a support reinforced with cable bolts do not corroborate the effectiveness of IR bolts in limiting the failure zone in the excavation roof [9]. Under similar geological and mining conditions, one would be hard-pressed to find any differences in the size of these zones in excavations with and without rock bolting using the bolts in question. This poses the question of why this form of rock bolting does not have a significant impact on the condition of the rock mass. This is a relevant question, since increasingly long cable bolts are being used for the safe maintenance of the longwall-gateroad junction with disconnected support arches. The effect is further augmented by the increasing depth of mining, the effect of mining-induced changes and the complex tectonics of Polish and Czech deposits. Making cable bolts longer (even to more than 10 m) results in a substantial increase in the cost of material and labour connected with the this process. The poor effects of reinforcing the roof rock mass are associated with the intensive convergence of excavations, which results in excessive limitation of their dimensions. The authors propose a thesis that the lack of interaction between IR cable bolts and the rock mass results from the lack of or negligible pre-tensioning and a small number of bolts per square meter of roof excavation. This, in turn, results from the design of the threaded bolt joint. Tests carried out in OKD 2 on IR-6 bolts prove the above. Taking into account the parameters of the IR-6 bolt head, we obtain a ratio of torque to tension of approximately 9.8. This means that a torque of 400 Nm on the nut of the bolt results in a pretension in the bolt strand of 40.82 kN only. At this point, it is worth highlighting that reaching this value of torque is impossible when using standard bolting rigs in Polish hard coal mines. In real conditions, it is required to use a wrench with an elongated arm and apply manual effort. Tests regarding the installations of IR-6 bolts, equipped with Glötzl KN 500 A50 dynamometer, were performed in OKD 2 and supervised by Ing. Pavel Dvořák. Of course, the objective of these tests was to determine not only the actual values of pretension in mining pit conditions, but also the course of changes in bolt load over the long time of excavation maintenance. In total, 6 test bolts with dynamometer were installed. The manner of their installation was always the same. The bolts were installed on 3 cartridges of Lokset resin using STB 1700/1500 2S devices, followed by mechanical and then manual tightening of the bolt nut to 400 Nm. The values indicated by the dynamometer immediately after applying this torque reached the level from 10 to 45 kN. As can be seen, the values obtained differ significantly from the theoretical value. The main reasons for this are probably the limitation of tension resulting from friction between the individual components – nuts and washers – and impurities in the bolt head thread. Whatever the cause, it should be noted that the IR bolts, commonly used in the Polish and Czech mining industry, make it impossible to reach a relevant level of pre-tensioning. As a result, the interaction between the bolts and the rock mass is significantly restricted, and their impact on the development of the failure zone in the roof is rather insignificant.

## 3 A proposed change to the system of reinforcing the support with cable bolts

The growing problems with maintaining the stability of excavations are caused by the increasingly difficult geological and mining conditions, and make it necessary to adopt a brand-new approach to the technique of reinforcing the support with roof bolting. We need cable bolt solutions whose active impact on the rock mass will result in increasing its load capacity. The utilisation of cable bolt only to "suspend" the elements of the support is an archaic solution which is poor in engineering terms, whose application can be justified only in very favourable rock mass conditions. According to the authors, for the Polish mining industry, it would suffice to adapt the solutions proven in many other countries, where roof boltings play a much greater role. The proposal described below is based on two bolting systems, Titan Bolt (by DSI) and Sumo Bolt (by Jennmar), which are very similar in terms of design. The presented solutions are among the systems characterised by the highest load capacity in the coal mining industry. The strands are composed of twisted wires with evenly spaced bulbs. In the middle of the strand, there is a steel tube which is used to pump the binder (mineral or resin). The bolt joint is a typical barrel and wedge system. The method of reinforcement bolting in excavations secured with a support proposed by the authors is described in detail below. First, in an excavation secured with steel arch yielding support, a hole with an appropriate length and diameter is bored several dozen metres from its face (Fig.10). At that point, the flexibility of the arch support and impact of the rock mass cause the roof surrounding the excavation is already cracked. The Titan or Sumo bolts are installed within the hole on two resin cartridges, with the barrel and wedge installed in place and pre-tensioning being applied using a hydraulic cable tensioner. Applying the pretension (usually 200 to 300 kN) directly to the roof results in the closure of some cracks and compression of rock strata in the immediate roof [10].

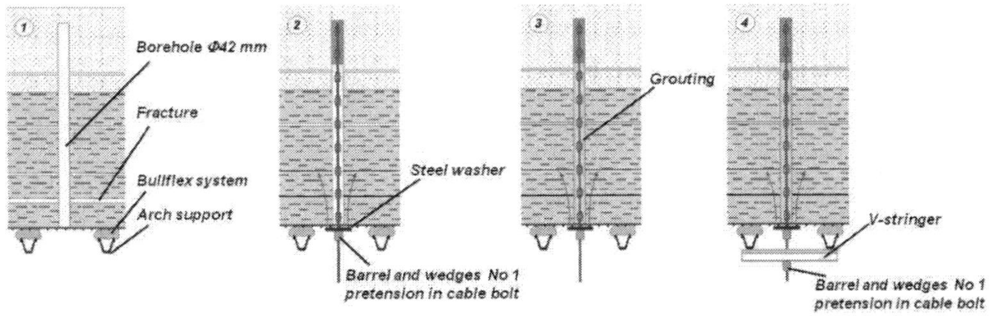


**Fig. 10.** Cable bolt installation.

A bolt constructed in this manner interacts with the rock mass from the very beginning, limiting the rate of cracking. The second stage usually takes place immediately after the first stage or after a slight delay. In this stage, mineral binder or liquid two-component synthetic resin are fed under pressure (Fig. 10). That way, an adhesive connection between the rock mass and the bolt, along its entire length, is obtained with cracks (not closed in stage 1) being filled and rocks becoming bonded. The last stage is the suspension of support arches on the protruding end of the bolt to facilitate the remove of several sidewall arches at the intersection with the wall. A steel V binding joist is built over the two adjacent arches and fixed with another barrel and wedge (Fig. 10).

## 4 Summary

The analyses and calculations presented in this article aimed at indicating the imperfections of the solutions used in Poland and the Czech Republic for reinforcing the yielding support with cable bolts. The impossibility of pre-tensioning and binding IR bolts along a short section results in their interaction with the rock mass being substantially limited. The increasing depth of mining (exceeding 1000 m) and complicated geological and mining conditions force mines to search for more effective ways of reinforcing excavation supports. The bolting techniques presented in Chapter 3 make it possible to eliminate the previous errors. According to the authors, they will result in a marked improvement of the stability of the support and roof of excavations. The method also facilitates disconnecting the sidewall arch at the longwall-gateroad junction without application of additional props.

*Article prepared as part of statutory work No. 11.11.100.005*

## References

1. A. Wardas, R. Bobek, T. Śledź, J. Twardokęs, A. Ratajczak, P. Głuch, *Ground control for ensuring the stabilisation of roadways located at considerable depths in the conditions of natural and mining risks,* Zeszyty Naukowe Instytutu Gospodarki Surowcami Mineralnymi i Energią Polskiej Akademii Nauk **94**, 53–66 (2016) (in Polish)

2. P. K. Kaiser, C. Ming, *Principle of rockbolting design*, Journal of Rock Mechanics and Geotechnical Engineering **4**, 215–227 (2012)
3. K. Skrzypkowski, E3S Web Conf. **29**, 00004 (2018) doi: 10.1051/e3sconf/20182900004
4. Z. Rak, J. Stasica, Z. Rawicki, *Selected problems in using roof bolting for the reinforcement of chock support in preparatory excavations,* Bezpieczeństwo Pracy i Ochrona Środowiska w Górnictwie **2**, 3–13 (2016) (in Polish)
5. P. Małkowski, Z. Rak, J. Stasica, *Elements of constructing mechanical lining in the light of previous experience*, Prace Naukowe GIG, Problemy Współczesnego Górnictwa **1/1**, 316–326 (2011) (in Polish)
6. P. Małkowski, Z. Rak, *Influence of mechanical lining on the state of stress around a gate drifted in weak carboniferous rocks*, Prace Naukowe GIG, Problemy Współczesnego Górnictwa **1/1**, 251–262 (2011) (in Polish)
7. M. Rotkegel, S. Stałęga, W. Węzik, W. Schopp, *Improving the work conditions of support through the application of mechanical lining*, Prace Naukowe GIG, seria Konferencje – Systemy obudowy wyrobisk korytarzowych dla zapewnienia stateczności i poprawy bezpieczeństwa pracy, Katowice, 151–158 (2007) (in Polish)
8. M. Cała, J. Flisiak, A. Tajduś, *Interaction between rock bolts and complex geology rock mass,* Biblioteka Szkoły Eksploatacji Podziemnej, Wydawnictwo Instytutu Gospodarki Surowcami Mineralnymi PAN, (2001) (in Polish)
9. W. Korzeniowski, Ł. Herezy, K. Krauze, Z. Rak, K. Skrzypkowski, *Rock mass monitoring based on the analysis of the work of the powered roof support*, Wydawnictwa AGH, (2013) (in Polish)
10. J. Stankus, X. Li, D. Giza, M. Rutyna, *Rock wall bolting and grouting: the application of SUMO bolts in Western U.S. Longwall Recovery,* Materiały konferencyjne: XXI Międzynarodowa Konferencja Naukowo-Techniczna z cyklu górnicze zagrożenia naturalne 2014, Zagrożenia naturalne a bezpieczne i efektywne Kopalnie, 35–37 (2014)

# Study on national activities and funding opportunities of furthering education programs for unemployed academics

*Susanne* Rahner[1,*], *Ilona* Winter[1] *Michael* Hartmann[2], *Frank* Wittich[2], *Aleksandra* Kasztelewicz[3], *Barbara* Tomaszewska[3], *Leszek* Pająk[3], *Marta* Dendys[3], *Agnieszka* Operacz[4], *Marta* Mraz[5], *Stelian* Nistor[6]

[1]Umweltbüro für Berlin-Brandenburg e.V., Berlin, Germany
[2]SRH Hochschule Berlin, Berlin, Germany
[3]Mineral Energy Economy Research Institute, Polish Academy of Sciences, Cracow, Poland
[4]University of Agriculture, Cracow, Poland
[5]Polip Youth Association, Szekszárd, Hungary
[6]University of Oradea, Oradea, Romania

**Abstract.** The joint German – Hungarian – Polish – Romanian Project EUBILD-UNAKLIM 2016-1-DE02-KA204-003254 in the ERASMUS + Scheme KA2 (Strategic Partnerships) designs an international adult-learning furthering education course. It took place in the participating countries Germany, Hungary, Poland and Romania. Environmental and cross-sector topics were addressed. A "green" job search and a questionnaire among employers, financing and educational institutions and NGOs was performed across the partner countries in 2017. The objectives were to find differences in employers' opinions on staff skills and the training topics provided by the furthering education organizations. In case differences were found, the course could close these gaps and will be adapted to the employers' needs, bringing more course participants into employment. Statistical evaluation compared countries, institution types, employers and furthering education- and financing institutions. The results were visualized in pie charts, histograms, and - where applicable - trend lines. Preliminary results show clear ranking of the new "green" job categories with differences in countries and institutions. The questionnaire results show differences in topic selection between employers and furthering education bodies. The results form the base for the course curriculum design closing the gaps between employers' expectations for "green" job topics and existing training curricula.

**Keywords:** furthering education, adult learning, environmental topics, green jobs, statistics

*Corresponding author: info@yggdrasil-dr-rahner.de

# 1 Introduction

In view of climate change and responding to political, economic and ecological challenges the number of "green" jobs is expected to rise. This will have an impact on education and training, and also on the long-life learning processes. In the joint German – Hungarian – Polish – Romanian project EUBILD-UNAKLIM 2016-1-DE02-KA204-003254 in the ERASMUS+ Scheme KA2 on Strategic Partnerships, five partners work on an international adult-learning furthering education course concept. Designed to take place in the four participating countries with training languages in the four national languages and in English. The course implementation is planned for the years 2019 – 2020.

**The goal** is a course with opportunities to attend in different countries, with online learning platforms, and internship and many practical, economy-driven activities and aspects.

**Target groups** are unemployed adults with an academic background, looking for new avenues in their careers, university drop-outs, migrants, single parents with children and other less privileged groups having difficulties and barriers to enter the employment market.

**The topics** are "green job" and environmental issues, i.e. nature protection, nature reserve management, European Fauna-Flora-Habitat (FFH) management, geology, water, emissions, sustainable energy, but also software topics such as Geographical Information Systems (GIS), and cross-sector topics such as project management, applied project work, team work, self-management and international environments. The course will take 6 months of theoretical input, followed by an internship period of 3 months in a country other than the one of residence.

**The outputs** are: a study on green jobs, a questionnaire on skill expectations for employees and course participants, a curriculum, training letter and e-learning platform, finally an implementation concept for the first course.

# 2 Definition of the "Green Jobs"

Green jobs are decent jobs that contribute to preserve or restore the environment, be they in traditional sectors such as manufacturing and construction, or in new, emerging green sectors such as renewable energy and energy efficiency [1–3].

The scope of Green jobs is:

- Improve energy and raw materials efficiency;
- Limit greenhouse gas emissions;
- Minimise waste and pollution;
- Protect and restore ecosystems;
- Support adaptation to the effects of climate change.

At the enterprise level, green jobs can produce goods or provide services that benefit the environment, for example green buildings or clean transportation. However, these green outputs (products and services) are not always based on green production processes and technologies. Therefore green jobs can also be distinguished by their contributions to more environmentally friendly processes. For example, green jobs can reduce water consumption or improve recycling systems. Green jobs defined through production processes do not necessarily produce environmental goods or services. Green jobs, are classified as, "jobs in business that produce goods or services that benefit the environment or conserve natural resources" or "jobs in which workers' duties involve making their establishment's production processes more environmentally friendly or use fewer natural resources" [1]. The Bureau of Labor Statistics, Washington [1] categorizes green jobs into: Water conservation, Sustainable forestry, Biofuels, Geothermal energy, Environmental Remediation, Sustainability, Energy auditors, Recycling, Electric Vehicles, Solar power, and Wind energy.

**Green jobs** or **green-collared jobs** are, according to the United Nations Environment Program [4], "work in agricultural, manufacturing, research and development (R&D), administrative, and service activities that contribute(s) substantially to preserving or restoring environmental quality. Specifically, but not exclusively, this includes jobs that help to protect ecosystems and biodiversity; reduce energy, materials, and water consumption through high efficiency strategies; decarbonize the economy; and minimize or altogether avoid generation of all forms of waste and pollution. The environmental sector has the dual benefit of mitigating environmental challenges as well as helping economic growth.

The definitions include jobs which seek to use or develop renewable forms of energy (i.e. wind, hydropower, geothermal, wind, landfill gas and municipal solid waste) as well as increase their efficiency. Under the green jobs domain education, training, and public awareness are also included. These jobs seek to enforce regulations, support education, and increase public influence for the benefit of the environment.

In 2008 the United Nations Environment Programme (UNEP) [4], the International Labour Organization ILO [5], the International Trade Union Confederation ITUC [6] and the International Employers Organization IEO [7] jointly launched the Green Jobs Initiative. The purpose is to bring a just transition to a green economy by providing space for workers, employers, and governments to negotiate on policy effective in providing equitable opportunity to green jobs.

## 3 Green Job Search in the Partner Countries

### 3.1 Green Jobs in Germany

The total 32 job portals were searched and about 16,778 green job ads were identified. Since an estimated of 10% of duplicates were identified, around 15,100 genuine job advertisements were identified. This equates to 2.29% of the annual total job offerings reported by the German Federal Statistics office. The main portal for green jobs is www.greenjobs.de. The research was performed between November 2016 and May 2017.

### 3.2 Green Jobs in Hungary

Green jobs is not a priority in job advertisements, it is difficult to decide what a green job is: any job done in a green economy (e.g. a driver) or a job is not a green economy, but it can be classified as a green job (e.g. environmental protection worker). The Hungarian research was conducted on the www.profession.hu website between 1 March and 30 April 2017, for 60 days, where the full offer will be published, and from 1 to 30 June 2017. Bids on a green jobs portal were search. According to the Ministry of National Economy, there is no reports –on green jobs at the national level. The green topic is not big enough to get into larger specifications or to make separate statistics, or to get attention. There is an aspiration to strike this strategy - but at the current labour market there is no focus on it. There is a tendency to increase in this area. The size of Hungarian job market is not comparable with the larger EU countries'.

### 3.3 Green Jobs in Poland

In Polish market the largest job portal is www.pracuj.pl, but only one portal (regularly updated), specialised in green jobs was found – portal www.teraz-srodowisko.pl. The total of 241 offers in the green topics were found in this portal between February and April 2017. Currently among all green job offers the most numerous were in the environmental management (21%) energy sector (20%), in industry and engineering sector (12%) and business management (11%). At present, the greatest demand on the market is on: environmental specialists (14 offers) and energy auditors (10 offers).

Preliminary conclusions:

- In Poland, development of the green job sector is slower than for e.g. in Germany;
- In Germany, thanks to the implementation of a series of activation and support measures, about 380,000 green jobs have already been generated. In Poland, this type of statistics has not been started yet, and the modernisation of the energy sector is proceeding very slowly (lack of support for pro-consumer and civic energy) [8];
- Lack of systematic analysis of the sectors and industries (indicators -> identify and remove barriers);
- No official definition of "green jobs", "green economy", "green sectors in the economy" in Poland.

### 3.4 Green Jobs in Romania

The market share has been taken into consideration, the number of registered companies, posted announces in the period of 60 days, number of registered CV (Curriculum Vitae) on the database and the monthly traffic of the website. The most important (negative) element is that there is no specific "green job" position mentioned on the top ranking. The search was made using two levels of search: first the search of national database and as second level the search of the offer for each county. At county level (there are 42 counties in Romania) the job offer is structured in two main directions. The first direction is represented by job offers posted at the website of County Agency of Labour Force Occupation, which are state companies, the main activity is collecting offers from different companies (state or private). The second direction is represented by job offers from private media (online, media a.s.o.) which also provide job offers from different economy sectors [9]. The total of 61 databases in Romania and abroad were searched. The search was performed between November 2016 and February 2017.

## 4 Questionnaire results on skill and knowledge expectations of potential employees

An online questionnaire was sent out in five languages: German, Hungarian, Polish, Romanian and in English. It was kept open from November 2016 until March 2017. A total of 1,866 targets were addressed, 400 in each consortium country was the goal. The total of 112 questions were asked in total, thereof 20 open questions, 23 closed questions (only "yes" and "no" as answer possible) and 69 questions that could be assessed by "1 = totally unimportant" until "6 = very important". The questionnaire was supported by structured interviews, five in each partner country. The structured interviews results were inserted into the statistics. The reflux quota was planned to be above 10%. We sent out several reminders, had lots of telephone calls, individual emails and also other activities throughout our networks. As a result we nearly achieved the total of 10%. Hungary had the best results; probably their networks are more efficient than the others. Table 1 shows the reflux results.

**Table 1.** Reflux results of the questionnaire in the countries and categories.

| Reflux | Germany | Poland | Rumania | Hungary | Total |
|---|---|---|---|---|---|
| Employers | 45 | 9 | 5 | 25 | 84 |
| | 9.57% | 4.54% | 3.67% | 13.03% | 8.9% |
| Furthering education and financing bodies | 15 | 31 | 10 | 23 | 79 |
| | 8.02% | 14.02% | 4.54% | 9.78% | 9.17% |
| Total | 60 | 40 | 15 | 48 | 163 |
| | 9.13% | 9.54% | 4.16% | 11.32% | 8.73% |

## 4.1 Statistical evaluation

The 84/70 questionnaires were completed. In some questions were empty answers, or people responded with "n.a=no answer", usually less than 4% "n.a." were given, in some delicate questions with high uncertainty, however, the rate rose up to 50%. Only the fully answered questions were statistically processed. In order to give quality estimates of the answers, the "n.a."-percentage is also given. We compared the absolute numbers, and the total percentages. However, countries with high feedback (Hungary, Germany and Poland) produce a high weighting for their answers and subsequently distort the comparison towards the German, Hungarian and Polish facts. In order to avoid this, internal country comparisons were carried out in addition. However, it should be kept in mind, that Germany has an absolute reflux of 45/15, Hungary 25/23, Poland 9/31 and Romania 5/10. The questions were categorized into countries and into institutional forms (i.e. private companies, public service, ministries, NGO, chambers, also Job centers). There were not enough data to categorize into countries and organizations. Comparisons of employers, furthering education institutions and institutions financing the furthering educations were performed.

It were performed state-of-the–art statistical data, i.e. average, mean, standard deviation, confidence interval at 95%, confidence minimum and maximum, skewness. It was visualized by pie charts, histograms, X/Y diagrams. In the case of the open questions, answers have been translated to English and categorized.

## 4.2 Results

In the employer respondents 33% were private enterprises, 30% NGOs, each 14% trade associations and public services. Of minor percentage were chambers, employers' associations and training and research institutions with 2-4%. The financing body respondents had 23% of NGOs, 39%, furthering education and financing institutions. Unemployment offices, municipal bodies and other institutions each 11%, and some had 1% (see Figures 1a and 1b). The employers were generally more than 10 years on the market. Romania here is the exception; here responded mainly young enterprises. The number of staff members among the employer respondents is significantly lower than of the financing and furthering education respondents. This phenomenon is visible across all countries. Most respondents were small and medium enterprises (SME) according to the EU-Definition (see Figure 2).

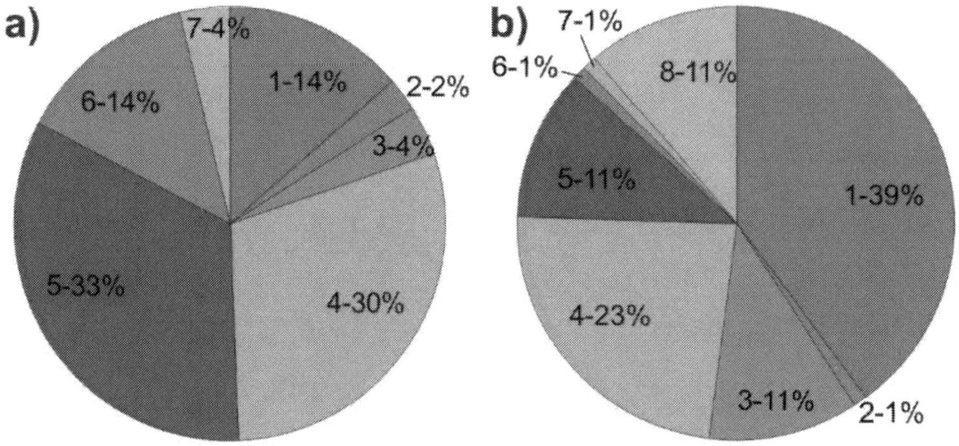


**Fig. 1.** a) Different institutions among the Employers (all countries). 1: Trade and association, 2: Chamber of Industry and Commerce, 3: Employers' Association, 4: NPO/NGO, 5: Private Enterprise, 6: Public Service Authority/Ministry, 7: Training and Research Institution. b) Different institutions among the furthering institutions and financing bodies (all countries). 1: Furthering education /Educational Institution, 2: Ministries/ Governmental Authority, 3: Municipal, communal, regional public institution, 4: NPO/ NGO, 5: Other, 6: Programme initiating institution, 7: Public Service/ Public Administration, 8: Unemployed office.

**Fig. 2.** Number of staff members for employers.

### 4.3 Hypotheses

Before sending out the questionnaire to obtain data, hypotheses was brainstormed to prove or reject (= Null-Hypotheses). The results are compiled in Figure 3.

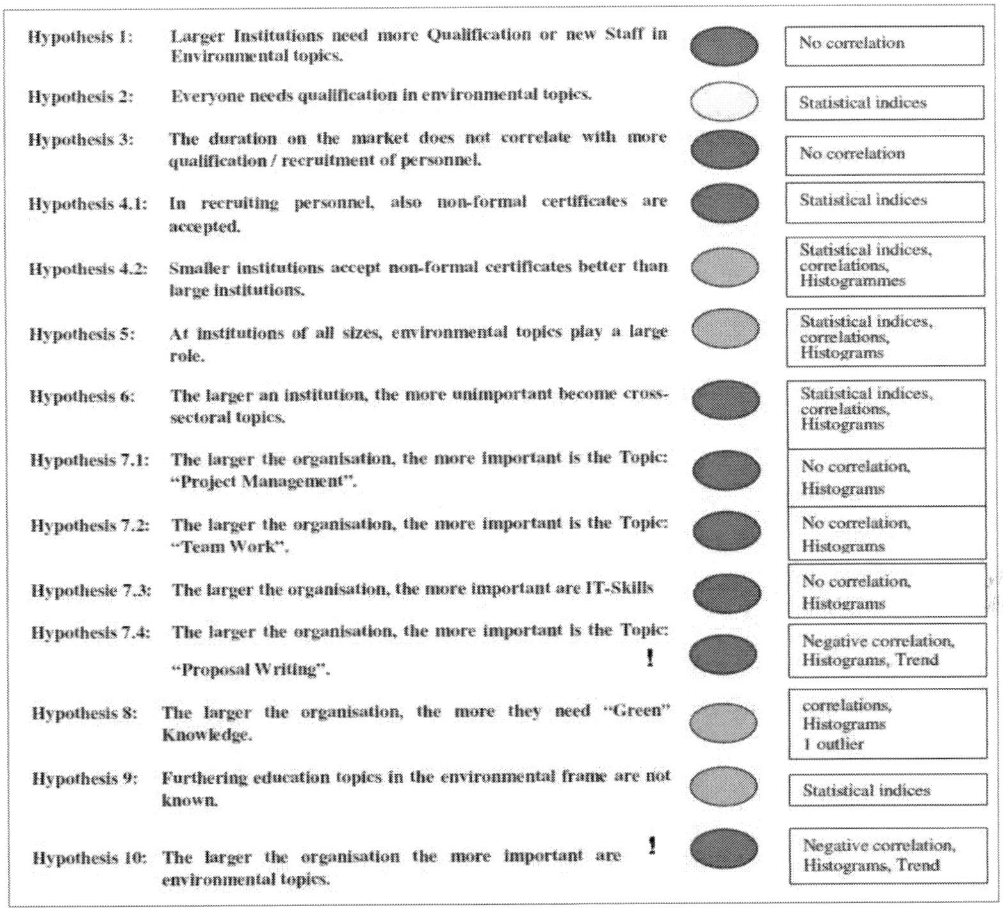


**Fig. 3.** Hypotheses situation after the statistical evaluation.

From the total of 14 hypotheses, 10 were rejected, one was indifferent and 4 were proved. The proved hypotheses:

- **Smaller institutions accept non-formal certificates better than large institutions (No. 4.2).** It was proved by all statistical indices, histograms and correlations.
- **At institution of all sizes environmental topics play a large role (No. 5).** It was proved by all statistical indices, histograms and correlations.
- **The larger the organisation, the more they need "Green" knowledge (No. 8).** It was proved by all statistical indices, histograms and correlations. There was one outlier.
- **Furthering education topics in the environmental frame are not known (No. 9).** It was proved by the statistical indices.

The following hypotheses proved the contrary. So it was proved that:

- **The smaller the institution, the more important is the topic: "Proposal Writing".** It was proved by the negative correlation of the Null-Hypothesis, the statistical indices and the histograms.
- **The smaller the organisation, the more important are environmental topics.** It was proved by negative correlations of the Null-Hypothesis, histograms and a trend line.

### 4.4 Gaps in Employers and Furthering education / Financing institutions expectations

The questionnaire had been designed in such a way that it should mirror the questions for employers and the furthering education/financing bodies as exactly as possible. It was assumed to prove if the employers' wishes and expectations for staff skills and of the furthering education courses offers of the furthering education bodies. The aim of the work is to provide a curriculum as closely as possible adapted to the commercial market. We consider that thus provide an uptake of our course participants by the labour market at its best possible rate. The analysis showed many gaps in these aspects (Fig. 4).

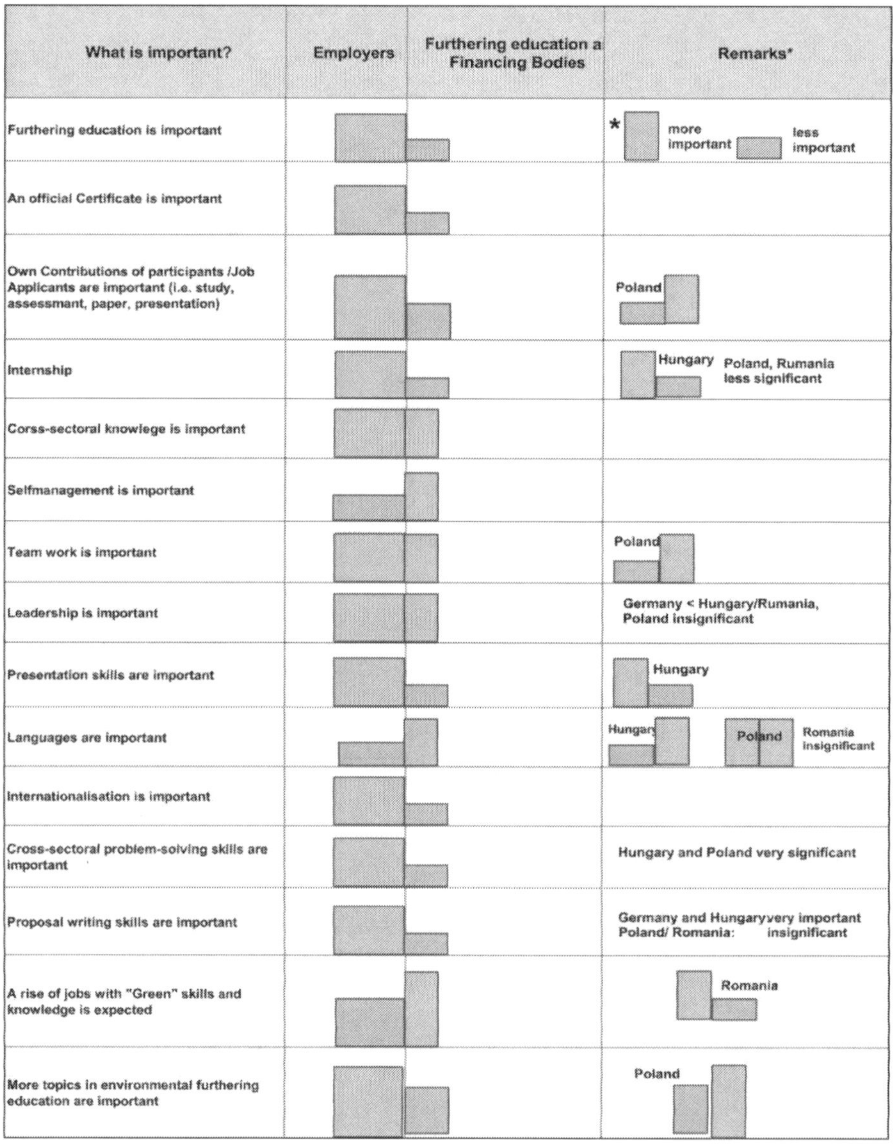


**Fig. 4.** Different assessments of employers and furthering education respondents.

## 4 Interpretation and future prospects

**The studies** revealed differences in the job topics of the "green jobs" in the four countries. Some countries do not define "green" in jobs yet. However, the topics are similar, even if the ranking is somewhat different.

**The questionnaire** questions addressed the opinions of employers and furthering education / financing bodies as closely as possible. The objective was to find out if the answers were as similar, or if mismatches in the opinions were detected. In that case, the curriculum would have to be adapted to the commercial and labour market environments. The results showed significant country-specific differences and gaps between employer and furthering education provider/financing body respondents. Employers rate the following topics higher than the Education/Finance providers:

- General more furthering education is needed, in Hungary this gap is significantly large;
- Official certificate should be handed out;
- Own participant contributions, i.e. presentation, practical project work, study, assessment are favourable, not tests;
- Internship should be part of the course, favourably abroad in order to expose the participants to an international environment, strengthen their internationalization and in consequence their resilience;
- Cross-sector knowledge and project management is important for both respondent groups;
- Self-management is less important for the employers;
- Team work is valued rather equally, in Poland significantly lower in employers' estimates. Here the curriculum has to be adapted to nation-specific needs;
- Leadership seems less important among German employers, however, it is significantly higher in value for Hungarian and Romanian employers;
- Presentation techniques are very high in value for employers. In Hungary, this trend is significantly sharper than in the other countries;
- Languages are not as important for the employers as the furthering institutes rated;
- Internationality, plays a big role for all the employers;
- Cross-sector problem solving skills are rated very important among employers, with even stronger trend in Poland and Hungary;
- Proposal writing is significantly important among the employers in Germany, in Hungary even more sharply visible.

Employers in Germany are rather indifferent in predicting a rising number of green jobs in the future. Romania is the opposite: here the employers expect a sharp rise in green jobs. The Authors can only speculate the causes: maybe in German the green job machine is very well established and has not as much growing potential any more (being on a high level already), whereas in Romania, the green jobs are in an earlier stage.

All employers like more furthering education topics in the "green" job environment, however only Poland has similar opinions among the both respondent groups.

**Further actions**

Actions will be taken at all topics analyzed that are very important for the employers. The course design will put more emphasis on them, when fine tuning the curricula, also by adding new modules, if necessary. Most of the topics can be taken up and added in a short-time activity, and there will be "quick wins" for the fine-tuned curriculum with the knowledge of now.

A mid-term or even a long-term goal will be establishing official certificates, acknowledged in the partner states or even on a European scale. It acquires quality management audits, working with accreditation bodies, disseminating the curriculum and the actual study and questionnaire findings in order to make the course and the certificate accepted. Stakeholder consultation also plays a large role, and this takes time.

The study of job titles in the job data bases revealed a job ranking. In many jobs our planned course will give a contribution. The potential thematic course contributions with an adapted curriculum will cover topics such as: geologist, hydrogeologists, waste management, climate protection commissioner, agricultural ecologist, environmental management and planning, CR specialist (energy, sustainability) or sustainable tourism.

When dealing with business data, the practical application of the findings has to be kept in mind. For many market decisions it is not necessary to follow such a strict significance levels as in scientific literature. It is advisable to combine the data results with good judgment and experience.

The course programme will be implemented in a follow-up ERASMUS project and put into life 2019 – 2021.

*This research has been funded by the European Union, ERASMUS Plus funding Schmes (EUBILD-UNAKLIM 2016-1-DE02-KA204-003254), which gratefully acknowledged.*

## References

1. Bureau of Labor Statistics, Washington, https://www.bls.gov/green/green_definition.htm (2013)
2. D.M. Kammen, & D. Engel, *Green Jobs and the Clean Economy; Copenhagen Climate Council*, (2009), https://antr.assembly.ca.gov/sites/antr.assembly.ca.gov/files/hearings/EngleKammen-CopenhagenClimateConcill%20-%202009.pdf (20.09.2018)
3. V. Éri, *Labor market impacts of environmentally conscious economic transformation, potentials of green jobs*. Center of Environmental Studies, Budapest (2010)
4. UNEP, http://www.unepfi.org/about/
5. ILO: *Guidelines for a just transition towards environmentally sustainable economies and societies for all* (2016), www.ilo.org
6. ITUC, https://www.etuc.org/en
7. IEO, https://www.ioe-emp.org/
8. *Biuletyn Mazowieckiego Obserwatorium Rynku Pracy* (2015), (in Polish)
9. http://hr.businesslive.ro/top-50-site-uri-de-job-uri-din-romania/

# European educational concept in environmental- nature- and climate protection to safeguard a cross border sustainable development

*Barbara* Tomaszewska[1,*], *Aleksandra* Kasztelewicz[1], *Marta* Dendys[1], *Wiesław* Bujakowski[1], *Susanne* Rahner[2], *Michael* Hartmann[3], *Jürgen* Weinreich[3]

[1]Mineral Energy Economy Research Institute, Polish Academy of Sciences, Kraków, Poland
[2]Umweltbüro für Berlin-Brandenburg e.V., Berlin, Germany
[3]SRH Hochschule Berlin, Germany

**Abstract.** In cooperation of 5 partners from 4 counties (Germany, Hungary, Poland, Romania), among them 1 public and 1 private university, 2 NGOs, 1 public research institute there are under realization concept to design a European educational in the ERASMUS Funding Scheme. The main objectives of the research are: 1) creation a curriculum concept with the participating countries in the professional advanced education area, in the topics: environmental -, nature-, and climate protection, 2) assessment of scope of realized research based on a research about existing furthering education programs, experiences, needs and requirements in the participating countries; 3) design of innovative training methods and combination of theoretical knowledge with practical key skills; 4) works in a complex environment that requires an interdisciplinary approach; 5) preparation of first test run of the course by exchanging potential trainers and trainees within the participating countries. The paper presents the research issues in the field of geoengineering and the use of renewable energy sources. The idea of the project is to open new chances on a future-oriented segment of the labor market for unemployed academic graduates, university dropouts and unemployed professionals, who have gained a multidisciplinary and yet very specific knowledge.

**Keywords:** geoengineering, "green" jobs, education, training

## 1 Introduction

In contrast to the background of climate change, scarcity of resources, reduction of biodiversity and increasing environmental pollution, it is necessary to find transboundary and integrated solutions to incorporate the interaction and synergy between all sectors and policies [1-3]. Therefore information, education and qualification in this topics are essential preconditions to forward assembly strategies and to reach broad basis [4]. Also the use of

* Corresponding author: tomaszewska@meeri.pl

existing human resources in Europe becomes an essential issue in this context. Underused human resources, affected by unemployment, can evolve into valuable protagonists of society and economy by qualification [5, 6].

Based on the fact of the development of a „Green Economy“ is expected to raises a new claim to an innovative oriented economy and effects employment with new requirements of qualification out of it is expected.

Project EUBILD-UNAKLIM (2016-1-DE02-KA204-003254) in the ERASMUS+ Scheme KA2 on Strategic Partnerships is implemented by a consortium of 5 partners from 4 countries (Germany, Poland, Hungry and Romania) since September 2016. Partners work on an international adult-learning furthering education course that will take place in the future, in the four participating countries with training languages in the four national languages and English.

The concept of the project is basing on the following objectives:

1. Create a curriculum concept with the participating countries in the professional advanced education area, in the topics: environmental-, nature-, and climate protection, based on a research about existing furthering education programs, experiences, needs and requirements in the participating countries.
2. Design of innovative training methods and the combination of theoretical knowledge with practical key skills.
3. Preparation of the first test run of the course by exchanging potential trainers and trainees with the participating countries.

The goals of the project are:

- To carry the curriculum concept into the institutions of the participating countries,
- To establish the curriculum in the countries' national educational structures in the long run,
- To opening new chances on a future-oriented segment of the labour market for unemployed academic graduates, university dropouts and unemployed professionals, who have gained a multidisciplinary and yet very specific knowledge.

The education course will be address to unemployed people with diverse background: long-term unemployed people of all ages, or working in inadequate positions compared to vocational training, people in orientation phase after vocational training, with different background grades, complementary work experience and interest for green jobs.

As part of the project, five tasks were carried out:

1. Study on national activities and funding opportunities of furthering education programs for unemployed academics in green jobs;
2. Curriculum to international course;
3. Model Training Letter on Topic - Geothermal Energy and Photovoltaic;
4. Online platform for e-learning tools and topics;
5. Experiments of model pilot plant to use as a field laboratory.

## 2 National studies on activities and funding opportunities in furthering education programs

Under the task comparative study of national activities and funding opportunities in furthering education programs for unemployed academics in green jobs have been conducted. Each partner conducted a labour market analysis in the field of green jobs, especially in field of geology, geophysics, environmental protection and renewable energy usage. Surveys were conducted using a specially created open-source platform run on SRH Hochschule Berlin surveys (Fig.1). The research included two main research methods to obtain necessary source data and information: quantitative (questionnaire surveys), and qualitative – open opinion, structural interviews. The questionnaire surveys enabled to gain required information from a large number of appointed employers and financial body & educational institutions in an efficient and standardized manner. The employer survey consisted of 27 questions, and the financial body & educational institution survey of 45 questions. Qualitative methods were used to emphasize experts' personal experiences and interpretation over quantification in an efforts to understand the meaning of the quantified social responses; to place and interpret the investigations and understand their results. The structural interview (five in each country) was to objectively assess the state of national activities and funding opportunities of furthering education programs in green jobs. The structured interviews results were inserted into the statistics. The achieved reflux quota was 10%.

An online questionnaire was sent out in five languages: German, Hungarian, Polish, Romanian and in English. It was kept open from November 2016 until March 2017. A total of 1,866 targets were addressed, 400 in each consortium country was the goal. The total 112 questions were asked, there of 20 open questions, 23 closed questions (only "yes" and "no" as answer possible) and 69 questions that could be assessed by "1- totally unimportant" until "6 - very important".

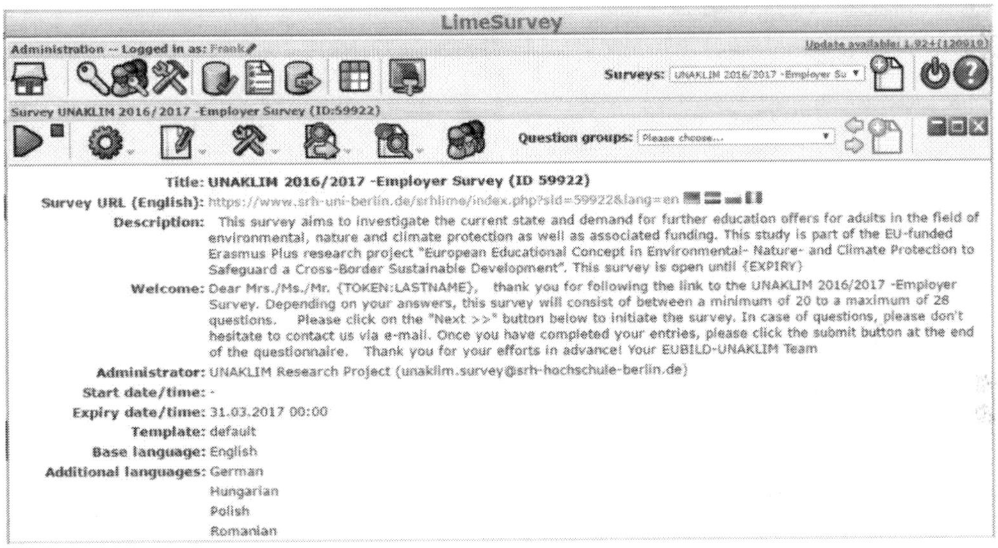


**Fig. 1.** SRH LimeSurvey Platform.

The objective of the questionnaires was to find if the curriculum would have to be adapted to the commercial and labour market environments. Significant country-specific differences and gaps between employer and furthering education provider / financing body respondents were found. According to that the course design could put more emphasis on the topics which were important for the employers also by adding new modules, if necessary. A long-term goal will be establishing official certificates, acknowledged in the partner states or on a European scale if possible.

The study of job titles in the job data bases revealed a job ranking in which planned course will give a contribution. Figure 2 shows the potential thematic course contributions with an adopted curriculum.

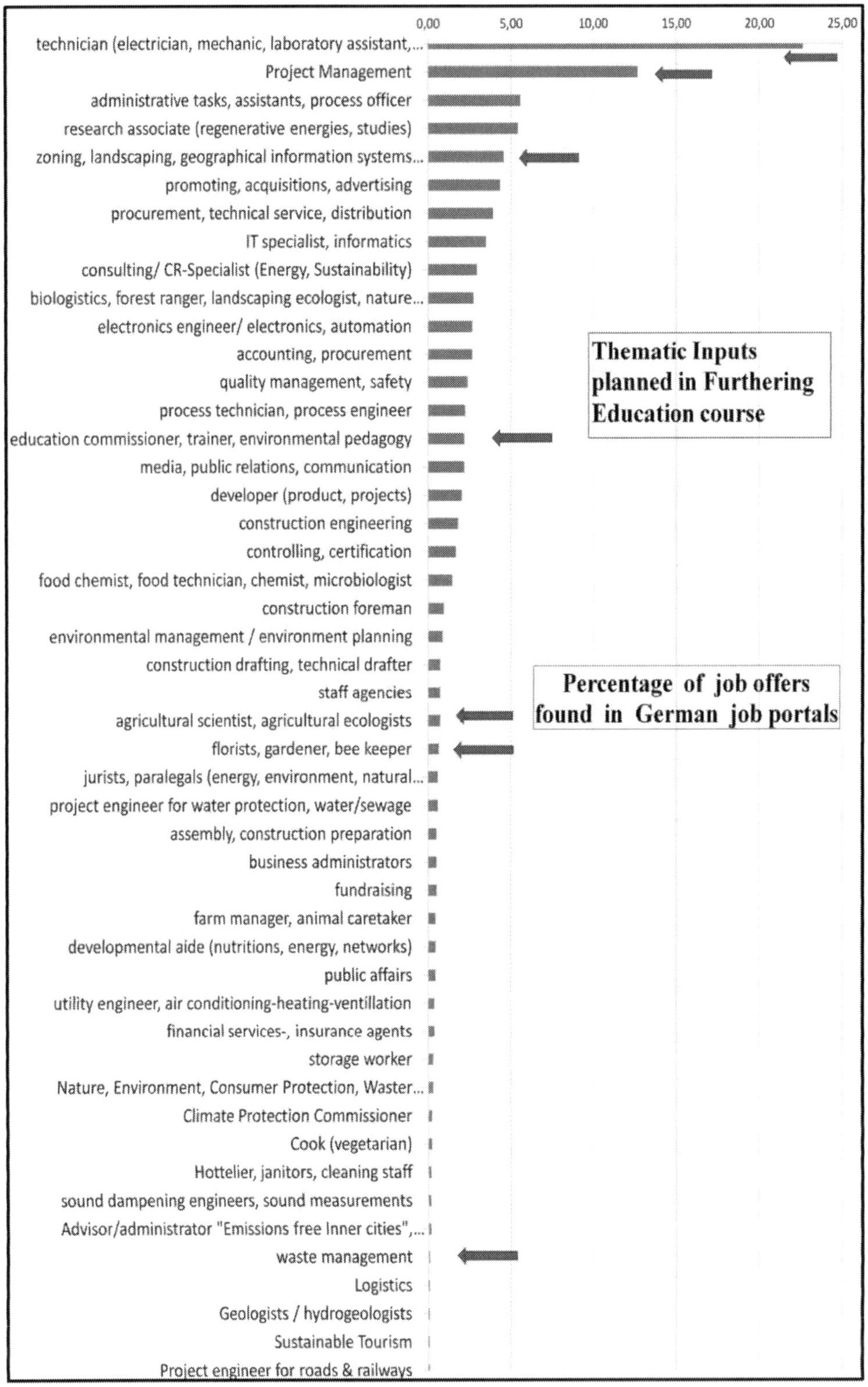


**Fig. 2.** Ranking of job offer titles found in the German data bases, and thematic input that the planned furthering education course will cover.

## 3 Curriculum to international course

The aim of the project is to develop a course to train and qualify professionals in clima-, environmental-, nature- protection and renewable energy are the charge of issues in the practical areas of these aspects. In assessment of plans in the field of environment, as well as strategic assessment of environmental impact to work in the area of environmental conservation, nature conservation and environmental protection in the respective of agencies, authorities and companies, research and educational institutions. As was suggested by Ocetkiewicz et al. [4] to let the world develop harmoniously, it is essential to propagate education for sustainable development. The role of teachers is crucial in this respect. Also wide aspect of environmental protection issues should is necessary to include in respected curricula.

The objective of the practically orientated specialty is also to diversify career options in the areas of policy, local, regional, national and international activities. Employment opportunities in the field include working for consultancies, local authorities, utility providers and contractors and organizations within the voluntary sector. Public and private organizations more and more are also looking for competent personnel capable of operating environmental management systems and skills that comply with national, EU-wide and international legislation, practical experience, and best-practice.

Elaborated under project curriculum concept based on research about existing furthering education programs, experiences, needs and requirements in the participating countries Germany, Hungary, Poland and Romania and respect to the future needs of an UE-wide approach. The design of the curriculum set up innovative training methods in combination of theoretical knowledge, case studies and practical key skills.

The Programme Structure is presented at Figure 3. The total length of the course will take 6 months. The course consist of 6 modules that covers 32 topics in total. The didactic methods provided in the courses includes lecturers, exercising (face-to-face to Learner), case studies and group working.

In the future curriculum will be tested and provide by exchanging potential European trainers and trainees within the participating countries.

## 4 Overview of the training letter on geothermal energy

As a part of task 3 the training letter on geothermal energy, district heating and cooling has been developed. The aim of the module is to introduce to the participants the fundamentals and potential of geothermal resources and system in the energy sectors and systems by building a general understanding of the resources use and technology approach participants are able to understand the economic and environmental impact of geothermal resources usage in projects with district heating and cooling (small, medium and big scale). The module covers information on the natural conditions, production, and utilization as well as the economic and environmental impact of geothermal energy. The purpose of the module is to provide a broad understanding of geothermal energy utilization which will prove useful in the practice. A detailed outline of the module include four specific topics:

1. Introduction to geothermal energy:
   - Geology and Earth heat
   - Basics of geothermal systems
   - Energy reservoirs and energy network
   - Investments, operating costs, costs of current production, efficiency
   - Case studies and reference installations
   - National and international utilization potentials

**EUBILD-UNAKLIM Course Overview**

**„Environmental-, Nature- and Clima Protection – Professional"**

Erasmus+

| Mod.No. | Module | ECTS | Month 21 days/m = 168 hrs/m | | | | | Workload = 1.512 hrs | | | | Workload 40 Credits (1 Credit = 30 hrs) | | | contribution to final certificate |
|---|---|---|---|---|---|---|---|---|---|---|---|---|---|---|---|
| | | | 1. | 2. | 3. | 4. | 5. | 6. | 7. | 8. | 9. | on campus hours | group work in projects | self-study hours | |
| Section I – Fundamentals, Methods and Tools | | | | | | | | | | | | | | | 21,7% |
| M 01 – 01 | Introduction to Sustainability in Environment, Renewable Energy and Grow | 0 | 16 | | | | | | | | | 12 | | 4 | 1,1% |
| M 01 – 02 | Urban Sustainability Assessment Framework | 1 | 32 | | | | | | | | | 20 | 4 | 8 | 2,1% |
| M 01 – 03 | Green Marketing and Services in Environment and Renewable Ennergy | 1 | 24 | | | | | | | | | 16 | 4 | 4 | 1,8% |
| M 01 – 04 | Career Opportunity in Green Economy / Job Coaching | | 8 | 4 | | 4 | | | | | | 8 | 4 | 4 | 1,1% |
| M 01 – 05 | Sustainability Finance and Funding in Urban, Environmental and Renewable Energy Projects | 1 | 24 | | | 8 | | | | | | 16 | 8 | 8 | 2,1% |
| M 02 – 01 | Project Management in Environmental and Renewable Energy Projects | 3 | 64 | 16 | | | | | | | | 32 | 32 | 16 | 5,3% |
| M 03 – 01 | GIS in Urban, Environmental and Renewable Energy Projects | 4 | | 64 | 32 | 32 | | | | | | 48 | 48 | 32 | 8,5% |
| Gesamt | | 10 | 168 | 84 | 32 | 44 | 0 | 0 | 0 | 0 | 0 | 152 | 100 | 76 | |
| Section II – Clima-, Environment and Nature Protection | | | | | | | | | | | | | | | 22,8% |
| M 04 - 01 | Facts in Environment and Nature Protection (EU- and National Framework) | 0 | | 8 | | | | | | | | 8 | | | 0,5% |
| M 04 – 02 | Circular Economy – Act and Law | 0,5 | | 16 | | | | | | | | 8 | 8 | | 1,1% |
| M 04 – 03 | Ecological footprint | 0 | | 8 | | | | | | | | 8 | | | 0,5% |
| M 04 – 04 | Water management, Protection and Law | 0,5 | | 16 | | | | | | | | 16 | | | 1,1% |
| M 04 - 05 | Waste Management, Separation and Recycling | 1 | | 32 | | | | | | | | 20 | 8 | 4 | 2,1% |
| M 04 – 06 | Soil Management, Protection and Law | 1 | | | 32 | | | | | | | 8 | 16 | 8 | 2,1% |
| M 04 – 07 | Emission and Pollutants in soil, air and water | 1 | | 4 | 28 | | | | | | | 24 | | 8 | 2,1% |
| M 04 – 08 | Environment in Logistic and Packaging | 0 | | | 4 | | | | | | | 4 | | | 0,3% |
| M 04 – 09 | Landfill – Management and Law | 0 | | | 4 | | | | | | | 4 | | | 0,3% |
| M 04 – 10 | Environmental criminal law | 1 | | | | 16 | | | | | | | 8 | 8 | 1,1% |
| M 04 – 11 | Nature protection – Management and Law | 1 | | | 32 | | | | | | | 16 | 8 | 8 | 2,1% |
| M 04 – 12 | NATURA 2000 Directive, Water Framework and Stakeholders | 2 | | | 36 | 28 | | | | | | 32 | 32 | | 4,2% |
| M 04 – 13 | Project work in Environment and Nature Protection | 2 | | | | 80 | | | | | | 8 | 40 | 32 | 5,3% |
| Gesamt | | 10 | 0 | 84 | 136 | 124 | 0 | 0 | 0 | 0 | 0 | 156 | 120 | 68 | |
| Section III – Sustainability in Renewable Energy | | | | | | | | | | | | | | | 21,2% |
| M 05 – 01 | Introduction to EnergyDistribution, SmartGrid and Future Mobility | 0 | | | | | 16 | 16 | | | | 20 | 8 | 4 | 2,1% |
| M 05 – 02 | Introduction to Photovoltaics and Storages | 1 | | | | | 40 | | | | | 24 | 12 | 4 | 2,6% |
| M 05 – 03 | Introduction to Solarthermal Energy and Storages | 1 | | | | | 24 | 8 | | | | 24 | 4 | 4 | 2,1% |
| M 05 – 04 | Introduction to Wind Energy and Power – to – Gas | 1 | | | | | 24 | 8 | | | | 20 | 8 | 4 | 2,1% |
| M 05 – 05 | Introduction to Bioenergy, Bio-Fuels and Storages | 1 | | | | | 24 | 8 | | | | 20 | 8 | 4 | 2,1% |
| M 05 – 06 | Introduction to Water flow energy, Hydro Power and Storages | 0 | | | | | | 8 | | | | 8 | | | 0,5% |
| M 05 – 07 | Introduction to Geothermal Energy, District Heating, Climatization and Cooling | 1 | | | | | 32 | 8 | | | | 20 | 12 | 8 | 2,6% |
| M 05 – 08 | Sustainability in green energy-efficient building | 1 | | | | | | 32 | | | | 24 | 4 | 4 | 2,1% |
| M 05 – 09 | Sustainability in energy-efficient production with energy management | 1 | | | | | | 32 | | | | 24 | 4 | 4 | 2,1% |
| M 05 – 10 | Study project in green energy and energy-efficiency | 3 | | | | | | 40 | | | | 8 | 24 | 8 | 2,6% |
| | | 10 | 0 | 0 | 0 | 0 | 160 | 160 | 0 | 0 | 0 | 132 | 84 | 44 | |
| Section IV - Profile and Practice – Internship | | | | | | | | | | | | | | | 33,3% |
| M 06 – 01 | Company Project / Internship | 10 | | | | | | | 168 | 168 | 152 | 16 | | 472 | 32,3% |
| M 06 – 02 | Course Final / Project Kolloquium | 0 | | | | | | | | | 16 | | 16 | | 1,1% |
| Gesamt | | 10 | 0 | 0 | 0 | 0 | 0 | 0 | 168 | 168 | 168 | 16 | 16 | 472 | |
| Workload / month | | | 168 | 168 | 168 | 168 | 168 | 168 | 168 | 168 | 168 | | | | 98,9% |

**Fig. 3.** Course overview.

2. Introduction to district heating:

- Concept and components of district heating and cooling systems
- Calculation of energy prices and cost effectiveness
- Case studies and reference installations
- Local and regional utilization potentials and impact

3. Introduction to air conditioning and cooling:
   - Heating pumps and air condition
   - Basics of air condition and cooling
   - Case studies and reference installations
   - Local and regional utilization potentials and impact
4. Best-practice and environmental impact of geothermal use in energy systems
   - Best practice
   - Environmental impact of geothermal exploration
   - Environmental impact of geothermal energy utilization
   - Environmental impact of district heating and cooling projects
   - Environmental impact of geothermal water utilization

Course participants will be expected to read and thing about the material outside class, and take part actively in the discussions. These discussions will enhance the learning process, allow sharing of experiences, and make this course more interesting.

The learning goal is to teach the participants to understand and implement current technical concepts of the geothermal use in energy concepts and solutions. They will be able to achieve commerciality while meeting legal, social and environmental challenges from the aspect of geothermal resources and use it in the local and regional matter. The lectures want to build an understanding of:

• Basic concepts of exploration and use of geothermal resources, the characterization of geothermal reservoirs and the production of heat and power;

• Integrated management techniques to deliver a geothermal energy project;

• Present and future potential of geothermal energy in the global and regional energy resource portfolio;

• District heating and cooling concepts, projects and best-practice;

• Environmental aspects of geothermal water and energy utilization.

The planned duration of the module will be 5 days with the workload of total 40 hours, respectively 20 hours for lectures, 12 hours for workgroup and 8 hours for self-work. As a part of group exercises, participants will consider regional and local examples of the use of geothermal energy as well as to prepare a short characteristic on given topics. Also a virtual trip to geothermal heating plant and laboratory of heating pumps will be provided.

It was pointed by Ocetkiewicz et al. [4] that the change of energy policy and the overall energy consumption model presumes a holistic approach. It should consider and manage a series of parameters such as technology, the economy, politics, society and education. That is why our proposed curricula contain such wide aspects to cover all important elements to introduce the fundamentals of geothermal resources system in the energy sectors and to better understanding of the resources use and technology, economic and environmental impact. Renewable energy sources, i.e. solar, wind, biomass, hydropower and especially in our case - geothermal, are nowadays widely being recognised as an effective response to the global issue of climate change. In this manner, the teacher's job is very responsible and demanding, and therefore teachers will be supported in the process of introducing into the curricula the issues related to the use of renewable energy sources and sustainable consumption. In education process it will be necessary to develop relevant methodological studies, meetings with experts and scientists, who will help teachers understand the issues of renewable energy and sustainable consumption. This aspects will be the aim of the second, planed in future Erasmus + project.

## 5 Conclusion

The sustainable development of the countries presumes a holistic approach which should consider and manage a series of parameters such as technology, the economy, politics, society and education. Education is an essential element and crucial to any nation's development. For adult-learning furthering education course stimulates scientific endeavor, enhances livelihoods, and injects a skilled workforce into the resultant modern job market. Investigating the knowledge, perceptions, as well as attitudes of the public concerning various aspects of environmental issues is of great importance to sustainable development. The conducted survey allowed to identify differences and gaps in the approach of employers and financial body and educational institutions, which are significant for individual countries. Based on the survey, topics important for education in "green jobs" were selected. Based on the research, a methodology was adopted, the effectiveness of which will be verified in the continuation of this project.

The Erasmus + project presented designing an international adult-learning furthering education course. It will take place in the participating countries: German, Hungarian, Polish and Romanian. Training languages will be the national languages and also in English. What is more, environmental and cross-sector topics will be addressed, so it can help in exchange of experiences, the use of good practice and bilateral exchange the participants from the project partners. These activities contribute to development of the knowledge and improve teachers' collaboration.

*This research has been funded by the European Union, ERASMUS Plus funding Schemes (EUBILD-UNAKLIM 2016-1-DE02-KA204-003254), which is gratefully acknowledged.*

## References

1. W. Bujakowski, A. Barbacki, M Miecznik, L. Pajak, R. Skrzypczak, *A structural-thermal model of the Karkonosze Pluton (Sudetes Mountains, SW Poland) for Hot Dry Rock (HDR) geothermal use*. Arch. Min. Sci. **61**, 4, 917–935 (2016)
2. M. Kaczmarczyk, E3S Web Conf. **24**, 02007 (2017) doi: 10.1051/e3sconf/20172402007
3. A. Operacz, Renew. Sust. Energ. Rev. **75**, 1453–1463 (2017) doi: 0.1016/j.rser.2016.11.141
4. I. Ocetkiewicz, B. Tomaszewska, A. Mróz, Renew. Sust. Energ. Rev. **80**, 92–97 (2017) doi: 10.1016/j.rser.2017.05.144
5. B. Tomaszewska, Adv. Intell. Syst. **599**, 60–71, (2018) doi: 10.1007/978-3-319-60204-2_7
6. W. Bujakowski, A. Barbacki, B. Czerwińska, L. Pająk, M. Prussak, M. Stefaniuk, Z. Trześniowski, Geothermics **39,** 78–93 (2010) doi: 10.1016/j.geothermics.2010.01.003

# AUTHOR INDEX